TOWARD
A MORE NATURAL
SCIENCE

TOWARD A MORE NATURAL SCIENCE

Biology and Human Affairs

LEON R. KASS, M.D.

THE FREE PRESS
A Division of Macmillan, Inc.
NEW YORK

The Free Press
A Division of Macmillan, Inc.
866 Third Avenue, New York, N.Y. 10022

Collier Macmillan Canada, Inc.

Printed in the United States of America

printing number

1 2 3 4 5 6 7 8 9 10

Library of Congress Cataloging in Publication Data

Kass, Leon.
    Toward a more natural science.

    Bibliography: p.
    Includes index.
    1. Medical ethics. 2. Bioethics. 3. Science and
ethics. I. Title.
R724.K318 1985   174′.2   85-1570
ISBN 0-02-918340-5

Credits

Grateful acknowledgment is made to the editors and publishers for permission to reprint the portions of this book that have previously been published.

"The New Biology: What Price Relieving Man's Estate?" was originally published, largely in its present form, in *Science 174*:779–788, November 19, 1971. Copyright 1971 by the AAAS.

"Making Babies: The New Biology and the 'Old' Morality," now considerably reworked, was originally published in *The Public Interest*, Winter 1972.

An earlier version of "Perfect Babies: Prenatal Diagnosis and the Equal Right to Life" appeared as "The Implications of Prenatal Diagnosis for the Human Right to Life" in *Ethical Issues in Human Genetics*, B. Hilton et al. (eds.), New York: Plenum, 1973.

An earlier version of "The Meaning of Life—in the Laboratory" was originally published as "'Making Babies' Revisited," in *The Public Interest*, Winter 1979.

"Patenting Life" was originally published, virtually in its present form, in *Commentary*, December 1981.

"The End of Medicine and the Pursuit of Health" was originally published as "Regarding the End of Medicine and the Pursuit of Health" in *The Public Interest*, Summer 1975.

An earlier version of "Practicing Prudently: Ethical Dilemmas in Caring for the Ill" was originally published as a two-part article, "Ethical Dilemmas in the

*To my mother
and the memory
of my father*

Contents

Preface

Nature is one thing, its scientific study another. Distinguishable from both is technology, providing power over nature made possible mainly by science. We human beings stand in the center of this triad: We belong to nature naturally, we place ourselves outside of nature to study it scientifically, in part so that we may be able to alter and control it technologically. Yet because we belong to the nature we study and seek to control, our power over nature eventually means power also over ourselves. We are not only agents but also and increasingly *patients* of our scientific project for the mastery of nature. Our self-conception, if not also our very being, lies upon the table science—biology, medicine, psychology—has prepared.

How shall we treat this patient? What standards of health and human flourishing shall guide our self-manipulations? We are, quite frankly, at a loss. For all our know-how, we know not whether or whither and why. Our knowledge of nature does not reach to its human import, to questions of meaning and goodness. This gap between nature studied scientifically and life lived naturally opens directly and necessarily because of the deliberate choice of modern science for "objectivity," for a stance outside of and removed from the world of our experience, from the world as it presents itself to us and as we engage it. Our natural science is, quite deliberately, most *un*natural, not only in what it enables us to do to one another, but even more in what it teaches us to think about who and what we are.

This volume, simply put, seeks to clarify and address this dilemma.

It does so by offering moral and philosophical reflections on the powers and teachings of modern biology and medicine. The themes of mastery and dehumanization provide its beginning, medicine and morality its center, man in nature its ground. The book moves from the moral challenges of the "new biology" to a search for a newer and more natural biology, one truer to our experience with room, perhaps, for morality and humanity within; the motion is accomplished across a bridge provided by that exemplary and inherently moral "art in the service of nature," medicine, which itself begs for such a more natural science. "Natural" and "more natural" mean here only "true (or 'truer') to life as found and lived." As used in this book, these terms imply no preconceived doctrine or school of thought: The careful reader will find here neither romanticism nor natural law, and no exhortations to eat organic foods. Yet although the author does not conflate "the natural" with "the good," he does insist that knowing the truth about nature, and our own nature, is crucial to thinking soundly about how we are to live.

Thoughtful people, scientists and laymen alike, are already perplexed and even disquieted by the meaning of the new biology for human affairs. But if we are truly to understand our present situation, we must be willing to reexamine and reevaluate its philosophical roots. To begin with, we must be open to looking afresh at phenomena made invisible by familiarity or inattention and to thinking anew about questions prematurely dismissed as settled or meaningless. This book aspires to assist such reconsiderations: to cast fresh light on the sensible and the familiar; to reopen foreclosed lines of inquiry; to reawaken the sense of awe and wonder, itself more human than even the desire for mastery—in short, to encourage and nurture the disposition of thoughtfulness about who we are and ought to be.

The author of this volume is by rearing a moralist, by education a generalist, by training a physician and a biochemist, by vocation a teacher—and student—of philosophical texts, and by choice a lover of serious conversation, who thinks best when sharing thoughts and speeches with another. Such a fellow incurs many debts—especially regarding a book written over fifteen years—which at this juncture he wishes gratefully to acknowledge. Thanks are owed first to my parents, Samuel and Anna Kass, who taught me by precept and example to put moral matters first and who pointed out, long before I read it in Rousseau, that learning and schooling are no substitutes for character. In the College of the University of Chicago, Joseph Schwab first woke me up and awakened, too, my interests in philosophy by showing me, painfully, that there were in fact questions where I had had only answers; it was he who first introduced me to the question of organism.

My father-in-law, Kalman Apfel, M.D., has exemplified for me that endangered species, the Hippocratic physician—for this and for much else, I am beholden to him. In the laboratory of Konrad Bloch I saw how a gentlemanly humanity and a love of natural beauty could flourish amidst centrifuges and scintillation counters and the clear-sighted pursuit of scientific discovery; while in the laboratory of Michael Yarmolinsky I was pushed to take seriously the connections between the beautiful and the true, and was encouraged by him and by Lee Rosner to biologize philosophically. Kit Mitchell initiated me into the delights of bird-watching, thereby awakening a beholder's wonder and admiration for living form and function, and thus providing a more natural access to nature.

From Paul Ramsey, I first learned that abortion is a moral question and, more important, that one could reason both carefully and profoundly about what is humanly at stake in the new biology; and from Hans Jonas I learned that one could think deeply and revealingly about living nature and man, against the reductionist and behavioralist tide. The example of these two men made it possible for me to shift my career in their directions; their encouragement and friendship has often nurtured me, even to the point of being able to take issue with some of their conclusions. Dan Callahan and Will Gaylin and my other colleagues at the Hastings Center have provided warm and lively collegiality and steady invitations to develop and present my own thinking. I am grateful especially to Alex Capron, Eric Cassell, Renée Fox, Harold P. Green, William F. May, Robert Morison, Ralph Potter, Barbara Rosenkrantz, Robert Stevenson, and Robert Veatch. My chairman, Milton Katz, gave me a free hand, much moral support, and invaluable lessons in tact and prudence with the Committee on Life Sciences and Social Policy of the National Research Council; and Tom Schelling often compelled me to think with and against notions that would never have occurred to me in many lifetimes. St. John's College, Annapolis, Maryland, in allowing me to serve part-time as tutor, introduced me to the vital reading of classic texts that informs many parts of this book and provided me the finest intellectual company I have yet enjoyed; conversations with Harvey and Mera Flaumenhaft, Robert Licht, and Tom McDonald have generated many of the better ideas in this book. My friend and patron, the late André Hellegers, and the generosity of Sargent and Eunice Shriver, brought me two productive years of research at the Kennedy Institute at Georgetown University, where André, Richard McCormick, and LeRoy Walters, among others, spurred me on and offered ample and constructive criticism. Godfrey Getz and the late Arnold Ravin made possible my return to the University of Chicago, where the freedom to teach whatever I wanted to learn to serious and thoughtful students has allowed me, alas, all too slowly, to begin to correct my ignorance of relevant philosophical and literary works; here friends—including Joel Beck, Martin Cook, John Cornell, Joseph Cropsey, John Gibbons, Ann Dudley Goldblatt, James Gustafson, Ralph Lerner, Nelson Lund, Rob McKay, Adam Schulman, Mark Schwehn, Mark Siegler, Jonathan Smith, Nathan Tarcov, and my colleagues on the Committee on Social Thought—have provided conversation, criticism of manuscripts, and a belief in the worthwhileness of my work, for all of which I am grateful. Were it not for my children, Sarah and Miriam, this book would have been written much sooner; were it not for what I have learned from and because of them, it would have been not worth writing.

Portions of the research for this work were conducted with support from the National Endowment for the Humanities and the John Simon Guggenheim Memorial Foundation; the generous patronage of the Henry R. Luce Foundation is also deeply appreciated. Curtis Black provided excellent help preparing the manuscript. Two outstanding editors, Irving Kristol and Joseph Epstein, improved several of the chapters; Celia Knight at The Free Press improved them all and ably shepherded the book through production. Faber & Faber Publishers kindly granted permission to use Sabine Baur's drawings for Adolf Portmann's *Animal Forms and Patterns,* which appear in my last chapter. Miriam Kass, age thirteen, has for several years helped me proofread every chapter, more times than either of us cares to remember, never missing a comma.

Finally, special thanks to three people who have worked with me and commented generously on the entire manuscript over its long period of gestation, preventing many a miscarriage: To Erwin Glikes, publisher and friend, who first urged me to write this book thirteen years ago and whose steady guidance and thoughtful reading have enabled me finally to see it to completion; to Harvey Flaumenhaft, favorite interlocutor and critic, editor and midwife, who first showed me what and how to read, and also how to write; and to my wife, Amy, who has shared every thought and labored with me over every paragraph, and who knows how indispensable she is to all that I have been and am.

November 1984

Introduction

The promise and the peril of our time are inextricably linked with the promise and the peril of modern science. On the one hand, the spread of knowledge has overcome superstition and reduced fear born of ignorance, and the application of science through technology has made life less poor, nasty, brutish, and short. As one of my colleagues puts it: Before the twentieth century, human life was simply impossible. Yet, on the other hand, new technologies have often brought with them complex and vexing moral and social difficulties, and the scientific discoveries themselves sometimes raise disquieting challenges to traditional notions of morality or of man's place in the world. Moreover, thanks to science's contributions to modern warfare, before the end of the twentieth century human life may become literally and permanently impossible. The age-old question of the relation between the tree of knowledge and the tree of life now acquires a special urgency.

The relation between the pursuit of knowledge and the conduct of life—between science and ethics, each broadly conceived—has in recent years been greatly complicated by developments in the sciences of life: biology, psychology, and medicine. Indeed, it is by now commonplace that the life sciences present new and imposing challenges, both to our practice and to our thought. New biomedical technologies (e.g., of contraception, abortion, and laboratory fertilization and embryo transfer; of genetic screening, DNA recombination, and genetic engineering; of transplanting organs and prolonging life by artificial means; of modifying behavior through drugs and brain surgery) pro-

1

vide vastly greater powers to alter directly and deliberately the bodies and minds of human beings, as well as many of the naturally given boundaries of human life. To be sure, many of these powers will be drafted for the battle against disease, somatic and psychic. But their possible and likely uses extend beyond the traditional medical goals of healing; they promise—or threaten—to encompass new meanings of health and wholeness, new modes of learning and acting, feeling and perceiving—ultimately, perhaps, new human beings and ways of being human.

The advent of these new powers is not an accident; they have been pursued since the beginnings of modern science, when its great founders, Francis Bacon and René Descartes, projected the vision of the mastery of nature. Indeed, such power over nature, including human nature, has been an explicit goal, perhaps the *primary* goal, of modern natural science for over three centuries, though the vision has materialized largely only in our century. By all accounts, what we have seen thus far is only the beginning of the biological revolution. Yet we have already seen enough to vindicate Aldous Huxley's predictions and concerns:

> It is only by means of the sciences of life that the quality of life can be radically changed. The sciences of matter can be applied in such a way that they will destroy life or make the living of it impossibly complex and uncomfortable; but, unless used as instruments by the biologists and psychologists, they can do nothing to modify the natural forms and expressions of life itself. The release of atomic energy marks a great revolution in human history, but not (unless we blow ourselves to bits and so put an end to history) the final and most searching revolution. This really revolutionary revolution is to be achieved, not in the external world, but in the souls and flesh of human beings.[1]

The practical problems—moral, legal, social, economic, and political—deriving from the new biomedical technologies have attracted widespread attention and concern. Over the past decade there has been much public discussion about such matters as the legality and morality of abortion, the definition of clinical death, the legitimacy of research on fetuses, the morality of "test-tube babies" and surrogate motherhood, the propriety of sperm banks, the right to refuse treatment, the rationale for psychosurgery, justice in the distribution of medical resources, the dangers and benefits of gene splicing, and the use and abuse of psychoactive drugs. Important practical challenges to individual freedom and dignity arise at every turn, most often as inescapable accompaniments of our ability to do good. On the one hand, freedom is challenged by the growing powers that increasingly permit some men to alter and control the behavior of others, as well as by the coming power to influence the genetic makeup of future generations. On the other hand, even the perfectly voluntary use of powers to prolong life, to initiate it in the laboratory, or to make it more colorful or less troublesome through chemistry carries dangers of degradation, depersonalization, and general enfeeblement of soul. Not only individuals, but many of our social and political institutions and practices may be affected: families, schools, law enforcement agencies, the military, and, especially, the profession of medicine, which already faces new dilemmas of practice and new challenges to the meaning of physicianship. None of these problems is easily resolved. Neither will they go away. On the contrary, we must expect them to persist and increase with the further growth of biomedical technologies.

But the biological revolution poses an even greater challenge, though one much less obvious and largely neglected. This challenge comes not so much from the technologies as from the scientific findings themselves. The spectacular advances in genetics and molecular biology, in evolutionary biology and ethology, and in neurophysiology and psychopharmacology, seem to force upon man a transformation— or at least a serious reconsideration—of his self-understanding and his view of his place in the whole. Even someone such as Jacques Monod, who helped usher in the new biology and who celebrates its triumphs, recognized the danger:

> There are far more grave and urgent dangers threatening modern societies already.
>
> Here, I am not referring to the population explosion, to the destruction of the natural environment, nor even to the stock pile of megatons of nuclear power; but to a more insidious and much more deep-seated evil: one which besets the spirit. One that was begot of the sharpest turning point ever taken in the evolution of ideas. An evolution, moreover, which continues and accelerates constantly in the same direction, ever increasing that bitter distress of the soul.
>
> The impact of his prodigious attainments in all areas of knowledge over the past three centuries is forcing man to make a heart-rending revision in his concept of himself and his relation to the world, a concept which had become rooted in him through tens of thousands of years.
>
> The whole of it, however—the spirit's disorder like our nuclear might—is the outcome of one simple idea: that nature is objective, that the systematic confronting of logic and experience is the sole source of true knowledge.[2]

Plainly, here is a challenge to our thinking that has potentially vast practical consequences, very possibly more profound and far-reaching than those of any given group of technologies. The technologies do indeed present troublesome ethical and political dilemmas; but the un-

derlying scientific notions and discoveries call into question the very
foundations of our ethics and the principles of our political way of life.

Now, it is an old but ill-remembered story, more or less forgotten
under the rosy optimism of the Enlightenment, that inquiry is, as such,
a risky business, in principle subversive of the authoritative beliefs
and practices of the community. When brought to trial by the city of
Athens on charges of not believing in the gods of the city, Socrates
understood that not he alone but philosophy itself was on trial. However much we are moved to take philosophy's side in its contest with
the city, the Socrates of Plato's dialogues shows us once and for all
how the activity of seeking knowledge undermines the rule of opinion,
and hence also, in principle, threatens the ruling opinions of one's time
and place. If philosophy was, and modern science is, the attempt to
replace opinion by knowledge, and if every society is rooted in certain
dominant opinions—whether about the gods or justice or the good life
or the equal rights of man—science essentially endangers society by
endangering the supremacy of its ruling beliefs. It is one thing to *hold
on trust* as true that one should honor one's father and mother or that
all men are created equal and endowed by their Creator with certain
unalienable rights; it is another thing to have to *prove* it.

Science—however much it contributes to health, wealth, and
safety—is neither in spirit nor in manner friendly to the concerns of
governance or the moral and civic education of human beings and citizens. Science fosters and encourages novelty; political society, governed by the rule of law, cannot do without stability. Science rejects
all authority save the truth, and prefers skepticism to trust and submission when truth is unavailing; the political community requires
trust in, submission to, and even reverence for its ruling beliefs and
practices. Science is universal and cosmopolitan; the political community is always particular and exclusive, resting on a distinction between who is in and who is out. The love of truth and the love of one's
own are not always reconcilable.

In the light of these fundamental and irreducible tensions, the
freedom accorded inquiry in liberal democratic regimes must be seen
as extraordinary, the exception rather than the rule. The remarkable
thing is not that democratic Athens executed Socrates, but that they
waited until he was seventy to do so; in Sparta—not to say Moscow—
he would not have been tolerated at all. These reflections should, by
right, make science and thought especially grateful to liberal democracy and eager to serve in its defense. Yet, paradoxically, as *science*
it is universalist and cosmopolitan, and does not take sides between
liberalism and totalitarianism. Worse, *modern* science, even modern
political science, by its own self-definition, declares that we cannot
know what we most need to know, namely, which way of life or form

of regime is better or best and why. The special character of modern science adds its own subversive elements to the iconoclasm of inquiry as such. And this brings us to the heart of the matter.

The pursuit of knowledge in our time differs radically from the Socratic pursuit of wisdom. When we say "knowledge," we mean *scientific* knowledge. The paradigm of our knowing, aped by the other sciences, is mathematical physics, a science that took its beginnings in the seventeenth century, in explicit opposition both to ordinary experience and to speculative philosophy. Most radically, it redefined what it means to *know* something, in terms of the standards of certainty and clarity possessed by symbolic mathematics and through the rigorous application of a universal method. Explicitly antiphilosophical in its spirit, it rejects as unworthy of its attention all questions that it cannot treat methodically and "objectively," and confines its attention to those problems that permit a scientific approach and solution. It is thus, at best, neutral to the large human and metaphysical questions that dominated ancient philosophy, and which human beings still ask and will always ask—questions about meaning, being, ultimate causes, the eternity or noneternity of the world, justice and injustice, the good, the true, and the beautiful. According to the scientific view there can be no knowledge, properly so-called, of these matters, no knowledge, strictly speaking, of theology or even about ethics: Opinions about good and bad, justice and injustice, virtue and vice have no cognitive status and are not subject to rational inquiry—they are, as we are fond of saying, values, and, therefore, merely subjective. As scientists we can, of course, determine more or less accurately what it is different people *believe* to be good, but we are, as scientists, impotent to judge between them. Even political science, once concerned with how men *ought* to live communally, now studies only how men *do* live and the circumstances that move them to change their ways. Man's political and moral life is studied scientifically not the way it is lived, but abstractly and amorally, like a mere physical phenomenon.

The sciences are not only methodologically indifferent to questions of better and worse. Seeking answers only in terms of their deliberately abstract questions, they find, not surprisingly, their own indifference substantively reflected in the nature of things. The scientific findings about nature and man are not congenial to human need, self-image, or aspiration. Nature, as seen by our physicists, proceeds deterministically, without purpose or direction, utterly silent on matters of better and worse, and without a hint of guidance as to how we are to live. According to our biological science, nature is indifferent even as between health and disease: Since both healthy and diseased processes obey equally and necessarily the same laws of physics and

chemistry, biologists conclude that disease is just as natural as health.
And concerning human longing, we are taught that everything humanly lovable is perishable, while all things truly eternal—like matter-
energy or space—are utterly unlovable. Indeed, many behavioral scientists and neurobiologists even explain away the existence of such
longings, and, *a fortiori*, of the human soul; they prefer instead an "objective" account, useful for predicting and controlling behavior, which
speaks in terms of stimulus and response, of input and output, and
neurotransmitters and end-plate potentials, and which relegates to
oblivion human inwardness, purposiveness, and consciousness. The
teachings of science, however gratifying as discoveries to the mind,
throw icy waters on the human spirit.

Now, one might justly say that there is no guarantee that the truth
will be edifying. Further, science, in its neutrality to matters moral
and metaphysical, can claim that it leaves to these separate domains
the care of the good and matters of ultimate concern. This division of
labor makes sense up to a point: Why should I cease to believe courage
is good or murder is bad just because science cannot corroborate these
opinions? In fact, many a pious man over the past century has thus
compartmentalized his beliefs, embracing Darwinism during the week
and biblical religion on the Sabbath. But this tolerant division of live
and let live is intellectually unsatisfying and finally impractical, because deep down we know that there cannot be incompatible truths
regarding the one universe, especially when one side claims to know
"objectively," i.e., "truly." Regardless of the *intent* of scientists, the
teachings of science, as they diffuse through the community, do not
stay quietly and innocently on the scientific side of the divide. They
challenge and embarrass the notions about man, nature, and the whole
that lie at the heart of our traditional self-understanding and our moral
and political teachings. The sciences not only fail to provide their own
standards for human conduct; their findings cause us also to doubt
the truth and the ground of those standards we have held and, more
or less, still tacitly hold.

The challenge goes much further than the notorious case of evolution versus biblical religion. Is there *any* elevated view of human life
and goodness that is proof against the belief that man is just a collection of molecules, an accident on the stage of evolution, a freakish
speck of mind in the mindless universe, fundamentally no different
from other living—or even nonliving—things? What chance have the
ideas of freedom and dignity, under even any high-minded *humanistic*
dispensation, against the teachings of strict determinism in behavior
and survival as the only natural concern of life? How fares the belief
in the self-evident truths of the Declaration of Independence and the

existence of unalienable rights to life, liberty, and the pursuit of happiness, to whose defense the signers pledged their lives, their fortunes, and their sacred honor? Does not the scientific worldview make us skeptical about the existence of *any* natural rights and therefore doubtful of the wisdom, and even suspicious of the motives, of those who risked their all to defend them? If survival is the only possible principle that nature does not seem to reject, does not all courage and devotion to honor look like folly?

The chickens are coming home to roost. Liberal democracy, founded on a doctrine of human freedom and dignity, has as its most respected body of thought a teaching that has no room for freedom and dignity. Liberal democracy has reached a point—thanks in no small part to the success of the arts and sciences to which it is wedded—where it can no longer defend *intellectually* its founding principles. Likewise also the Enlightenment: It has brought forth a science that can initiate human life in the laboratory but is without embarrassment incompetent to say what it means either by life or by the distinctively human, and, therefore, whose teachings about man cannot even begin to support its own premise that enlightenment enriches life.

What I have said does not arise from hostility to science. I think I properly appreciate its accomplishments. I intend no aid or comfort to the enemies of science or the friends of ignorance. My intention, rather, is to point out that the teachings and discoveries of science are at best partial—indeed, partial in *principle*. They are necessarily incomplete, hence in need of being supplemented. Our current difficulties call for more and better thought, not less, albeit also thought of a somewhat different kind. They beckon us to seek deeper knowledge, precisely about the adequacy of what we already know—or think we know—and also about the possible knowability of what we have declared to be unknowable. A new intellectual challenge presents itself: to study and think through—much more thoroughly, precisely, and deeply—the questions about science and ethics I have but touched on here: (1) questions about the reasonableness of divorcing science from ordinary experience, on the one hand, and from philosophy, on the other—and also questions about the relation between knowledge and wisdom; (2) questions about the proper relation between the universalist character of science and the necessarily particularistic demands of human institutions and polities—in particular, the connection between free thought and liberal democracy; (3) questions about the correctness of the claims that reason is impotent and nature is silent in matters of ethics and morality; (4) questions about the relations among the sciences, in search of a more coherent understanding of the whole.

In short, we must ponder the full range of questions raised by the relation between knowledge and human life, or between science and the broader community.

In conducting these reexaminations we are not seeking a pie in the sky. For despite the tremendous achievements of our nonteleological and mechanistic natural and human sciences, there is ample reason to believe that the fundamental questions about the nature of nature and the being of man are far from closed. For example, should not the remarkable powers of self-healing, present in all living things, make us suspect that dumb nature in fact inclines purposively toward wholeness and is not simply neutral between health and disease? Should we suppose that the lower can properly account for the higher, that animals—never mind human beings—can be finally understood in terms of inorganic matter and motion? Should the reductionists persuade us that a chicken is just an egg's way of making another egg or, more precisely, a gene's way of making more genes? Can biochemistry and neurophysiology ever do justice to what we know first and best: our inward experience of ourselves as passionate, purposeful, and thoughtful beings? Is a colleague of mine right when he claims that human love will soon have a biochemical explanation? Finally, is nature itself (as distinguished from science) really "objective"? Do not the deterministic and "objective" accounts of behaviorists or neurophysiologists utterly fail to account for their own activities, most especially their own passionate and spontaneous quest for the truth, never mind the thoroughly mysterious arrival of their flashes of insight? Is there not something finally defective about objective thinking if it is in principle blind to the mind of the thinker who thinks?

On the other side, we must seriously also reexamine, in the light of the genuine discoveries science has made, the traditional notions of freedom and virtue, choice and responsibility, and man's place in nature. Until we do this, carefully and thoroughly, we do not know if they need to be reaffirmed, abandoned, or revised. Ultimately, our goal is a richer, more comprehensive "new science" of man in relation to the whole. This must be compatible with the findings—if not necessarily the interpretations—of the natural, psychological, and social sciences. But it must also do justice to the full range and complexity of human powers and activities, and it might thus provide some standards for addressing moral and political questions.

There can, of course, be no guarantee that such a unified science is possible. But even if the quest for it fails, the search will make us more keenly aware of what we can and cannot know—and why. In the process, we will have gained self-knowledge, including worthwhile knowledge of our ignorance. We may also recover a lost sense of awe and wonder regarding the natural world of which we are both the scientific and the ethical part.

This volume is an invitation to reflection on these matters. Although the essays collected here were originally written on separate occasions over the past dozen years, and most of them previously published in widely different journals, they all nevertheless have the same intention: to search out the human significance of the presently new biology, and to search for a yet newer and richer biology that will do justice to matters of human significance. They are informed by the intuition that a science of nature that aspires to give an account of the human must be able to account for the natural ground of human aspiration, including the aspiration to give an account. Most observers who recognize and deplore the gap between the modern scientific view of the world and common experience's view of the world try to bridge the gap by shoring up the humanistic side. Some even call for a new humanism or a new ethic based upon, or simply harmonious with, the scientific worldview. In contrast, this book is informed by the belief that the so-called two cultures can properly be bridged, if at all, only by a philosophical reconstruction of the scientific side of the divide.

The order of topics largely reflects both the development of my own thinking and the logic of the subject: from the ethical dilemmas raised by new biomedical technologies to the philosophical notions—such as purpose, embodiment, finitude, and form—that are crucial for a richer science of life and, thereby, possibly important for the foundations of ethics. Reflections on the art of medicine link the discussion of ethics and new technologies to the search for an ethically relevant philosophical biology. The following remarks should help the reader in following the thread that connects the separate chapters, each of which, I might add, has been revised, in part to make the connections more evident.

The first part, entitled "Eroding the Limits," concentrates on the new biomedical technologies. The first chapter, "The New Biology: What Price Relieving Man's Estate?," provides an overview of the field and sets forth the direction of the entire inquiry. It surveys the new powers to alter the boundaries of birth and death and to manipulate human capacities and activities; it articulates certain neglected ethical questions made urgent by these new powers, especially questions of voluntary self-degradation and dehumanization; it shows how these ethical issues point to fundamental philosophical questions about human nature and human good, man's place in the world, and the meaning and purposes of knowledge; and it concludes with some thoughts regarding policy. Some of these practical and theoretical questions are probed more thoroughly in the subsequent chapters in Part I, which concentrate on those technological powers connected with reproduction and genetics: *in vitro* fertilization and cloning ("Making Babies: The New Biology and the 'Old' Morality"); screening for genetic disease and abortion of the genetically defective ("Perfect Babies: Prenatal Diagnosis and the Equal Right to Life"); laboratory culture of and experimentation on human embryos ("The Meaning of Life—in the Laboratory"); and the creation, exploitation, and ownership of new forms of life through the techniques of DNA recombination and genetic engineering ("Patenting Life: Science, Politics, and the Limits of Mastering Nature"). Though one could argue for the greater human significance of other biomedical and psychobiological technologies (behavior modifying drugs, for example, or a perfected technology of pleasure), nonetheless the topics discussed suffice to demonstrate the main points. First, and most obviously, the use of the new technological powers brings vexing and even sometimes truly novel challenges to human freedom, dignity, equality, individuality, bodily integrity, and privacy. For example, how does one respect and preserve the freedom and dignity of those human beings who are used instrumentally as subjects of biomedical research? Is the practice of aborting the genetically handicapped compatible with our professed commitment to the equal worth of every life? How are we to judge the merits of efforts to prolong life by means that render the beneficiary more helpless, dependent, and degraded? Is our growing dominion over living nature compatible with respecting our own given nature?

Second, these technologies and their scientific underpinnings threaten to erode the existence or at least the meaning and human significance of many of the naturally given boundaries, attributes, and relations that frame and structure human life—birth, father, mother, child, gender, lineage, embodiment, selfhood and identity, health and normality, aging, and death. What, for example, does "mother" mean—and what can and should it mean for human affairs—if one woman donates the egg, another houses it for insemination, a third hosts the transferred embryo and gives birth to the baby, a fourth nurses it, a fifth rears it, and a sixth has legal custody? And how is male distinguished from female: Is it by genotype (XX or XY), or external genitalia, or psychological outlook, or sexual preference, or even none of the above because gender can be "reassigned" through reconstructive surgery? What does organ transplantation or surrogate motherhood imply about the relation between a person and his or her body? And, at both ends of a human life, what constitutes the clear and distinct boundary between alive and dead?

Third, the erosion of these natural boundaries and definitions is both cause and effect of the much broader erosion of limits: the absence of any clear standards to guide the use of the enormous new powers. Everything is in principle open to intervention; because all is alterable, nothing is deemed either respectably natural or unwelcomely unnatural, nothing in principle better or worse. Here lies the deepest danger of the new biology: limitless power—both unlimited in its extent and without clear limits or standards to guide its use.

To some extent this danger has been hidden from view because the new biomedical technologies have entered our society largely through the benevolent offices of the medical profession, whose minions have traditionally practiced an art with relatively clearly defined ends and norms of conduct. To conserve health and to cure disease— the traditionally understood goals of medicine—implicitly carry a natural reference: the healthy, normal human being, fit both in body and in soul. Yet the advent of the new technological powers and the many attending ethical dilemmas have raised profound questions about the nature and purpose of the medical profession. The traditional art of healing is a "normative" art constituted by a view of the naturally determined goal of health. Can this be reconciled with the limitless new powers and the so-called value-free science to which it is now wed?

The chapters in Part II, "Holding the Center," are devoted to a reconsideration of the nature of medicine, in the light of its contemporary predicament. Conceding that the boundaries of medicine may now be less clear, these chapters attempt to locate and defend its center. "The End of Medicine and the Pursuit of Health" argues that, despite all the changes of modern times, health remains the true goal of medicine, that health is a naturally given although precarious standard or norm, characterized by "wholeness" and "well-working," toward which the body aspires on its own, and that the pursuit of health depends far more than we realize on cultivating habits of living that assist the body in its efforts toward wholeness. This approach is, admittedly, too sanguine and simple; the next two chapters ("Practicing Prudently: Ethical Dilemmas in Caring for the Ill" and "Professing Medically: The Place of Ethics in Defining Medicine") show why, and provide the necessary qualifications. Though health remains a high goal, tacitly sought and explicitly desired, it is difficult to attain and preserve. Following all the rules can guarantee neither good health nor good physicianship. Even if health were the only or the highest goal that we naturally seek, we can attain it but provisionally and temporarily, for we are finite and frail. Even if health is an enduring idea, each person's particular embodiment of it is not. Medicine thus finds itself in-between: the physician is called to serve the high and universal goal of health while also ministering to the needs and sufferings of the frail and particular. The task of weighing the claims of the high and the needy is the work of prudence, the cardinal virtue of the true physician, as it is the cardinal virtue of all practical men. Indeed, medicine, properly understood, turns out to be the very model of human moral activity: activity in pursuit of a genuine and worthy good, in the face of unavoidable impediments to its attainment, requiring the virtues of firmness in support of aspiration for the good and (as in the tempering of justice with mercy) patient understanding and sympathy in support of the ones who fall short. Medicine, properly understood, also provides an attractive model for the moral relation between knowledge or expertise and the concerns of life: The doctor, because of his knowledge of what is best and how to attain it, cannot be made the mere servant of the patient's wishes. But neither does his expertise entitle him to be a despot, for he is himself in the service of the natural powers of healing and the goal of health. Moreover, it is the particularized needs and concerns of his patients, whose mortal condition he also shares, that restrain and moderate any possible claims of supremacy through knowledge. In the best case, the doctor and patient are a partnership, albeit asymmetric, with shared goals. The patient is more than stuff; the doctor is less than God—or, rather, more than just mind and art and power. Both share in the vitality and aspirations, as well as the frailty and disappointments, of our mindfully embodied life.

The last chapter in Part II, "Is There a Medical Ethic?: The Hippocratic Oath and the Sources of Ethical Medicine," begins from the intuition that medicine is necessarily a moral activity and explores the question of whether or not medical ethics has its origins in medicine, or, more precisely, in the understanding of nature, man, and healing that provides the philosophical basis of medicine. In its form largely a commentary on and a defense of the venerable Hippocratic Oath, the chapter argues that medicine once did and still can understand that its ethical principles grow out of medical (or natural) roots. In fact, once again, the traditional understanding of the healing activity and the healing relation is seen to illuminate much of our moral life.

If the theory of medicine can withstand the theoretical challenges posed by the new biology, if medicine still is nature served rather than nature mastered, and if medicine tacitly knows things about our nature and our life that our biology cannot support, we are invited to seek for a richer and more adequate biology (and psychology and anthropology) that will serve as the ground for the medical—and perhaps also the moral. If medicine knows, but biology denies, that nature in living bodies is not neutral to the difference between health and disease, then we should try to find a more adequate account of nature, one that will affirm about life what life knows about itself. The last group of chapters, "Deepening the Ground: Nature Reconsidered," begins an effort toward that end.

A proper philosophy of nature would seek and explore the biological ground of those attributes, capacities, and activities that seem to the unprejudiced observer to characterize all or some living things: wholeness, self-maintenance, purposiveness, organic form, finitude, inwardness, identity, neediness, aspiration, locomotion, individuality, sociality, awareness, display, beauty, perpetuation, lineage, freedom, morality, artfulness and playfulness, and the concern with what is true and good. It would also consider man in relation to the rest of living nature, taking account of both his continuity of descent from and hence his kinship to the animals, as well as his irreducible difference. The four chapters in Part III offer no such comprehensive presentation, but they do consider certain carefully selected and important themes, and suggest ways to fruitful and more thoroughgoing inquiries. The first, "Teleology, Darwinism, and the Place of Man: Beyond Chance and Necessity?," explores the question of purpose in living nature in the light of the fact of evolution and the Darwinian account of its workings. The ascent and place of man is also considered in the course of examining the question of hierarchy in nature—that is, whether "higher" animals are really higher. As in the beginning of Part II, the conclusions of this chapter are provocatively, if too unqualifiedly, cheerful: Organisms are indeed naturally teleological; nature is hierarchic and tends, at least in part, toward the emergence and growth of higher powers of freedom, awareness, and self-awareness, now culminating in man.

Man is the explicit subject of the next chapter, "Thinking About the Body," a meditation on the nature and meaning of the human bodily form. It seeks to supplement, and thereby correct, science's reductionistic understanding of bodily life by showing how certain apparently superficial aspects of the body are in fact deeply revealing of human being—and its irreducible perplexities. Man's high standing among living things, as the thinking animal, is shown to be implicit in his upright posture and the associated changes it entails. Yet though separate as thoughtful, in need and dependence man is also no higher than equal, as a reflection on his nakedness discloses. Thinking about the body, and its dual intimations, also induces wonder about the relation between thought and embodiment, and also invites consideration of the proper treatment of the body, both living and dead.

But if life naturally tends toward its own fullnesses and fulfillments, what are we to make of death? Mortality has long been regarded as a blot on the dignity of life, as evidence of nature's indifference to the needs and desires of living things. According to some views, death is the wage of human transgression, or at least a contradiction of the goodness that life is and seeks. Indeed, modern science has inherited this view of death; yet it responds neither with resignation nor hope of divine redemption, but instead with a call to arms, ultimately in pursuit of bodily immortality. "Mortality and Morality: The Virtues of Finitude" calls this project into question and makes a case for the benefits of mortality. Necessity—or the recognition of necessity—turns out to be the mother of aspiration toward the beautiful, the good, the transcendent.

How ought we to conduct ourselves, poised as we are between the high and the urgent, between the good we seek and the necessities that call? What kind of aspirations lead most toward self-fulfillment? How is self-fulfillment related to the needs and concerns of our community, those with whom we are compelled by necessity to live? What can nature itself teach us? The last chapter, "Looking Good: Nature and Nobility," suggests that nature is not silent on this subject. By examining the meaning of animal appearance, then the manifestly human phenomenon of blushing, and finally the complex and rich meaning of shame, it argues that nature itself invites us to a concern with looking—and being—good, for ourselves and toward others. This aspiration to nobility is one response (dare I say, a naturally sanctioned response?) to our being precariously "in-between." So, too, is "looking well," that goal of clear-sighted, far-seeing understanding of this wonderful and mysterious world of which we are lucky enough to be the self-conscious part.

PART I

Eroding the Limits
Troubles with the Mastery of Nature

CHAPTER ONE

The New Biology
What Price Relieving Man's Estate?

*Science bestowed immense new powers on man and at the same time
created conditions which were largely beyond his control. While he
nursed the illusion of growing mastery and exulted in his new trappings he became the sport and presently the victim of tides and cur-
rents, whirlpools and tornadoes, amid which he was far more helpless
than he had been for a long time.*

—Winston Churchill

May you live in interesting times.

—Ancient Chinese curse

Recent advances in biology and medicine suggest that we may be rapidly acquiring the power to modify and control the capacities and activities of men by direct intervention and manipulation of their bodies
and minds. Certain means are already in use or at hand, others await
the solution of relatively minor technical problems, while yet others,
those offering perhaps the most precise kind of control, depend upon
further basic research. Biologists who have considered these matters
disagree on the question of how much how soon, but all agree that the
power for "human engineering," to borrow from the jargon, is coming
and that it will probably have profound social consequences.

These developments have been viewed both with enthusiasm and
with alarm, but only recently have they started to receive serious attention. Several biologists have undertaken to inform the public about
the technical possibilities, present and future. Practitioners of social
science "futurology" have begun to predict and describe the likely social consequences of and public responses to the new technologies. New
institutions, public and private, have been established to assess the
new technologies. All of these activities are based upon the hope that
we can harness the new technology of man for the betterment of mankind.

Yet this commendable aspiration points to another set of ques-

tions that are in my view sorely neglected—questions that inquire into
the meaning of such phrases as the "betterment of mankind." A *full*
understanding of the new technology of man requires an exploration
of ends, principles, and standards. What ends will or should the new
techniques serve? What principles should guide society's adjustments? By what standards of better and worse should the assessment
agencies assess? Behind these questions are others: what is a good
man; what is a good life for man; what is a good community?

While these questions about ends and ultimate ends are never un-
important or irrelevant, they have rarely been more important or more
relevant than they are now. That this is so can be seen once we recognize that we are here dealing with a group of technologies that are
in a decisive respect unique in that the object upon which they operate
is man himself. The technologies of energy or food production, of communication, of manufacture, and of motion greatly alter the imple-
ments available to man and the conditions under which he uses them.
In contrast, the biomedical technology works to change the user himself. To be sure, the printing press, the automobile, the television, and
the jet airplane have greatly altered the conditions under and the way
in which men live, but men as biological beings have remained largely
unchanged. They have been and remain able to accept or reject, to use
and abuse these technologies; they choose, whether wisely or foolishly,
the ends to which these technologies are means. Biomedical technology may make it possible to change the inherent capacity for choice
itself. Indeed, both those who welcome and those who fear the advent
of human engineering ground their hopes and fears in the same prospect, that *man can, for the first time, re-create himself.*

Engineering the engineer seems to differ in kind from engineering
his engine. Some have argued, however, that biomedical engineering
does not differ qualitatively from toilet training, education, government, law, and moral teachings—all of which are forms of so-called
social engineering with man as their object, used by one generation to
mold the next. In reply, it must at least be said that the techniques
that have hitherto been employed are feeble and inefficient when compared to those on the horizon. This quantitative difference rests in
part on a qualitative difference in the means of intervention. The traditional influences operate by speech or by symbolic deeds. They pay
tribute to man as the animal who lives by speech and who understands
the meanings of actions. Also, their effects are, in general, reversible,
or at least subject to attempts at reversal. Each person has greater
or lesser power to accept or reject or abandon them. Biomedical engineering, on the other hand, circumvents the human context of speech
and meaning, bypasses choice, and goes directly to work to modify

the human material itself, and the changes wrought may be irreversible.

There is also an important practical reason for considering biomedical technology apart from the other technologies. The advances we shall examine are fruits of a large, humane project dedicated to the conquest of disease and the relief of human suffering. The biologist and physician, regardless of their private motives, are seen with justification to be the well-wishers and benefactors of mankind. Thus, in a time in which technological advance is more carefully scrutinized and increasingly criticized, biomedical developments are still viewed largely as benefits without qualification. The price we pay for these developments is thus more likely to go unrecognized. For this reason, I shall consider only the dangers and costs of biomedical advance. As the benefits are well known, there is no need to dwell upon them here. *My discussion is, in this regard, deliberately partial.*

Yet it is, in another sense, also comprehensive. Though we are compelled in practice to consider the implications of each technological innovation as it arises—and I myself do so in the next four chapters—we can understand its full human significance only if we attend also to the whole biomedical project of which it is a part. In this opening chapter I seek such a synoptic view. I begin with a survey of the pertinent technologies. Next, I consider some of the basic ethical and social problems in the use of these technologies. Then, I briefly raise some fundamental questions to which these problems point. Finally, I offer some very general reflections on what is to be done.

The Biomedical Technologies

Biomedical technologies can be usefully organized into three groups according to their major purpose: (1) control of death and life; (2) control of human potentialities; and (3) control of human achievement. The corresponding technologies are: medicine, or those parts of medicine engaged in prolonging life and controlling reproduction; genetic engineering; and neurological and psychological manipulation. I shall briefly summarize each group of techniques.

Control of Death and Life

Previous medical and public health triumphs have greatly increased man's average life expectancy. Yet other developments, such as organ transplantation or replacement and research into aging, hold forth the promise of increasing not just the average but also the maximum life expectancy. Indeed, medicine seems to be sharpening its tools to do battle with death itself, treating death as if it were just one more disease.

More immediately and concretely, available techniques of life prolongation—respirators, cardiac pacemakers, artificial kidneys—are already in use in the fight against death—though, ironically, the success of these devices has introduced confusion in determining when death has indeed occurred. The traditional signs of life—heartbeat and respiration—can now be maintained entirely by machines. As a result, most physicians have adopted so-called new definitions of death, although others more radically maintain that the technical advances have shown that death is not a concrete event at all, but rather a gradual process—like twilight—incapable of precise temporal localization.[1]

The real challenge to death will come from research into aging and senescence, a field just entering its own puberty. Recent studies suggest that aging is a manipulable process, distinct from disease, under biological control but alterable by diet or drugs. Extrapolating from animal studies, some scientists have suggested that a decrease in the rate of aging might also be achieved simply by effecting a very small decrease in human body temperature. According to some optimistic estimates, it may soon be technically possible to add from twenty to forty extra years to the human life span.*

Medicine's success in extending life is already a major cause of excessive population growth: death control points to birth control. Although we are already technically competent, new techniques for lowering fertility and chemical agents for inducing abortion will greatly enhance our powers over conception and gestation. But problems of definition have been raised here as well. The need to determine when individuals acquire enforceable legal rights gives society an interest in the definition of human life and the time when it begins. These matters are too familiar to need elaboration.

Technologies to conquer infertility proceed alongside those to promote it. The first successful laboratory fertilization of a human egg by human sperm was reported in 1969,[2] and by a year later human embryos could be grown *in vitro* up to at least the blastocyst stage (that is, to the age of one week).[3] In 1978, after many failures, such a laboratory-grown embryo was successfully reimplanted into a woman previously infertile because of oviduct disease, and the first "test-tube baby" was born, an achievement now repeated many times in many places.

---

*A discussion of this prospect is the point of departure for Chapter 12, "Mortality and Morality: The Virtues of Finitude."

Work continues on techniques to support and sustain embryonic and fetal growth in artificial environments. The development of an artificial placenta, now under investigation, will make possible full laboratory control of fertilization and gestation. In addition, sophisticated biochemical and cytological techniques for monitoring the "quality" of the fetus have been developed and are being used.* These developments not only give us more power over the generation of human life, they also make it possible to manipulate and modify the quality of the human material.

Control of Human Potentialities

Genetic engineering, when fully developed, will wield two powers not shared by ordinary medical practice. Medicine treats existing individuals and seeks to correct deviations from a norm of health. Genetic engineering, in contrast, will be able to make changes that are transmittable into succeeding generations and may be able to create new capacities and, hence, new norms of health and fitness.

For now, however, the primary interest in human genetic manipulation remains strictly medical: to develop treatments for individuals with inherited diseases. Genetic disease is prevalent and increasing, thanks partly to medical advances that have enabled those affected to survive and perpetuate their mutant genes. The hope is that normal copies of the appropriate gene, obtained biologically or synthesized chemically, can be introduced into defective individuals to correct their deficiencies. While this *therapeutic* use of genetic technology is still probably some years away, astounding progress has been made in the past decade in sequencing, synthesizing, and transferring genetic material—thanks to new techniques for DNA recombination (or gene-splicing). Yet, there is some doubt that gene therapy will ever be practical because of difficulties in delivering the therapeutic gene precisely and specifically to the desired bodily target and because the same end might be more easily achieved by simply transplanting cells or organs that could compensate for the missing or defective gene product.

Far less remote are technologies that could serve *eugenic* ends. Their development has been endorsed by those concerned about a general deterioration of the human gene pool and by others who believe that even an undeteriorated human gene pool needs upgrading. Artificial insemination with selected donors, the eugenic proposal of Herman Muller,[4] has been possible for several years due to the perfection

---

*For a discussion of these developments, see Chapter 3, "Perfect Babies: Prenatal Diagnosis and the Equal Right to Life."

of methods for long-term storage of human spermatozoa. At least one commercial sperm bank (in California) now overtly boasts a eugenic intention, offering interested women the semen of several Nobel Laureates and other high achievers. The successful maturation of human oocytes in the laboratory and their subsequent fertilization now make it possible to select donors of ova as well. But a far more suitable technique for eugenic purposes may soon be upon us: nuclear transplantation, or cloning.

Bypassing the lottery of sexual recombination, nuclear transplantation permits the asexual reproduction or copying of an already developed individual. The nucleus of a mature but unfertilized egg is replaced by a nucleus obtained from a specialized cell of an adult organism or embryo (e.g., an intestinal cell, a skin cell). The egg, with its transplanted nucleus, develops as if it had been fertilized and, barring complications, may give rise to a normal adult organism. Since almost all the hereditary material (DNA) of a cell is contained within its nucleus, the renucleated egg and the individual into which it develops are genetically identical to the adult organism that was the source of the donor nucleus. Cloning could be used to produce sets of unlimited numbers of genetically identical individuals, each set derived from a single parent. Cloning has been successful in amphibians and is now being tried in mice; its extension to man merely requires the solution of certain technical problems.*

Production of man-animal chimeras by the introduction of selected nonhuman material into developing human embryos is also expected. Fusion of human and nonhuman cells in tissue culture has already been achieved. The scientific grapevine also reports attempts (thus far unsuccessful), using artificial fertilization, to cross human egg or sperm with sperm or egg of other primates.

Other, less direct means for influencing the gene pool are already available, thanks to our increasing ability to identify and diagnose genetic diseases. Genetic counselors can now detect biochemically and cytologically a variety of severe genetic defects (e.g., Down's syndrome, Tay-Sachs disease) while the individual is still a fetus *in utero*. Since treatments are at present largely unavailable, diagnosis is often followed by abortion of the affected fetus. With some diseases, more sensitive tests will also permit the detection of heterozygotes (i.e., unaffected individuals who carry a single dose of a given deleterious gene). The eradication of a given genetic disease might then be attempted by aborting all such carriers. In fact, it has been suggested

---

*The best-selling book, published a few years back, that reported the first successful human cloning was and is regarded as a hoax. For a fuller treatment of the ethical issues, see Chapter 2, "Making Babies: The New Biology and the 'Old' Morality."

that cystic fibrosis, a fairly common genetic disease, could be completely eliminated over the next forty years by screening all pregnancies and aborting the 17 million unaffected fetuses that will carry a single gene for this disease. Such zealots need to be reminded of the consequences should each geneticist be allowed an equal assault on his favorite genetic disorder, given that each human being is a carrier for some four to eight such recessive lethal genetic diseases.

Control of Human Achievement

Although human achievement depends at least in part upon genetic endowment, heredity determines only the material upon which experience and education impose the form. The limits of many capacities and powers of an individual are indeed genetically determined, but the nurturing and perfection of these capacities depends upon other influences. Neurological and psychological manipulation hold forth the promise of controlling the development of human capacities—in particular, those long considered most distinctively human: speech, thought, choice, desire, emotion, memory, and imagination.

These techniques are in a rather primitive state at present because we understand little about the brain and mind. Nevertheless, we have already seen the use of electrical stimulation of the human brain to produce sensations of intense pleasure and to control rage, the use of brain surgery (e.g., frontal lobotomy) for the relief of severe anxiety, and the use of aversive conditioning with electric shock to treat sexual perversion. Operant-conditioning techniques are widely used, apparently with success, in schools and mental hospitals. The use of so-called consciousness-expanding, euphoriant, and hallucinogenic drugs is widespread, to say nothing of tranquilizers and stimulants. We are promised drugs to modify memory, intelligence, libido, and aggressiveness.

From its inception, modern science has been especially interested in finding reliable biological means—means more effective than exhortation or praise and blame—to attain the ends of sensible, decent human conduct and peace of mind. Insisting on the mind's acute "dependence on the temperament and disposition of the bodily organs," René Descartes projected a new medicine, based upon what one might call "psychophysics," that would render mankind more prudent and capable than ever before. This dream—or nightmare—may soon come true, as the following passages from a book by neurophysiologist José Delgado—a book instructively entitled *Physical Control of the Mind: Toward a Psychocivilized Society*—make evident. In the early 1950s, it was discovered that with electrodes placed in certain discrete regions of their brains, animals would repeatedly and indefatigably press levers to stimulate their own brains with obvious resultant enjoyment. Even starving animals preferred stimulating these so-called pleasure centers to eating. Delgado comments on the electrical stimulation of a similar center in a human subject:

> [T]he patient reported a pleasant tingling sensation in the left side of her body "from my face down to the bottom of my legs." She started giggling and making funny comments, stating that she enjoyed the sensation "very much." Repetition of these stimulations made the patient more communicative and flirtatious, and she ended by openly expressing her desire to marry the therapist.[5]

Delgado sees no reason to be alarmed or troubled by these electrifying prospects:

> Leaving wires inside of a thinking brain may appear unpleasant or dangerous, but actually the many patients who have undergone this experience have not been concerned about the fact of being wired, nor have they felt any discomfort due to the presence of conductors in their heads. Some women have shown their feminine adaptability to circumstances by wearing attractive hats or wigs to conceal their electrical headgear, and many people have been able to enjoy a normal life as outpatients, returning to the clinic periodically for examination and stimulation. In a few cases in which contacts were located in pleasurable areas, patients have had the opportunity to stimulate their own brains by pressing the button of a portable instrument, and this procedure is reported to have therapeutic benefits.[6]

Progress on the pharmacological front, though yielding less spectacular images, has been even more impressive—and worrisome. Ritalin and other drugs have been widely used in schools to calm hyperactive (and other, just troublesome?) children. Anabolic steroids (and numerous other so-called performance-enhancing drugs) are taken by athletes (as well as other performers). The discovery of a large number of natural psychoactive substances in the brain—including the encephalins, peptides with opiumlike properties—brings us much closer to a perfected pharmacology of pleasure and other precise desire and mood modifying powers. And new immunochemical methods for delivering drugs selectively to precise target areas in the brain promise a sophisticated technology to put these new chemically based powers to work in the proper place, without the need for brain surgery. It bears repeating that the sciences of neurophysiology and psychopharmacology are in their infancy. The techniques that are now available are crude, imprecise, weak, and unpredictable compared to those that may flow from a more mature neurobiology.

Basic Ethical and Social Problems in the Use
of Biomedical Technology

After this cursory review of the powers now and soon to be at our disposal, I turn to the questions concerning the use of these powers. First, we must recognize that questions of *use* of science and technology are also moral and political, never simply technical. All private or public decisions to develop or to use biomedical technology, as well as the decisions *not* to do so, inevitably contain judgments of good and bad, better and worse—what our jargon aseptically calls "values." This is true even if the values guiding those decisions are not articulated or made clear, as indeed they often are not. Second, the standards of better and worse cannot be derived from biomedical science. This is true even if scientists themselves make the decisions.

These important points are often overlooked for at least three reasons. (1) They are obscured by those who like to speak of "the control of nature by science." It is men who are in control, not that abstraction "science." Science may provide the means, but men choose the ends, and the choice of ends comes from beyond science. (2) Introduction of new technologies often appears to be the result of no decision whatsoever, or of the culmination of decisions too small or unconscious to be recognized as such. Fate seems to hold the reins: What can be done is done. But technological advance is *not* automatic. Someone is deciding on the basis of some notions of desirability, no matter how self-serving or altruistic. (3) Desires to gain or keep money and power—allegedly unreasoning and subrational—are seen to influence much of what happens. But these desires can in fact be formulated as reasons, and then discussed and debated.

Insofar as our society has tried to deliberate about questions of use, how has it done so? Pragmatists that we are, we prefer a utilitarian calculus: We weigh "benefits" against "risks" and we weigh them both for the individual and for society. We often ignore the fact that the very definitions of a benefit and a risk are themselves open to dispute and, in every case, based upon judgments and notions of "good" and "bad." In the biomedical areas just reviewed the benefits are considered to be self-evident: prolongation of life; control of fertility and population size; treatment and prevention of genetic disease; the reduction of anxiety and aggressiveness; the enhancement of memory, intelligence, and pleasure. The assessment of risk is in general simply pragmatic: Will the technique work effectively and reliably; how much will it cost; will it do detectable bodily harm; who will complain if we proceed with development? As these questions are familiar and congenial, there is no need to belabor them.

But the very pragmatism that makes us sensitive to considerations of economic cost often blinds us to the larger social costs exacted by biomedical advances. We seem to be unaware that we may not be able to maximize all the benefits, that several of the goals we are promoting conflict with each other. On the one hand, we seek to control population growth by lowering fertility; on the other, we develop techniques to enable every infertile woman to bear a child. On the one hand, we try to extend the lives of individuals with genetic disease; on the other, we wish to eliminate deleterious genes from the human population. I am not urging that we resolve these conflicts in favor of one side or the other, but simply that we recognize that such conflicts exist. Once we do, we are more likely to appreciate that most "progress" is heavily paid for in terms not generally included in the simple utilitarian calculus.

To become sensitive to the larger costs of biomedical progress, we must attend to several serious ethical and social questions. I will briefly discuss three of them: questions of distributive justice; questions of the use and abuse of power; and questions of self-degradation and dehumanization.

Distributive Justice

The introduction of any biomedical technology presents a new instance of an old problem: How to distribute scarce resources justly. We should assume that demand will usually exceed supply. Which people should receive a kidney transplant or an artificial heart? Who should get the benefits of genetic therapy or of brain stimulation? Is "first-come first-served" the fairest principle? Or are certain people "more worthy," and if so, on what grounds?

It is unlikely that we will arrive at answers to these questions in the form of deliberate decisions. More likely, the problem of distribution will continue to be decided *ad hoc* and locally. If this is so, the consequence will probably be a sharp increase in the already far too great inequality of medical care. The extreme case will be longevity, in the beginning obtainable probably only at great expense. Who is likely to be able to buy it? Do conscience and prudence permit us to enlarge the gap between rich and poor, especially with respect to something as fundamental as life itself?

Questions of distributive justice also arise in the prior decisions to acquire new knowledge and to develop new techniques. Personnel and facilities for medical research and treatment are scarce resources. Is the development of a new technology the best use of the limited resources given current circumstances? How should we balance efforts aimed at cure against those aimed at prevention, or either of these against efforts to redesign the species? How should we balance the delivery of available levels of care against further basic research? More fundamentally, how should we balance efforts in biology and medicine against efforts to reduce poverty, pollution, urban decay, discrimination, and poor education? This last question is perhaps the most profound. We should seriously reflect upon the social consequences of seducing many of our brightest young people into spending their lives locating the biochemical defects in rare genetic diseases while our more serious problems go begging. The current squeeze on money for research provides us with an opportunity to rethink and reorder our priorities.

Problems of distributive justice are frequently mentioned and discussed, but they are hard to resolve in a rational manner. *We* find them especially difficult because of the enormous range of conflicting "values" and interests that characterizes our pluralistic society. We cannot agree—we usually do not even try to agree—on standards for just distribution. Rather, decisions tend to be made largely out of a clash of competing interests.* Thus, regrettably, the question of *"how* to distribute *justly"* often gets reduced to *"who* shall *decide* how to distribute." Our question about justice has led us to the question about power.

Use and Abuse of Power

We have difficulty recognizing the problems of the exercise of power in the biomedical enterprise because of our delight with the wondrous fruits it has yielded. This is ironic because of the cardinal place of power in the modern conception of science. The ancients conceived of science as the *understanding* of nature, pursued for its own sake. We moderns view science as power, as *control* over nature; the conquest of nature "for the relief of man's estate" was the charge issued by Francis Bacon, one of the leading architects of the modern scientific project.[7]

Another source of our difficulty is our fondness for speaking of the abstraction "man." I suspect that we prefer to speak figuratively about "Man's power over Nature" because it obscures an unpleasant reality about human affairs. It is in fact particular men who wield power, not man. What we really mean by "Man's power over Nature"

---

*This is, of course, the way the American political system is intended to work, institutional arrangements (e.g., balance of power, checks and balances, multiplication of factions) taking the place of the difficult-to-find superior wisdom of men.

is that some men, with knowledge of nature as their instrument, exercise power over other men.

While applicable to technology in general, these reflections are especially pertinent to the technologies of human engineering with which men deliberately exercise power over future generations. An excellent discussion of this question is found in *The Abolition of Man* by C. S. Lewis:

> It is, of course, a commonplace to complain that men have hitherto used badly, and against their fellows, the powers that science has given them. But that is not the point I am trying to make. I am not speaking of particular corruptions and abuses which an increase of moral virtue would cure: I am considering what the thing called 'Man's power over Nature' must always and essentially be. . . .
>
> In reality, of course, if any one age really attains, by eugenics and scientific education, the power to make its descendants what it pleases, all men who live after it are the patients of that power. They are weaker, not stronger: for though we may have put wonderful machines in their hands we have pre-ordained how they are to use them. . . . The real picture is that of one dominant age . . . which resists all previous ages most successfully and dominates all subsequent ages most irresistibly, and thus is the real master of the human species. But even within this master generation (itself an infinitesimal minority of the species) the power will be exercised by a minority smaller still. Man's conquest of Nature, if the dreams of some scientific planners are realized, means the rule of a few hundreds of men over billions upon billions of men. There neither is nor can be any simple increase of power on Man's side. Each new power won *by* man is a power *over* man as well. Each advance leaves him weaker as well as stronger. In every victory, besides being the general who triumphs, he is also the prisoner who follows the triumphal car.[8]

Please note that I am not yet speaking about the problem of the misuse or abuse of power. The point is rather that the power which grows is unavoidably the power of only some men, and that the number of these powerful men tends to grow fewer and fewer as the power increases.

Specific problems of abuse and misuse of specific powers must not, however, be overlooked. Some have voiced the fear that the technologies of genetic engineering and behavior control, though developed for good purposes, will be put to evil uses. These fears are perhaps somewhat exaggerated, if only because biomedical technologies would add very little to our highly developed arsenal for mischief, destruction, and stultification. Nevertheless, any proposal for large-scale human engineering should make us wary. Consider a program of positive eugenics based upon the widespread practice of asexual reproduction. Who shall decide what constitutes a superior individual worthy of replication? Who shall decide which individuals may or must reproduce, and by which method? These are questions easily answered only for a tyrannical regime.

And about tyrannical regimes our century cannot exaggerate. The desire of men to dominate and subjugate other men is, no doubt, coeval with the race. But tyranny in our time has been, shall we say, elevated to a science, thanks in no small part to the marriage of science and politics, as well as to the utopian dreams unleashed by the project for the mastery of nature. Here the most relevant sciences concern communication, surveillance, and the susceptibility of the human mind to manipulation by terror, propaganda, pain, drugs, sensory deprivation, and other devices for controlling memory, thought, and feeling.

The conditions that prepare modern tyranny are notoriously complex; but it seems fair to say that Hitler's grip on the minds of the German people would not have been possible without loudspeakers and the insights, however primitive, of mass psychology and propaganda, not to mention the sophisticated techniques of torture and intimidation. But it is the Soviets and their compatriots who have achieved a stable and efficient tyranny, seemingly impregnable and thus unrivaled in human history, not least because they are armed with sophisticated psychological and organizational techniques and elaborate devices for controlling the flow of information. Imagine what they will do with a really developed science of human psychology or neurobiology.

Concern about the use of power is equally necessary in the selection of means for desirable or agreed-upon ends. Consider the desired end of limiting population growth. An effective program of fertility control is likely to be coercive, if not by directly applying brute force then by imposing powerful sanctions (as in contemporary China) or by making offers that cannot be refused. Who should decide the choice of means? Will the program penalize conscientious objectors?

Serious problems arise simply from obtaining and disseminating information, as in mass screening programs now being proposed for detection of genetic diseases. For what kinds of disorders is compulsory screening justified? Who shall have access to the data obtained, and for what purposes? To whom does information about a person's genotype belong? In ordinary medical practice, the patient's privacy is protected by the doctor's adherence to the principle of confidentiality. What will protect his privacy under conditions of mass screening?

More than privacy is at stake if screening is undertaken to detect psychological or behavioral abnormalities. Some years ago, a highly placed psychologist-advisor to the Nixon administration called for the psychological testing of all six-year-olds to detect future criminals and misfits. The proposal was rejected because current tests lacked the requisite predictive powers. But would such a proposal have been rejected if reliable tests had been available? Will they be in the future? What if certain genetic disorders, diagnosable in childhood, can be shown to correlate with subsequent antisocial behavior? For what degree of correlation and for what kinds of behavior can mandatory screening be justified? What use should be made of the data? Might not the dissemination of the information itself undermine the individual's chance for a worthy life and, therefore, contribute to his so-called antisocial tendencies?

Consider the seemingly harmless effort to redefine clinical death. The primary stimulus to seek a new definition of death was the need for organs for transplantation.[9] Was it not possible for this concern to influence the definition at the expense of the dying? One physician, in fact, calls the revised criteria for declaring a patient dead a "new definition of heart donor eligibility."[10]

Problems of abuse of power arise even in the acquisition of basic knowledge. The securing of a voluntary and informed consent is an abiding difficulty in the use of human subjects in experimentation. Gross coercion and deception are now rarely seen; the pressures are generally subtle, often related to an intrinsic power imbalance in favor of the experimentalist.

A special problem arises in experiments on or manipulations of the unborn. Here it is impossible to obtain consent of the human subject. If the purpose of the intervention is therapeutic (e.g., to correct a known genetic abnormality), consent can reasonably be implied. But can anyone ethically consent for nontherapeutic interventions in which parents or scientists work their wills or their eugenic visions on the child-to-be? Would not such manipulation represent in itself an abuse of power, independent of consequences?

There are many clinical situations that already permit, if not invite, the manipulative or arbitrary use of power provided by biomedical technology: obtaining organs for transplantation; refusing to let a person die with dignity; giving genetic counseling to a frightened couple; recommending eugenic sterilization for a mental retardate; ordering electric shock for a homosexual. In each situation there is an opportunity to violate the will of the patient or subject. Such opportunities have generally existed in medical practice, but the dangers are becoming increasingly serious. With the growing complexity of the technologies, the technician gains in authority because he alone understands what he is doing. The patient's lack of knowledge makes him deferential and often inhibits him from speaking up when he feels threatened. Physicians *are* sometimes troubled by their increasing power, yet they feel they cannot avoid its exercise. "Reluctantly," one commented to me, "we shall have to play God." With what guidance and to what ends I shall consider later. For the moment, I merely ask: By whose authority?

While these questions about power are pertinent, and I think important, they are in one sense misleading. They imply an inherent conflict of purpose between physician and patient, between scientist and citizen. The discussion conjures up images of master and slave, of oppressor and oppressed. Yet it must be remembered that conflict of purpose is largely absent, especially with regard to general goals. To be sure, the purposes of medical scientists are not always the same as those of the experimental subjects. Nevertheless, the basic sponsors and partisans of biomedical technology are precisely those upon whom the technology will operate. The will of the scientist and physician is the offspring of the desire of all of us for better health, longer life, and peace of mind.

Most future biomedical technologies will probably be welcomed, as have those of the past, and their use will require little or no coercion. Some developments, such as pills to improve memory, control moods, or induce pleasure, are likely to need no promotion. Thus, even if we should escape from the dangers of coercive manipulation, we shall still face large problems posed by the voluntary use of biomedical technology, problems to which I now turn.

Voluntary Self-Degradation and Dehumanization

Modern liberal opinion is sensitive to problems of restriction of freedom and abuse of power. Indeed, many hold that a man can be injured only by violating his will, but this view is much too narrow. It fails to recognize the great dangers we shall face in the uses of biomedical technology that stem from an excess of freedom, from the uninhibited exercise of will. In my view, our greatest problem—and one that will continue to grow in importance—will be voluntary self-degradation, or willing dehumanization—dehumanization not directly chosen, to be sure, but dehumanization nonetheless—as the unintended yet often inescapable consequence of relentlessly and *successfully* pursuing our humanitarian goals.

Certain desired and perfected medical technologies have already had some dehumanizing consequences. Improved methods of resuscitation have made possible heroic efforts to "save" the severely ill and injured. Yet these efforts are sometimes only partly successful: They may succeed in salvaging individuals, but these individuals may have severe brain damage and be capable of only a less-than-human, vegetating existence. Such patients, found with increasing frequency in the intensive care units of university hospitals, have been denied a death with dignity. Families are forced to suffer seeing their loved ones so reduced, and are made to bear the burden of a protracted "death watch."

Even the *ordinary* methods of treating disease and prolonging life have changed the context in which men die. Fewer and fewer people die in the familiar surroundings of home or in the company of family and friends. At that time of life when there is perhaps the greatest need for human warmth and comfort, the dying patient is kept company by cardiac pacemakers and defibrillators, respirators, aspirators, oxygenators, catheters, and his intravenous drip. Ties to the community of men are replaced by attachments to an assemblage of machines.

This loneliness, however, is not confined to the dying patient in the hospital bed. Consider the increasing number of old people still alive thanks to medical progress. As a group, the elderly are the most alienated members of our society: Not yet ready for the world of the dead, not deemed fit for the world of the living, they are shunted aside. More and more of them spend the extra years medicine has given them in "homes for senior citizens," in hospitals for chronic diseases, and in nursing homes—waiting for the end. We have learned how to increase their years, but we have not learned how to help them enjoy their days. Yet we continue to bravely and relentlessly push back the frontiers against death.

Paradoxically, even the young and vigorous may be suffering because of medicine's success in removing death from their personal experience. Those born since the discovery of penicillin represent the first generation ever to grow up without experience or fear of probable death at an early age. They look around and see that virtually all their friends are alive. A thoughtful physician, Dr. Eric Cassell, writes:

> [W]hile the gift of time must surely be marked as a great blessing, the *perception* of time, as stretching out endlessly before us, is somewhat threatening. Many of us function best under deadlines, and tend to procrastinate when time limits are not set. . . . Thus, this unquestioned boon, the extension of life, and the removal of the threat of premature death, carries with it an unexpected anxiety: the anxiety of an unlimited future.
>
> In the young, the sense of limitless time has apparently imparted not a feeling of limitless opportunity, but increased stress and anxiety, in addition to the anxiety which results from other modern freedoms: personal mobility, a wide range of occupational choice, and independence from the limitations of class and familial patterns of work. . . . A certain aimlessness (often ringed around with great social consciousness) characterizes discussions about their own aspirations. The future is endless, and

their inner demands seem minimal. Although it may appear uncharitable to say so, they seem to be acting in a way best described as "childish"— particularly in their lack of a time sense. They behave as though there were no tomorrow, or as though the time limits imposed by the biological facts of life had become so vague for them as to be nonexistent.[11]

Consider next the coming power over reproduction and genotype. We endorse the project that will enable us to control numbers and to treat individuals who have genetic diseases. But our desires outrun these defensible goals. Many would welcome the chance to become parents without the inconvenience of pregnancy, others would wish to know in advance the characteristics of their offspring (sex, height, eye color, intelligence), still others would wish to design these characteristics to suit their tastes. Some scientists have called for the use of the new technologies to assure the "quality" of all new babies.[12] As one obstetrician put it: "The business of obstetrics is to produce *optimum* babies." But the price to be paid for the optimum baby is the transfer of procreation from the home to the laboratory and its coincident transformation into manufacture. Increasing control over the product can only be purchased by the increasing depersonalization of the process. The complete depersonalization of procreation (possible with the development of an artificial placenta) shall be in itself seriously dehumanizing, no matter how optimum the product. It should not be forgotten that human procreation not only issues new human beings, but is also in itself a human activity. Would the laboratory production of human beings still be *human* procreation? Or would not the practice of making babies in laboratories—even perfect babies— mean a degradation of parenthood?*

The dehumanizing consequences of programmed reproduction extend beyond the mere acts and processes of giving life. Transfer of procreation to the laboratory will no doubt weaken what is for many people the best remaining justification and support for the existence of marriage and the family. Sex is now comfortably at home outside of marriage; child-rearing is progressively being given over to the state, the schools, the mass media, the child-care centers. Some have argued that the family, long the nursery of humanity, has outlived its usefulness. To be sure, laboratory and governmental alternatives might be designed for procreation and child-rearing. But at what cost?

This is not the place to conduct a full evaluation of the biological family. Nevertheless, some of its important virtues are, nowadays, too often overlooked. The family is rapidly becoming the only institution in an increasingly impersonal world where each person is loved not for

---

*These matters of dehumanization of generation and parenthood are discussed more fully in Chapter 2, "Making Babies: The New Biology and the 'Old' Morality."

what he does or makes, but simply because he is. The family is also the institution where most of us, both as children and as parents, acquire a sense of continuity with the past and a sense of commitment to the future. Without the family, we would have little incentive to take an interest in anything after our own deaths. These observations suggest that the elimination of the family would weaken ties to past and future, and would throw us, even more than we are now, on the mercy of an impersonal, lonely present.

Neurobiology and psychobiology probe most directly into the distinctively human. The technological fruit of these sciences is likely to be both more tempting than Eve's apple and more momentous in its result.* One need only consider contemporary drug use to see what people are willing to risk or sacrifice for novel experiences, heightened perceptions, or just "kicks." The possibility of drug-induced instant and effortless gratification will be welcomed—and one must not forget the possibilities of voluntary self-stimulation of the brain to reduce anxiety, to heighten pleasure, or to create visual and auditory sensations unavailable through the peripheral sense organs. Once these techniques are perfected and safe, is there much doubt that they will be desired, demanded, and used?

What ends will these techniques serve? Most likely, only the most elemental, those most tied to the bodily pleasures. What will happen to thought, to love, to friendship, to art, to judgment, to public-spiritedness in a society with a perfected technology of pleasure? What kinds of creatures will we become if we obtain our pleasure by drug or electrical stimulation without the usual kind of human efforts and frustrations? What kind of society will we have?

We need only consult Aldous Huxley's prophetic novel *Brave New World* for a likely answer to these questions.† There we encounter a

---

*Curiously, the implicit goal of biomedical technology—indeed, of the entire project for the conquest of nature—could well be said to be the reversal of the Fall, and a return of man to the hedonic and immortal existence of the Garden of Eden. Yet we can point to at least two difficulties. First, the new Garden of Eden will probably have no gardens; the received, splendid world of nature will be buried beneath asphalt, concrete, and other human fabrications, a transformation that is already far along. (Recall that in Aldous Huxley's *Brave New World*, elaborate consumption-oriented, mechanical amusement parks—featuring, e.g., centrifugal bumble-puppy—have supplanted wilderness and even ordinary gardens.) Second, the new inhabitant of the new "Garden" will have to be a creature for whom we have no precedent, a creature as difficult to imagine as to bring into existence. He will have to be simultaneously an innocent like Adam and a technological wizard who keeps the Garden running. (I am indebted to Robert Goldwin for this last insight.)

†Especially for those of us in the West, Huxley's novel must be regarded as more profound than its justly famous counterpart, George Orwell's *1984*. Orwell captured the prospect of tyranny made permanent thanks to the powers that science and technology help provide. He understood the enormous difficulty in restraining the appetite for power, once that appetite has been swollen by the gifts, present and promised, of the project for mastery over nature—especially when that project is fueled by utopian vi-

society dedicated to homogeneity and stability, administered by means of instant gratifications, and peopled by creatures of human shape but of stunted humanity. They consume, fornicate, take "soma," and operate the machinery that makes it all possible. They do not read, write, think, love, or govern themselves. Creativity and curiosity, reason and passion, exist only in a rudimentary and mutilated form. In short, they are not men at all.

True, our techniques, like theirs, may enable us to treat schizophrenia, to alleviate anxiety, and to curb aggressiveness. And we, like they, may be able to save mankind from itself, but it will probably be at the cost of our humanness. In the end, the price of relieving man's estate might well be the abolition of man.*

There are, of course, many other routes to the abolition of man, and there are many other and better-known causes of dehumanization. Disease, starvation, mental retardation, slavery, and brutality—to name just a few—have long prevented many, if not most, people from living a fully human life. We should work to reduce and, where possible, eliminate these evils. But their existence should not prevent us from appreciating the fact that the use of the technology of man, uninformed by wisdom concerning proper human ends, and untempered by an appropriate humility and awe, can unwittingly render us all irreversibly less than human. Unlike the man reduced by disease or slavery, the people dehumanized à la *Brave New World* are not miserable, don't know that they are dehumanized, and, what is worse, would not care if they knew. They are, indeed, happy slaves with a slavish happiness.

---

sions for the remaking of mankind. In contrast, Huxley deals with a far more interesting danger, ineradicably connected to the benevolent and humanitarian goals of relieving man's estate, *even in the absence of any misuse or abuse of power.* He shows us the soft and dehumanizing despotism that is the full meaning of that pursuit of peace, health, sanity, and contentment that nearly all of us seek. Orwellian evils are easily seen to be evils. Yet harder to combat, *because* harder to see *as* evil, are the evils imbedded in inadequate notions of good. These Huxley makes vivid, as he exposes the evils posed by the ostensibly generous well-wishers of mankind.

*Some scientists naively believe that an engineered increase in human intelligence will steer us in the right direction. It is surely true that intelligence often seems to be in short supply, and frequently in our complex world we find persons in authority who oversimplify questions, often grossly, to make them compatible with their own limited understanding. But, surely, we have also learned by now that intelligence, whatever it is and however measured, is not synonymous with wisdom and that, if harnessed to the wrong ends, it can cleverly perpetrate great folly and evil. Given the activities in which many, if not most, of our best minds are now engaged, should we simply rejoice in the prospect of enhancing I.Q.? On what would this increased intelligence operate? At best, the programming of further increases in I.Q. It would design and operate techniques for prolonging life, for engineering reproduction, for delivering gratifications. With no gain in wisdom, our gain in intelligence can only enhance the rate of our dehumanization. This will especially be true under conditions of modern thought, which regards reason or intellect as but a tool or instrument in the service of subrational ends (survival, pleasure, will to power), and itself incapable of knowing or discovering the good.

In fairness, to conclude this part of the discussion, we must concede immediately that Huxley presents us with an extreme, with a
fictional and futuristic picture—albeit a picture of the likely consequences of certain present tendencies and aspirations carried forward
to their logical conclusion. In our world, at least for now, necessity is
too much with us—including the necessity of opposing the danger of
harsh tyranny—for us to succumb quickly to the more fundamental
but soft despotism into which we might otherwise contentedly slide.
And, in any case, we are still far enough from that humanitarian hell
of *Brave New World* that most of us still comfortably hold the humanitarian vision of the technological project. Most of us, I am sure,
are still attached to an optimistic view: Technology, and the power it
provides, is, we concede, morally neutral, and the twin dangers of tyranny and voluntary degradation cannot be denied. But we can have
the benefits without suffering the harms if we exercise *moral* and *political control* over technology, if we guide our use of technology by
sound notions of humanity, justice, and human good. Sure, there will
always be practical difficulties, but with compassion, sincerity, good-
will, and charity we will in the long run and for the most part be able
to succeed. We can have our technological cake and eat it, freely and
with dignity.

I am sorry to disappoint you. Moral and political control requires
sound and firm moral and political beliefs and practices. But these
beliefs and practices are insecure. They, too, are threatened, as I soon
will show, but not so much from the biomedical technologies as from
the scientific teachings themselves. At stake is the very status and
possibility of morality altogether.

Some Fundamental Questions

The practical problems of distributing scarce resources, of curbing the
abuses of power, and of preventing voluntary dehumanization point
beyond themselves to some large, enduring, and most difficult questions: the nature of justice and the good community; the nature of man
and the good for man. My appreciation of the profundity of these questions and my own ignorance before them makes me hesitant to say
any more about them. Nevertheless, previous failures to find a short-
cut around them have led me to believe that these questions must be
faced if we are to have any hope of understanding where biology is
taking us. Therefore, I shall try to show in outline how I think some
of the larger questions arise from my discussion of dehumanization
and self-degradation.

My remarks on dehumanization can hardly fail to arouse argument. It might be said, correctly, that to speak about dehumanization presupposes a concept of the "distinctively human." It might also be said, correctly, that to speak about wisdom concerning proper human ends presupposes that such ends do in fact exist and that they may be more or less accessible to human understanding, or at least to rational inquiry. It is true that neither presupposition is at home in modern thought.

The notion of the distinctively human has been seriously challenged by modern scientists. Darwinists hold that man is, at least in origin, tied to the subhuman; his seeming distinctiveness is an illusion, or at most not very important. Biochemists and molecular biologists extend the challenge by blurring the distinction between the living and the nonliving. The laws of physics and chemistry are found to be valid and are held to be sufficient for explaining biological systems. Man is a collection of molecules, an accident on the stage of evolution, endowed by chance with the power to change himself, but only along determined lines.

Psychoanalysts have also debunked the distinctively human. To them, the essence of man is located in those drives he shares with other animals: pursuit of pleasure and avoidance of pain. The so-called higher functions are understood to be servants of the more elementary, the more base. Any distinctiveness or "dignity" that man has consists in his superior capacity for gratifying his animal needs.

The idea of "human good" fares no better. In the social sciences, historicists and existentialists have helped drive this question underground. The former hold all notions of human good to be culturally and historically bound, and hence mutable. The latter hold that values are subjective: each man makes his own, and ethics becomes simply the cataloguing of personal tastes.

Such appear to be the prevailing opinions. Yet there is nothing novel about reductionism, hedonism, and relativism; these are doctrines with which Socrates contended. What is new is that these doctrines seem to be vindicated by scientific advance. Not only do the scientific notions of nature and of man flower into verifiable predictions, but they yield marvelous fruit. The technological triumphs are held to validate their scientific foundations. Here, perhaps, is the most pernicious result of technological progress—more dehumanizing than any actual manipulation or technique, present or future. We are witnessing the erosion, perhaps the final erosion, of the idea of man as something splendid or divine, and its replacement with a view that sees man, no less than nature, simply as more raw material for manipulation and homogenization. Hence, our peculiar moral crisis. We are in turbulent seas without a landmark precisely because we adhere more and more to a view of nature and of man that both gives us enormous power and, at the same time, denies all possibility of standards to guide its use. Though well-equipped, we know not who we are nor where we are going. We are left to the accidents of our hasty, biased, and ephemeral judgments.

Let us not fail to note a painful irony: Our conquest of nature has made us the slaves of blind chance. We triumph over nature's unpredictabilities only to subject ourselves to the still greater unpredictability of our capricious wills and fickle opinions. That we have a method is no proof against our madness. Thus, engineering the engineer as well as the engine, we race our train we know not where.*

While the disastrous consequences of ethical nihilism are insufficient to refute it, they invite and make urgent a reinvestigation of the ancient and enduring questions of what is a proper life for a human being, what is a good community, and how is each to be achieved.† We must not be deterred from these questions simply because the best minds in human history have failed to settle them. Shouldn't we, rather, be encouraged by the fact that *they* considered them to be the most important questions?

As I have suggested before, our ethical dilemma is, at bottom, caused by the victory of modern natural science with its nonteleological view of man. We ought, therefore, to reexamine with great care

---

*The philosopher, Hans Jonas, has made the identical point:[13]

> Thus the slow-working accidents of nature . . . are to be replaced by the fast-working accidents of man's hasty and biased decisions, not exposed to the long test of the ages. His uncertain ideas are to set the goals of generations, with a certainty borrowed from the presumptive certainty of the means. The latter presumption is doubtful enough, but this doubtfulness becomes secondary to the prime question that arises when man indeed undertakes to "make himself": in what image of his own devising shall he do so, even granted that he can be sure of the means? In fact, of course, he can be sure of neither, not of the end, nor of the means, once he enters the realm where he plays with the roots of life. Of one thing only can he be sure:-of his power to move the foundations and to cause incalculable and irreversible consequences. Never was so much power coupled with so little guidance for its use.

Contrary to popular belief, we are not even on the right road toward a rational control over human nature and human life. It is indeed the height of irrationality triumphantly to pursue rationalized technique, while at the same time insisting that questions of ends, purposes, and better or worse lie beyond rational discourse.

†It is encouraging to note that these questions are seriously being raised in other quarters, say, by persons concerned with the decay of cities or the pollution of nature. There is a growing dissatisfaction with ethical nihilism. In fact, its tenets are unwittingly abandoned, by even its staunchest adherents, in any discussion of "what to do." For example, in the biomedical area, everyone, including the most unreconstructed and technocratic reductionist, finds himself speaking about the use of powers for human betterment. Yet he has wandered unawares onto ethical ground, for one cannot speak of human betterment without considering what is meant by the *human* and by the related notion of the *good for man.* These questions can be avoided only by asserting that practical matters reduce to tastes and power, and confessing that one's use of the phrase "human betterment"—and, indeed, all practical speech—is a deception to cloak one's own will to power. In other words, these questions can be avoided only by ceasing seriously to discuss.

the modern notions of nature and of man that undermine those earlier notions that provide a basis for ethics. If we consult our common experience, we are likely to discover some grounds for believing that the questions about man and human good are far from closed. Our common experience suggests many difficulties for the modern scientific view of man. For example, this view fails to account for the concern for justice and freedom that appears to be characteristic of all human societies. Consider the widespread acceptance, in a variety of cultures and legal systems, of the principle and practice of third-party adjudication of disputes. Also consider why no slave-holder has preferred his own enslavement to his own freedom. Some notions of justice and freedom, as well as right and truthfulness, it seems, are constitutive for any society, and a concern for these values may be a fundamental characteristic of human nature.

The scientific view of man also fails to account for or explain the fact that men have speech and not merely voice, that men can choose and act and not merely move or react. It fails to explain why men engage in moral discourse, or for that matter, why they speak at all. Finally, the scientific view of man cannot account for scientific inquiry itself, for why men seek to know. Might there not be something the matter with a knowledge of man that does not explain or take account of his most distinctive activities, aspirations, and concerns?*

Having gone this far, let me offer one suggestion as to where the difficulty might lie: in the modern understanding of knowledge. Since Bacon, as I have mentioned earlier, technology has increasingly come to be the basic justification for scientific inquiry. The end is power, and not knowledge for its own sake. And yet, power is not simply the end. It is also an important *validation* of knowledge. One definitely knows that one knows only if one can make. Synthesis is held to be the ultimate proof of understanding.† A more radical formulation holds that one knows only what one makes: knowing *equals* making.

Yet therein lies a difficulty. If truth be power to change or to make the object studied, then of what do we have knowledge? If there are no fixed realities but only material upon which we may work our wills, will not science be merely the knowledge of the transient and the manipulable? We might indeed have knowledge of the laws by which things

---

*Scientists may, of course, continue to believe in righteousness or justice or truth, but these beliefs are not grounded in their scientific knowledge of man. They rest instead upon the receding wisdom of an earlier age.

†This belief, silently shared by many contemporary biologists, has been given the following clear expression:[14]

> One of the acid tests of understanding an object is the ability to put it together from its component parts. Ultimately, molecular biologists will attempt to subject their understanding of all structure and function to this sort of test by trying to synthesize a cell. It is of some interest to see how close we are to this goal.

change and the rules for their manipulation, but no knowledge of the things themselves. Can such a view of science yield any knowledge about the nature of man, or indeed about the nature of anything? Our questions appear to lead back to the most basic of questions: What does it mean to know? What is it that is knowable?*

We have seen that the practical problems point toward and make urgent certain enduring fundamental questions.† Yet while pursuing these questions, we cannot afford to neglect the practical problems as such. Let us not forget Dr. Delgado and the psychocivilized society. The philosophical inquiry could be rendered moot by our blind confident efforts to dissect and redesign ourselves. While awaiting a reconstruction of theory, we must act as best we can.

What Is to Be Done?

First, we sorely need to recover some humility in the face of our awesome powers. The arguments I have presented should make apparent the folly of arrogance, of the presumption that we are wise enough to remake ourselves. Because we lack wisdom, caution is our urgent need.

---

*When an earlier version of this essay was presented publicly, it was criticized by one questioner as being "antiscientific." He suggested that my remarks "were the kind that gave science a bad name." He went on to argue that, far from being the enemy of morality, the pursuit of truth was itself a highly moral activity, perhaps the highest. The relation of science and morals is a long and difficult question with an illustrious history, and deserves a more extensive discussion than space permits. However, because some readers may share the questioner's response, I offer a brief reply. First, on the matter of reputation, we should recall that the pursuit of truth may be in tension with keeping a good name (witness Oedipus, Socrates, Galileo, Spinoza, Solzhenitsyn). For most of human history, the pursuit of truth (including science) was not a reputable activity among the many, and was in fact highly suspect. Even today, it is doubtful whether more than a few appreciate knowledge as an end in itself. Science has acquired a good name in recent times largely because of its technological fruit; it is, therefore, to be expected that a disenchantment with technology will reflect badly upon science. Second, my own attack has not been directed against science, but against the use of some technologies, and even more, against the unexamined belief—indeed, I would say, superstition—that all biomedical technology is an unmixed blessing. I share the questioner's belief that the pursuit of truth is a highly moral activity. In fact, I am in my paper inviting him and others to join in a pursuit of the truth about whether all these new technologies are really good for us. This is a question that merits and is susceptible of serious intellectual inquiry. Finally, we must ask whether what we call science has a monopoly on the pursuit of truth. What is truth? What is knowable, and what does it mean to know? Surely, these are also questions that can be examined. Unless we do so, we shall remain ignorant about what science is and about what it discovers. Yet science—modern natural science—cannot begin to answer them; they are philosophical questions, the very ones I am trying to raise at this point in the text.

†Some of these fundamental questions of natural philosophy—purposiveness in living things, the place of man, the relation of mind and body, the meaning of mortality, the meaning of the looks of whole organisms and its bearing on ethics—are the main themes of the four chapters in Part III of this volume.

Or to put it another way, in the absence of that "ultimate wisdom," we can be wise enough to know that we are not wise enough. When we lack sufficient wisdom to do, wisdom consists in not doing. Caution, restraint, delay, abstention are what this second-best (and, perhaps, only) wisdom dictates with respect to the technology for human engineering.

If we can recognize that biomedical advances carry significant social costs, we may be willing to adopt a less permissive, more critical stance toward new developments. We need to reexamine our prejudice not only that all biomedical innovation is progress, but also that it is inevitable. Precedent certainly favors the view that what can be done will be done, but is this necessarily so? Ought we not to be suspicious when technologists speak of coming developments as automatic, not subject to human control? Is there not something contradictory in the notion that we have the power to control all the untoward consequences of a technology but lack the power to determine whether it should be developed in the first place?

What will be the likely consequences of the perpetuation of our permissive and fatalistic attitude toward human engineering? How will the large decisions be made? Technocratically and self-servingly, if our experience with previous technologies is any guide. Under conditions of *laissez-faire*, most technologists will pursue techniques, and most private industries will pursue profits. We are fortunate that, apart from the drug manufacturers, there are at present in the biomedical area few large industries that influence public policy. Once these appear, the voice of the public interest will have to shout very loudly to be heard above their whisperings in the halls of Congress. These reflections point to the need for institutional controls.

Scientists understandably balk at the notion of the regulation of science and technology. Censorship is ugly and often based upon ignorant fear; bureaucratic regulation is often stupid and inefficient. Yet there is something disingenuous about a scientist who professes concern about the social consequences of science, but who responds to every suggestion of regulation with one or both of the following: "no restrictions on scientific research," and "technological progress should not be curtailed." Surely, to suggest that *certain* technologies ought to be regulated or forestalled is not to call for the halt of *all* technological progress (and says nothing at all about basic research). Each development should be considered on its own merits. Although the dangers of regulation cannot be dismissed, who, for example, would still object to efforts to obtain an effective, complete global prohibition on the development, testing, and use, of biological and nuclear weapons?

The proponents of *laissez-faire* ignore two fundamental points.

They ignore the fact that not to regulate is as much a policy decision as the opposite, and that it merely postpones the time of regulation. Controls will eventually be called for—as indeed they have been, say, in demands to end environmental pollution. If attempts are not made early to detect and diminish the social costs of biomedical advances by intelligent institutional regulation, the society is likely to react later with more sweeping, immoderate, and throttling controls.

The proponents of *laissez-faire* also ignore that fact that much of technology is already regulated. The Federal government is already deep in research and development (e.g., space, electronics, weapons) and is the principal sponsor of biomedical research. One may well question the wisdom of the direction given, but one would be wrong in arguing that technology cannot survive social control. Clearly, the question is not control versus no control, but rather what kind of control, when, by whom, and for what purpose.

Means for achieving international regulation and control need to be devised. Biomedical technology can be no nation's monopoly. The need for international agreements and supervision can readily be understood if we consider the likely American response to the successful asexual reproduction of 10,000 Mao Zedongs.

To repeat, the basic short term need is caution. Practically, this means that we should shift the burden of proof to the *proponents* of a new biomedical technology—at least those that directly challenge fundamental aspects of human life. Concepts of risk and cost need to be broadened to include some of the social and ethical consequences discussed earlier. The probable or possible harmful effects of the *widespread* use of a new technique should be anticipated, and introduced as costs to be weighed in deciding about the *first* use. The regulatory institutions should be encouraged to exercise restraint, and to formulate the grounds for saying no. We must all get used to the idea that biomedical technology makes possible many things we should never do.

But caution is not enough, nor are clever institutional arrangements. Institutions can be little better than the people who make them work. However worthy our intentions, we are deficient in understanding. In the long run, our hope can only lie in education: in a public educated about the meanings and limits of science and enlightened in its use of technology; in scientists better educated to understand the relationships between science and technology on the one hand and ethics and politics on the other; in human beings who are as wise in the latter as they are clever in the former.

CHAPTER TWO

Making Babies
The New Biology
and the "Old" Morality

One egg, one embryo, one adult—normality. But a bokanovskified egg will bud, will proliferate, will divide. From eight to ninety-six buds, and every bud will grow into a perfectly formed embryo, and every embryo into a full-sized adult. Making ninety-six human beings grow where only one grew before. Progress.

—Aldous Huxley, *Brave New World*

Good afternoon ladies and gentlemen. This is your pilot speaking. We are flying at an altitude of 35,000 feet and a speed of 700 miles an hour. I have two pieces of news to report, one good and one bad. The bad news is that we are lost. The good news is that we are making very good time.

—Anonymous

Thoughtful men have long known that the campaign for the technological conquest of nature, conducted under the banner of modern science, would someday train its guns against the commanding officer, man himself. That day is fast approaching, if not already here. New biomedical technologies are challenging many of the formulations that have served since ancient times to define the specifically human—to demarcate human beings from the beasts on the one hand, and from the gods on the other. Birth and death, the boundaries of an individual human life, are already subject to considerable manipulation. The perfection of organ transplantation and especially of mechanical organs will make possible wholesale reconstructions of the human body. Genetic engineering, a prospect already visible on the horizon, holds forth the promise of a refined control over human capacities and powers. Finally, technologies springing from the neurological and psychological sciences (e.g., electrical and chemical stimulation of the brain) will permit the manipulation and alteration of the higher human functions and activities: thought, speech, memory, choice, feeling, appetite, imagination, and love.

The advent of these new powers for human engineering means that some men may be destined to play God, to re-create other men in their own image. This Promethean prospect has captured the imagination of scientists and laymen alike, and is being hailed in some quarters as the final solution to the miseries of the human condition. But this optimism (not to say hubris) has been tempered by the dim but growing recognition that the use of these new powers will raise profound and difficult moral and political questions precisely because the objects on which they are to operate are human beings. An overview of these questions, in relation to the new biology in general, was presented in Chapter 1. In this chapter, I consider at greater length some of these moral and political questions in connection with a particular group of new technologies: the technologies for making babies.*

Why Make Babies?

Why would anyone want to provide new methods for making babies? A major reason is that, in many instances, the "old"† method is not possible. Despite greatly increased abilities to diagnose and treat the causes of infertility in recent years, some couples still remain involuntarily childless. Thus, paradoxically, while the need to limit fertility becomes ever more apparent, some scientists and physicians have taken it as their duty to satisfy the natural desire of every couple to have a child, by any means, natural or artificial.

Some rather large questions arise here. Physicians have a duty to treat infertility by whatever means only if patients have a right to have children by whatever means. But the right to procreate is an ambiguous right, and certainly not an unqualified one. Whose right is it, a woman's or a couple's? Is it a right to carry and deliver (i.e., only

---

*The original version of this chapter was written well before the birth of the first "test-tube baby."[1] Changes have been made to bring the account of the state of the art more up to date, and some of the arguments have been recast in light of recent events. However, because the direction and shape of the project for making babies is so clear, the larger moral and political questions are independent of the precise state of the present technology. In fact, the successful developments in recent years make the arguments about the overall project all the more urgent—especially given our tendency to let ourselves be overwhelmed by events.

†This awkward use of "old" calls attention to the subtle traps laid for us by the abuse of language. In a time and place where novelty and originality are considered cardinal virtues, and when trendiness has replaced tuberculosis as the scourge of the intellectual classes, one should vigorously resist the tendency to make things attractive simply by emphasizing their newness. Is the old way of beginning life merely *old*, and simply traditional or conventional?

a woman's right) or is it a right to nurture and rear? Is it a right to have your own biological child? Even if involuntary sterilization imposed by a government would violate such a right, however defined, is the right also violated or denied by sterility not imposed from without but due to disease? Is the inability to conceive a disease? Whose disease is it? Can a couple have a disease? Does infertility demand treatment wherever found? In women over seventy? In virgin girls? In men? Can these persons claim either a natural desire or natural right to have a child, which the new technologies might or must provide them? Does infertility demand treatment by any and all available means? By artificial insemination? By *in vitro* fertilization? By extracorporeal gestation? By parthenogenesis? By cloning—"xeroxing" of existing individuals by asexual reproduction?*

Simply posing these questions suggests that both the language of rights and the language of disease can lead to great difficulties in thinking about infertility. Both point to possessions or properties of single individuals, for it is an individual who bears rights and diseases. Yet infertility is a relationship as much as a condition—a relationship between husband and wife, and also between generations. More is involved than the interests of any single individual. Ultimately, to consider infertility (or procreation) solely from the perspective of individual rights can only undermine—in thought and in practice—the meaning of childbearing and its bond to the covenant of marriage. And in a technological age, to view infertility as a disease, one demanding treatment by physicians, automatically fosters the development and encourages the use of all the new technologies mentioned above.

A second reason given for seeking new methods for making babies is that sometimes the old method is thought to be undesirable or inadequate, primarily on eugenic grounds. A diverse—and ultimately incompatible—collection of champions are presently in bed together under this rationale: patient-centered physicians and genetic counselors seeking to prevent the transmission of inherited diseases by carrier parents to prospective children; species-centered pessimists interested in combating the alleged deterioration of the human gene pool; and zealous optimists eager to engineer so-called improvements in the human species. The new methods include the growth of early embryos in the laboratory with selective destruction of those that do not pass genetic muster; directed mating with eugenically selected

---

*Those who seek to submerge the distinction between natural and unnatural means would do well to ponder these questions and reflect on what they themselves mean when they speak of "a natural desire to have children" or "a [natural human] right to have children." One cannot speak of natural desires or natural human rights or, indeed, about disease without some notion of what would be normal, natural, and healthy for human beings. (See Chapters 3, "Perfect Babies: Prenatal Diagnosis and the Equal Right to Life," and 6, "The End of Medicine and the Pursuit of Health," as well as Part III.)

eggs, sperm, or both; and asexual replication of existing "superior" individuals. But serious questions can be raised with respect to these ends as well. For example, we may know which disease we would wish not to have inflicted upon ourselves and our offspring, but is it clearly wise to act upon these desires? What are the social implications of efforts to prevent the birth of the genetically handicapped?* And, social consequences aside, in view of our ignorance concerning why certain genes survive in our populations, can we be sure that the eradication of genetic disease (or of any single genetic disease) is even biologically a sensible goal? Might it not have unanticipated genetic consequences?

The species-centered goals are even more problematic. Do we know what constitutes a deterioration or an improvement in the human gene pool? Might one not argue that, at least under present conditions, the crusaders against the deterioration of the species are worried about the wrong genes? After all, how many architects of the Vietnam War or the suppression of Solidarity suffered from Down's syndrome? Who uses up more of our irreplaceable natural resources and who produces more pollution: the inmates of an institution for the retarded or the graduates of Harvard College? And which of our genetic mutants display more vanity, self-indulgence, and the will-to-power, or less courage, reverence, and love of country than many of our so-called best and brightest? It seems indisputable that the world suffers more from the morally and spiritually defective than from the genetically defective. Thus, it is sad that our best minds are busy fighting our genetic shortcomings while our more serious vices are allowed to multiply unmolested.

Perhaps this is too harsh a judgment. Certainly, our genetic inheritance is entrusted to us for safekeeping and not for abuse or neglect. Perhaps a case could be made for the desirability and wisdom of certain negative or even positive eugenic goals. Still, as in the treatment of infertility, we shall also have to consider which means, if any, can be justified in the service of *any* reasonable goals.

Third, there are scientific and technological goals—distinct from making babies—that themselves generate new beginnings in life. For example, there is a limit to what can be learned about the nature and regulation of fertilization, embryonic development, or gene action from lives begun in the old, undisturbed, natural manner. This is no doubt true. But if the goal is scientific knowledge of these processes for its own sake, there is, for now, little need to develop new beginnings in *human* life. Embryological experimentation in a wide range of mam-

*This is the theme of Chapter 3, "Perfect Babies: Prenatal Diagnosis and the Equal Right to Life."

mals, employing all the new technological possibilities, would yield the basic understanding. There is at present no reason to believe that the fundamental mechanisms of differentiation differ in monkeys and in man. Until extensive animal studies show otherwise, the human experiments can only be given a technological and not a purely scientific justification. (Indeed, it is the philanthropic foundations interested in finding new drugs for abortion or contraception who supported much of the pioneering work on the laboratory growth of human embryos. For example, the work of Robert G. Edwards and his colleagues in Cambridge, England, has been supported by the Ford Foundation.) These technological purposes and activities (and others such as the use of early embryos in culture to test for mutation- or cancer-producing chemicals and drugs, or to work out techniques for genetic manipulation) may well be desirable, but they need to be so identified and distinguished from the pure quest for knowledge. To adequately assess the desirability of any specific means and to properly weigh alternative means requires a clear understanding of which ends are being served.

Finally, new methods for making babies are being sought precisely because they are new and because they can be sought. While not praiseworthy reasons, they certainly are important, and all-too-human, ones. Drawn by the promise of fame and glory, driven by the hot breath of competitors, men do what can be done. Biomedical scientists are no less human than anyone else. Some of them are unable to resist the lure of immortality promised, say, to the scientific father of the first test-tube baby. Moreover, regardless of their private motives, they are encouraged to pursue the novel because of the widespread and not unjustified belief that their new findings will probably help to alleviate one form or another of human suffering. They are even encouraged by that curious new breed of technotheologians, who after having pronounced God dead, disclosed that God's dying command was that mankind should undertake its limitless no-holds-barred self-modification by all feasible means.

So much then for reasons why some have called for and helped to promote new beginnings for human life. But what precisely is new about these new beginnings? Such life will still come from preexisting life, no new formation from the dust of the ground is being contemplated, nothing as new—or as old—as that first genesis of life from nonliving matter is in the immediate future. What is new is nothing more radical than the divorce of the generation of new human life from human sexuality, and ultimately, from the confines of the human body, a separation that began with artificial insemination and which will finish with ectogenesis, the full laboratory growth of a baby from sperm to term. What is new is that sexual intercourse will no longer be needed for generating new life. (The new technologies provide the corollary to the pill: babies without sex.) This piece of novelty leads to two others: There is a new co-progenitor, the embryologist-geneticist-physician, and there is a new home for generation, the laboratory. The mysterious and intimate processes of generation are to be moved from the native darkness of the womb to the bright (fluorescent) light of the laboratory, and beyond the shadow of a single doubt.

But this movement from natural darkness to artificial light has the most profound implications. What we are considering, really, are not merely new ways of beginning individual human lives but also, and this is far more important, new ways of life and new ways of viewing life and the nature of man. Man is defined partly by his origins and his lineage; to be bound up with parents, siblings, ancestors, and descendants is part of what we mean by human. By tampering with and confounding these origins and linkages, we are involved in nothing less than creating a new conception of what it means to be human.

Consider the views of life and the world reflected in the following different expressions to describe the process of generating new life. Ancient Israel, impressed with the phenomenon of transmission of life from father to son, used a word we translate as "begetting" or "siring." The Greeks, impressed with the springing forth of new life in the cyclical processes of generation and decay, called it *genesis*, from a root meaning "to come into being." (It was the Greek translators who gave this name to the first book of the Hebrew Bible.) The premodern Christian English-speaking world, impressed with the world as given by a Creator, used the term "pro-creation." We, impressed with the machine and the gross national product (our own work of creation), employ a metaphor of the factory, "re-production." And Aldous Huxley has provided "decantation" for that technology-worshipping Brave New World of tomorrow.

In Vitro Fertilization—State of the Art

From the point of view of making babies, the most significant technological achievements in recent decades involve *in vitro* fertilization. In 1969, Dr. Robert G. Edwards in Cambridge, England, reported the first successful fertilization, in the test tube, of a human egg by human sperm and, in 1970, the subsequent laboratory culture of the young embryo. To surmount the difficulty of obtaining mature functional eggs, Edwards and his obstetrician colleague, Dr. Patrick C. Steptoe, had devised a surgical method, known as laparoscopy, to obtain matured eggs directly from the ovaries prior to ovulation. From a single woman as many as three or four eggs can be recovered at one operation. Upon addition of sperm, fertilization occurs with a small but significant fraction of these eggs. Kept in culture medium, a majority of the fertilized eggs begin to divide, and a small fraction reach the blastocyst stage (i.e., age of about seven to eight days), the stage at which the early embryo normally implants itself in the wall of the uterus. Successful implantation of laboratory-grown embryos had been reported in rabbits and in mice, and Edwards and Steptoe—and others as well—began to attempt the transfer in women. The physical transfer of the embryo into the uterine cavity posed no problem, but implantation was at first difficult to achieve. Because the uterine lining is receptive to implantation only for a short portion of the menstrual cycle, and because only embryos at a certain stage are capable of implantation, the timing of transfer is critical. After many failures, changes in the hormonal treatment of the woman brought success, and in the summer of 1978, Louise Brown, the first identified human child conceived *in vitro,* was born. At least several dozen more live births have been achieved since then, and numerous *in vitro* fertilization laboratories and clinics have been or are being started all over the world. The age of making babies has arrived.

This summarizes the current state of the art on new beginnings in human life. But there is much work being done with other mammals that will provide knowledge and techniques someday applicable to humans. Some of these developments deserve mention, although a detailed treatment is beyond the scope of this essay. Considerable progress has been made in growing older mammalian embryos in the laboratory. Dr. D. A. T. New and his colleagues in Cambridge, England, have been successful in growing rat and mice embryos bathed in a blood-serum medium for about one-third (the middle third) of the whole gestation period. As the embryo approaches term size, it can no longer be maintained bathed in media, but requires more efficient circulation and exchange of nutriments, gases, and wastes. Various artificial pump and perfusion techniques, analogous to the artificial kidney machine, are being studied in an effort to design an artificial placenta. Finally, a long-standing early barrier to extensive laboratory culture, located just after the blastocyst stage, has fallen with the successful culture of mouse embryos from the blastocyst stage to a stage having a differentiated and beating heart. Thus, from both ends and from the middle, researchers are gradually closing in on the possibility of complete extracorporeal gestation. It should be stressed that these techniques are being pursued primarily to make possible a better understanding of the full scope of embryonic development. However, even though no scientists at present appear to be interested in going from fertilization to birth entirely in the laboratory,* the technology to do so is gradually being worked out piece by piece.

Techniques to predetermine the sex of unborn children may be just around the corner. Since gender is determined solely by the X or Y chromosome content of the sperm, techniques for physical separation of X-carrying from Y-carrying sperm would make sex control possible through artificial insemination. Attempts to effect such a separation have not yet succeeded, but new efforts can be expected, partly because of recently discovered methods for detecting the Y-chromosome in cells, methods that may serve as an assay for successful separation. A second method of sex control, already successfully demonstrated in rabbits, involves the sexing of embryos (prior to implantation) by cell-staining techniques or chromosome analysis. Embryos of the desired sex could then be transferred to the recipient females. Though accurate, this sexing technique is not without its problems, since embryonic tissue needs to be removed for testing. In the rabbit work, only about one in five embryos sexed by these methods developed to full term, and there was at least one monstrous birth. Even if these technical difficulties were ironed out, it is doubtful that many will accept the costs and inconvenience of *in vitro* fertilization or even artificial insemination to control the sex of their offspring unless perhaps they are known carriers of a sex-linked genetic disease such as hemophilia. Less cumbersome methods (e.g., a chemical method that would selectively destroy either the X- or the Y-carrying sperm in the man's body or soon after intercourse) are conceivable but as yet unreported even in animals. A negative form of sex-selection is, of course, already feasible, using amniocentesis to disclose gender and selective abortion if the finding is displeasing.

Finally, the generation of man-animal hybrids or chimeras has been predicted by some reputable scientists. These might be produced by the introduction of selected nonhuman genetic material into the developing human embryos. Fusion of human and nonhuman cells in tissue culture has already been achieved, and so has the transfer of functional genes, from one species into another, by means of the new techniques of DNA recombination.

Experimenting on the Unborn and the Unconceived

As in so many other cases of technological innovation, the ability to initiate human life in the laboratory and to make babies using labo-

---

*Although no scientist *appears* to be, at least one *is* interested, according to a quotation in Albert Rosenfeld's book, *The Second Genesis:* " 'If I can carry a baby all the way through to birth *in vitro*,' says an American scientist who wants his anonymity protected, 'I certainly plan to do it—though obviously, I am not going to succeed on the first attempt, or even the twentieth.' "[2]

ratory fertilization arrived on the scene before we had given thoughtful consideration either to the meaning or the desirability of doing so. Once again, scientists—acting entirely on their own, under a liberal polity that supports their freedom to do so—have produced the means before the community has even endorsed the end, let alone this particular choice of means. The *in vitro* fertilizers were somehow aware of this problem, but it did not trouble them much. At the end of a popular review article in *Scientific American* summarizing the early work on human embryos, Drs. Robert Edwards and Ruth Fowler offered the following conclusion: "We are well aware that this work presents challenges to a number of established social and ethical concepts. In our opinion the emphasis should be on the rewards that the work promises in fundamental knowledge and in medicine."[3] Here, not surprisingly, we are told we *should* emphasize promised rewards in knowledge and power at some future time at the expense of established (don't they really mean "establishment" or "conventional"?) ideas of right. But this is itself a moral judgment—untenable as a general proposition for experimentation on human subjects—whose soundness in this particular case cannot be determined until the full range of ethical, social, legal, and political implications is carefully studied and understood.

Let us consider first what is probably the least controversial and, thus far, the most popular use for *in vitro* fertilization: the provision of their own child to a childless couple, where oviduct disease in the woman obstructs the free passage of egg and sperm, and hence fertilization.*

At first glance, the intramarital use of artificial fertilization resembles, ethically, artificial insemination (husband). The procedure simply provides for the union of the wife's egg and the husband's sperm, circumventing the pathological obstruction to that union. But there is at least this difference: There is an alternative treatment for infertility due to tubal obstruction, namely surgical reconstruction of the oviduct, which, if successful, permanently removes the cause of infertility (i.e., it treats the underlying disease, not merely the desire to have a child). At present, the success rate for oviduct reconstruction is only fair, but with effort and practice, this is bound to improve. This therapeutic surgery for women is without possible moral objection or adverse social consequences. Therefore, should both options be

*There are many infertile women so affected, crude estimates suggesting as many as 1 percent of all women. Approximately 10 to 15 percent of couples are infertile, and in more than half of these cases, the cause is in the female. Blocked or abnormal oviducts account for perhaps 20 percent of the female causes of infertility. However, not all such women are suitable candidates for the intramarital use of *in vitro* fertilization. Some women who have blocked tubes also have associated disordered ovaries, making it difficult or impossible for a doctor to obtain eggs from them. For this reason, only some of these women are likely to be able to provide their own eggs for fertilization.

feasible and available, oviduct repair is to be preferred over artificial fertilization both in principle (namely, one should use the least objectionable means to achieve the same unobjectionable end) and in practice.

Why this preference? Because the attempt to generate a child with the aid of *in vitro* fertilization constitutes an experiment upon the prospective child—beyond any loose sense in which we might call any generation of a child experimental. The use of *in vitro* fertilization to initiate a new human life—unlike oviduct repair, artificial insemination, and, of course, sexual union—involves the necessary and deliberate manipulation of the human embryo itself, conceived and nurtured, at least for a time, in an artificial environment. Serious questions can be raised about the effects of the manipulations on the child who is eventually produced by this procedure. These medical questions about safety and normality lead to a perplexing moral question: Does the parents' desire for a child (or the obstetrician's desire to help them get one) entitle them to have it by methods which carry for that child an unknown and untested risk of deformity or malformation?

How known are the risks? How well tested were they before Edwards and Steptoe undertook to make a baby for Mr. and Mrs. Brown? The truth is that the risks were very much *un*known. Although there had been no reports of gross deformities at birth following successful transfer in mice and in rabbits, the number of animals produced in this way was much too small to exclude even a moderate risk of such deformities. In none of the research had the question of abnormalities been systematically investigated. No prospective studies had been made to detect defects that might appear at later times or lesser abnormalities apparent even at birth. In species more closely related to humans (e.g., in primates), successful *in vitro* fertilization had yet to be accomplished. Many respected researchers in the field believed—even after the birth of Louise Brown—that the ability regularly to produce normal monkeys by this method was a minimum prerequisite for using the procedure in humans.*

But, it must, in fairness, be added that even after normal young were produced in monkeys, one would not yet be fully certain that normal young would be produced in humans. There might be species differences in sensitivity to the physical manipulations or to possible teratogenic agents in a culture medium. Also, monkey experiments

---

*One such individual is Dr. Luigi Mastroianni, chairman of obstetrics and gynecology at the University of Pennsylvania. Dr. Mastroianni believes that scientists must be sure that normal monkeys can be produced with regularity before the procedure is tried on humans, and that scientists must also know all the risks involved in order to obtain "truly informed consent." He says: "We must be very careful to use patients well and not be presumptuous with human lives. We must not be just biologic technicians."

could neither rule out nor establish the risk of mental retardation for children resulting from experiments in humans. Unfortunately, as is often true, only humans could provide the test system for fully assessing the risks of using the procedure in humans.

Laboratory testing of the human embryos themselves, prior to transfer, could not provide enough information about normality, and might itself do damage. Ordinary microscopic observation of the early embryo can disclose gross abnormalities, but the measure of normality that it does provide is too crude. Not only can genetic tests not be done on a given embryo without damaging it, but there are few genetic tests presently available for the doing. Furthermore, damages could be introduced during the transfer procedure, after the last inspection is completed.

In sum, there is still no way of finding out *in advance* whether or not the viable progeny of the procedures of *in vitro* fertilization, culture, and transfer of human embryos will be deformed, sterile, or retarded. Only the made babies, after their birth and growth, can show whether it was safe to have made them in the first place.

This conclusion could lend support to two opposite courses of action: to proceed or to abstain, permanently. Unconcerned about the risks, and confident that there was no prior way to come to know them, Edwards and Steptoe threw caution to the winds and proceeded. Those of us who worried about the prospective children must now be pleased and relieved that the experience to date has been so encouraging, with no (published) reports of severe abnormalities—though the number of human births remains small and insufficient time has elapsed to assess the development of the children. There is, perhaps, no quarreling with success, as the pioneering baby-makers no doubt will now argue. However, when it comes to human experimentation, the proof of the pudding is not only in the eating. Besides, the next innovators may not be so lucky.

Against the stream, Paul Ramsey has almost single-handedly upheld the opposite conclusion: permanent abstention. In these prospective experiments on the unborn, it is not enough not to know of any grave defects; one needs to know with confidence *that there will be no such defects*—or at least no more than there are without the procedure. In his book *Fabricated Man,* Ramsey puts the matter quite forcefully (and, I used to think, correctly): "The decisive moral verdict must be that we cannot rightfully *get to know* how to do this without conducting unethical experiments upon the unborn who must be the 'mishaps' (the dead and the retarded ones) through whom we learn how."*[4]

---

*The matter of mishaps was completely ignored by Edwards and David Sharpe (the latter, an American professor of law), in their article in *Nature* surveying some of the social and ethical issues attending research in human embryology. Feigning neutrality,

One might object that all new medical technologies are risky, and that the kind of ethical scrupulosity Ramsey advocates would put a halt to medical progress. But such an objection ignores an important distinction. It is one thing to accept voluntarily the risk of a dangerous procedure for yourself (or to consent on behalf of your child) if the purpose is therapeutic. Some might say this is not only permissible, but obligatory, in line with a duty to preserve one's own health. It is quite a different thing deliberately to submit a child, born or unborn, to hazardous procedures that can in no way be considered therapeutic for him. This argument against nontherapeutic experimentation on children applies with even greater force against experimentation "on" a hypothetical child (whose conception is as yet only intellectual). One cannot ethically choose for him the unknown hazards he must face and simultaneously choose to give him life in which to face them.*

Whether we agree with him or not, we must all be grateful to Professor Ramsey for his insistent raising of the question. For he forces us to remember that there are real human beings who suffer the consequences of our decisions to reproduce (or to alter their genotypes,

---

these authors reveal their prejudices—and their misconception of medical ethics—when they invoke the principle "Do no harm" to justify, rather than oppose, the use of *in vitro* techniques for making babies: "The beginning of medical ethics, however, is *primum non nocere;* this permits alleviation of infertility, and has been stretched to cover destruction of foetuses with hereditary defects, but would it permit the more remote techniques like modifying embryos?"[5] "Do no harm" is a principle that can only justify *omitting* a medical intervention; it does not permit, let alone justify, committing an intervention. Rightly understood, it can enter medico-moral arguments only in opposition to risk-filled technologies for making babies.

*Some readers may be perplexed to see this matter of prospective children treated under the principles that govern experimentation on human subjects. Who is the human subject of the fertilization, culture, and implantation procedures? Do we mean to call a blastocyst a human subject? The questions as raised here can easily be distinguished from a similar question raised concerning abortion. With abortion, the moral question is when and whether one can justify killing the fetus; one of the underlying issues is whether the fetus is a human being, or a potential human being, worthy of protection. Here, we are concerned with the possible harm inflicted upon the live, breathing children who come to be born after getting their start via *in vitro* fertilization and laboratory culture. The underlying issue is whether one can speak of such children and their ontogenetic precursors as "human subjects of experimentation," especially when they themselves are the products of such experiments.

The issue can be clarified and possibly resolved, if analogous harmful manipulations of the unborn are considered. A useful, though not perfect, analogy would be the deliberate administration of thalidomide or some other known teratogen to a pregnant woman. Whether the fetus is then human or not, the child it becomes would be an unwilling and unjustly injured victim of such unethical practice. More analogous would be the generation of a child through artificial insemination with sperm that had been deliberately irradiated or mutagenized. These examples, though hypothetical, serve to show that a child can be deliberately injured before its birth, even coincidentally with his conception. And if the child-to-be can be deliberately injured, he can be negligently injured, as he might be if he were the product of a new baby-making technique employed before it were shown to be free of risk for serious harm.

etc.). More important, in this way he forces us to consider the meaning of our desire to have a child. If having children is regarded primarily as the satisfaction of parental desires, to attain our own fulfillment and happiness, then Ramsey's argument seems to me decisive. But if we have children not primarily for ourselves but for our children, if procreation means to pass on the gift of life to the next generation, if the gift of life can be held to be morally at least the equal of therapy, then this clear benefit to a child-to-be—even to a child-at-risk, as all of our children are—could justify the risks taken because they are taken in the child's behalf—provided, of course, that the risks are not excessive. This is now my view.

There is, then, it seems to me, good reason for insisting that risk of incidence and the likely extent of possible harm be very, very low— lower, say, than in therapeutic experimentation in children or adults. But I do not think that the risk of harm must be positively excluded (as it certainly cannot be). It would suffice if those risks were roughly equivalent to the risks to the child from normal procreation. To insist on more rigorous standards, especially when we permit known carriers of genetic disease to reproduce, would seem a denial of equal treatment to infertile couples contemplating *in vitro* assistance. It also gives undue weight to the importance of bodily harm over risks of poor nurture and rearing after birth, or, to repeat, against the goodness of bodily life itself. Wouldn't the couple's great eagerness for the child count, in the promise of increased parental affection, toward offsetting even a slightly higher but unknown risk of mental retardation? It should suffice that the risks be comparable to those for ordinary procreation, not much greater but no less.

Important as these matters are, there is a danger in belaboring them. To devote so much attention to the subject of risks in laboratory-assisted reproduction is probably to attach too much of one's concern to the wrong issue. True, everyone understands about harming children, while very few worry about problems of confounding lineage or dehumanizing life and procreation. But it is these neglected matters, not the risk of bodily harm to offspring, that are the distinctive and deepest dangers of laboratory-assisted reproduction. We go forward, soon to consider them.

Experiments with Childless Couples

Before I leave the subject of the ethics of experimentation, however, let me add a few comments concerning the adult participants. During the roughly ten years of *in vitro* research leading up to the first successful test-tube baby, most of the scientific reports on human embryo experimentation were strangely silent on the nature of the egg donors, on their understanding of what was to be done with their eggs, and on the manner of obtaining their consent. This silence was surprising considering the growing sensitivity of the medical and scientific communities to the requirement of informed consent, and especially surprising given the kind of experiments being performed. Who were these women and how did they come to "volunteer"? In the *Scientific American* article by Edwards and Fowler there was this solitary comment: "Our patients were childless couples who hoped our research might enable them to have children."[6] From the report that they had hopes, we can surmise that they considered themselves to be patients. But so far as those experiments were concerned, they were in fact only experimental subjects. One wonders if they were told this.

Only one of the many scientific articles, the one which described the use of laparoscopic surgery to recover human oocytes, told anything more about the persons used as experimental subjects (in this case, for perfecting the laparoscopy technique), and about how they were informed: "The object of the investigations was fully discussed with the patients, including the possible clinical applications to relieve *their* infertility."[7] Though welcome, this statement left many questions unanswered. Were the couples also told that the much more likely possibility was that it would be future infertile women, rather than they themselves, whose infertility might be relieved? Were they told about alternative possibilities, such as surgery on the blocked oviduct or adoption? Since the same article told us that three out of forty-six "infertile" women became pregnant—by the old, customary method—during the first month after laparoscopy, we had to wonder about the criteria used for subject selection. Were all other possibilities exhausted before bringing these couples into this uncertain program of experimentation? Finally, we were left to wonder how the discussions were conducted, especially in the light of the following quotation attributed (in 1969) to R. G. Edwards: "We tell women with blocked oviducts, 'Your only hope of having a child is to help us. Then maybe we can help you.' "[8]

It is altogether too easy to exploit, even unwittingly, the desires of a childless couple. It is cruel to generate false hopes by inflated publicity of the sort that some of these researchers promoted. It is both cruel and unethical falsely to generate hope—for example, by telling women that they themselves, rather than future infertile women, might be helped to have a child—in order to secure participation in experiments. Injuries done to the original experimental subjects are in no sense rectified or excused by the latter-day successes.

Experimentation was also done with eggs obtained from women undergoing ovarian surgery for clinical reasons. Did these women know what was to be done with their eggs? To whom belong the rights governing ordinary tissue removed at surgery? Is reproductive tissue a special case? Surely, if the eggs were going to be implanted in another woman, one would think that the donor's permission should be obtained. Then what about their simple fertilization? If a woman from whom eggs are taken has religious or other objections against *in vitro* fertilization that would lead her to refuse permission if asked, is she wronged by not being asked or informed? No matter how worthy the research and how well-intentioned the investigator, ethical experimentation on human subjects and their tissues requires that persons be treated as ends and not merely as means, and hence that their wishes and beliefs in these matters be considered and respected.

The Case of the Surplus Embryos*

So far, I have discussed questions about the ethics of experimentation connected with attempts to generate a normal child by transferring a laboratory-grown human embryo into the uterus of an infertile woman. But what about all the embryos that are not so implanted? Dr. Donald Gould, editor of the British journal, *The New Scientist*, has asked: "What happens to the embryos which are discarded at the end of the day—washed down the sink? There would necessarily be many. Would this amount to abortion—or to murder? We have no law to cope with this kind of situation."[9]

I don't wish to mislead anyone. At this state of the art, the largest embryo we are talking about is a blastocyst, barely visible to the naked eye. But the moral question does not turn on visibility, any more than it does in the case of murder committed by a blind man. The embryos are clearly biologically alive, even at the blastocyst stage. At some future date, improved techniques will permit their growth to later stages, someday, perhaps, even viable stages. Before then, however, the question of discarding will have to be faced. Since there is a continuity of development between the early and the later stages, we had better face the question now and draw whatever lines need to be drawn.

When, in the course of human development, does a living human embryo acquire "protectable humanity"? This is a familiar question that I shall not belabor. But the situation here, though similar, is not identical to that of abortion. For one thing, we don't start with a fetus already *in utero*, which one destroys reluctantly, and, one hopes, only for good reasons. Here, nascent lives are being deliberately created

*The status of the human embryo, here a somewhat subordinate issue in the context of making babies, becomes a central issue in the context of growing and experimenting with life in the laboratory. Accordingly, it receives a more thorough treatment in Chapter 4, "The Meaning of Life—in the Laboratory."

despite certain knowledge that many of them will be destroyed or discarded. (Several eggs are taken for fertilization from each woman, the extra ones being available for experimentation.) Unlike the unwanted fetuses killed in abortion, the embryos discarded here are wanted, at least for a while; they are deliberately created, used for a time, and then deliberately destroyed. Even if there is no wrong done by discarding at the blastocyst stage—and I am undecided on this question—there certainly would be at later stages. (Those who disagree should at least be concerned about the effects on the attitude toward and respect for human life engendered in persons who are engaged in these practices, a topic to which I shall return.)

There is a second, related difference between abortion and discarding laboratory-grown embryos. Who decides what are the grounds for discard? What if there is another recipient available who wishes to have the otherwise unwanted embryo? Whose embryos are they? The woman's? The couple's? The geneticist's? The obstetrician's? The Ford Foundation's? If one justifies abortion on a paramount right of the woman to decide about her family size and spacing, or even on that unbelievable ground that a woman has a right to do what she wishes with her body,* whose rights are paramount here? Shall we say that discarding laboratory-grown embryos is a matter solely between a doctor and his plumber?†

One Thing Leads to Another

Having discussed so far only one serious moral objection to implantation of embryos and having raised some questions about the discarding and perpetuation of unimplanted ones (and deliberately neglecting the ethics of creating the embryos in the first place), I suspect that I have not persuaded anyone not already opposed to, or at least doubtful about, *in vitro* fertilization. Furthermore, the first objection to future implantation is losing its practical force. The experimentation has proceeded, deaf to this objection, and, as already noted,

---

*Even if such a right exists, it does not govern actions involving the fetus, because the fetus is simply not a mere part of a woman's body. One need only consider whether a woman can ethically take thalidomide while pregnant to see that this is so. It is distressing that seemingly intelligent people would sincerely look upon a fetus whose heart was beating, and which had its own EEG (five weeks), as indistinguishable from a tumor of the uterus, a wart on the nose, or a hamburger in the stomach. On the general question of our relations to our bodies, see Chapter 11, "Thinking About the Body."

†Dr. Steptoe, when once asked about the problem of "discarding," rejected the use of the term, preferring to refer to the embryos as "*not* chosen for implantation." These were, he said, fixed on slides and studied, to learn what could be learned. No embryos were discarded.

the results to date show no evidence of severe harm to the newborn children. The discarding issue will hardly get a fresh hearing in a society that has so recently converted to feticide.* Moreover, apart from these questions I can find no intrinsic moral reason to reject the intramarital use of *in vitro* fertilization and implantation—at least no reason that would not also rule against artificial insemination (husband). But the argument does not stop here, for we must consider both the likely other uses and abuses of this procedure, and also the other and more objectionable procedures that this one makes possible.

Some may object to my making an argument based upon likely or possible misuses and abuses. After all, there are few if any powers, technological and nontechnological, that cannot be abused. Many of us would prefer arguments from principle concerning intrinsic rightness or wrongness, arguments that abstract from the difficult task of predicting and weighing consequences, often quite remote and intangible ones. Nevertheless, we can ill afford such intellectual purism, especially for technologies that touch the foundations of man's biological nature. No technology exists autonomously or in isolation; each arises in the context of other technologies and, more importantly, in a complex and heterogeneous world of men whose proclivity for mischief and folly we cannot in good conscience ignore. We would ourselves display such folly were we to justify the introduction of each new technology simply because *some* good use can be found for it.†

Once introduced for the purpose of treating intramarital infertility, *in vitro* fertilization can now be used for any purpose. There is no reason why the embryo need be implanted in the same woman from whom the egg was obtained. An egg taken from one woman (the biological mother) could be donated to another woman (the gestational mother), either before or after fertilization—the former has been termed "artificial inovulation (donor)," the latter, "embryo transfer (donor)" or "prenatal adoption." Since obtaining eggs for donation is more difficult than obtaining sperm, and requires surgery on the donor, this might not seem a likely occurrence except on a small scale. However, recently developed procedures for freezing and storing eggs or young embryos will circumvent this problem. Egg and embryo banks will

---

*Some may object to my use of the term "feticide," but I am opposed to hiding behind euphemisms. If we are going to be brave enough to practice abortion, let us at least not be cowardly in describing it. Even if feticide were made legal everywhere, and even if it were morally justified, it would still be killing (though not necessarily murder, which is "wrongful killing").

†I doubt if many readers would find acceptable as a reason sufficient to justify the development of chemical and biological weaponry the fact that it may be more humane to use nonlethal gas on an enemy soldier than to kill him without first asking whether we can foreclose the further consequences to human civilization and the human race of introducing such military technology.

almost certainly be established—as are sperm banks today—partly to avoid having to do repeated operations on the same woman, and also because there is money to be made.

There are enough women whose infertility is due to reasons other than blocked oviducts to make it extremely likely that donation of eggs and embryos will be attempted, and that the technique will not be confined to those intramarital cases in which it was first used. Clinical use of *in vitro* fertilization will probably rely mainly on donor eggs in a manner closely analogous to donor insemination. And why stop at couples? What about single women, widows, or lesbians? If adoption agencies now permit these women to adopt, are they likely to be denied a chance to bear and deliver?

The converse possibility will also follow, namely the use of one woman simply to incubate and deliver another woman's child. If the previous practice might lead to new business ventures, advertised under "eggs for sale," this practice might lead to one advertised under "wombs for rent." Women with uterine abnormalities that preclude normal pregnancy may seek surrogate gestational mothers, as may women who don't want pregnancy to interfere with the skiing season. There are certainly enough poor women—and not only poor women—available to form a caste of childbearers, especially for good pay.

In fact, in the past five years, the practice of surrogate motherhood has become a thriving business throughout the nation, far beyond anyone's expectation (though, for now, the practice uses artificial insemination [donor] rather than embryos from *in vitro* fertilization). The cost for this service, performed under contract, is between $20,000 and $25,000: roughly $10,000 for the surrogate, the same for the broker (lawyer), plus medical costs for the pregnancy.

The more sentimentally minded will point out that these twin forms of foster pregnancy can be humanely and respectfully practiced. A woman may wish to donate an egg or an embryo to her sister, or may agree out of generosity to gestate a friend's embryo. It can be argued that no one should stand in the way of such acts of love,* but it is simply naive to think the practice would be limited to these more innocent cases. Moreover, there are psychological and ethical reasons

---

*Even if this argument were correct, it surely cannot be correct on the basis that these are acts of love. The current sentimentality that endorses all acts done lovingly because lovingly done leads to some strange judgments. A teacher friend not long ago asked a student, who was having difficulty appreciating the crimes of Oedipus, what she would think if she discovered that her sister was having an affair with their father. The girl replied that, although she was personally disgusted by the prospect, she thought that there was probably nothing wrong with it "provided that they [sister and father] had a good relationship." It is unlikely that an individual or a whole society that is unable to find reasons (other than genetic ones) for rejecting incest will be willing to appreciate any of the questions raised in this chapter.

for thinking that these cases may not be so innocent. What are the psychological consequences for the womb-lending sister (and others) of giving birth to her nephew? Will she feel like giving him up? Confusion and conflict would seem almost inevitable. If the donor of sperm has no claim over a child born after artificial insemination, why should the donor of ova, especially if there is a later dispute between the two women? Conversely, what happens if the surrogate mother delivers a defective child? Can she be forced to keep it? These acts of love cannot be kept anonymous; the visible and continued presence of the progeny might chronically stimulate such conflict. Also, might not female relatives be under intolerable pressures to donate eggs or to lend wombs, once the first such acts of "kindness" are well-publicized (just as relatives now often feel constrained to serve as transplantation donors of kidneys)? And is a person morally justified in allowing her body to be used as a "hot house," as a human incubator? Indeed, is not a decisive objection to the extramarital use of these techniques that it requires and fosters, both in thought and in deed, the exploitation of women and their bodies? And is not a second decisive objection the fact that it fosters the notion that children are property, and encourages the practice of child buying and selling?* (The further question concerning the separation of procreation from sexual love and marriage will be treated later in this chapter.)†

Use of these technologies need not be confined—nor is it likely to be confined—to the scale of individual couples making private decisions, nor to treatment of infertility. Indeed, several proposals for additional uses have already been placed before the public. As suggested in these proposals, and as I noted at the start, these techniques could serve eugenic purposes. Artificial insemination with semen from selected donors, the (positive) eugenic proposal of Herman Muller, has now been put into practice with the opening of the Repository for Germinal Choice in California, the so-called Nobel Sperm Bank. It may soon be possible to provide selection of ova as well. For many people, these prospects raise the fear of directed breeding programs under the dictates of a totalitarian regime. Such programs need not be coercive, since the desired donors of egg and sperm, as well as the foster moth-

---

*The British blue ribbon Committee of Inquiry into Human Fertilization and Embryology (chaired by Dame Mary Warnock), which endorsed *in vitro* fertilization and even some embryo research, vigorously opposed surrogate motherhood. Its report (Summer, 1984) recommended that it be made a crime to organize or operate agencies, whether profit-making or nonprofit, to recruit surrogate mothers or couples who might use them and that surrogacy agreements be declared both illegal and unenforceable in court. The British understand the gravity of this matter better than we do.

†Questions of lineage and the meaning of our embodiment, again in connection with the new reproductive technologies, will be discussed in Chapter 4: "The Meaning of Life—in the Laboratory."

ers of the regime, might be handsomely paid and highly honored. But, perhaps perversely in a time when suspicion and fear of governmental abuse of power are high and growing, I am not very worried about government-directed breeding by these methods. The eugenic advantages of this method, if there are any, are also available—and more cheaply, too—simply by directed sexual intercourse. A regime could more easily compel or induce this less troublesome, more enjoyable practice. While those who hold to the demonic theory of politics may think me naive, I expect that artificial fertilization and embryo culture would add very little to the already large arsenal of those who would practice mischief and evil.

We stand in much greater danger from the well-wishers of mankind, for folly is much harder to detect than wickedness. The most serious danger from the widespread use of these techniques will stem not from desires to breed a super-race, but rather from the growing campaign to prevent the birth of all defective children, and in the name of population control, quality of life, and the "right of every child to be born with a sound physical and mental constitution, based on a sound genotype."[10] Thus continues the former president of the American Association for the Advancement of Science, geneticist Bentley Glass, in his presidential address (1971): "No parents will in that future time have a right to burden society with a malformed or a mentally incompetent child."[11] These are not the words of a dictator, but of a gentle biologist. Even granting the desirability of his end—optimum children of no burden to society (except from their "perfection")—just consider what it would require in the way of means. This perfect condition is to be accomplished not by infanticide, not just by prenatal diagnosis and abortion of defectives, but by the laboratory growth and implantation of human embryos.

> The way is thus clear to performing what I have called "prenatal adoption," for not only might the selected embryos be implanted in the uterus of the woman who supplied the oocytes, but in that of any woman at the appropriate time of her menstrual cycle. Edwards cautiously limits the application of his developing techniques to the provision of a healthy embryo for a woman whose oviducts are blocked and prevent descent of the egg. It should be obvious that the technique can be quickly and widely extended. The embryos produced in the laboratory might come from selected genotypes, both male and female. Preservation of spermatozoa in deep frozen condition could permit a high degree of selectivity among the sperm donors, who so far have been limited to the husbands of the women donors of the oocytes. Sex determination of the embryos is possible before implantation; and embryos with abnormal chromosome constitutions can be discarded. By checking the sperm and egg donors with a battery of biochemical tests, matching of carriers of the same defective genes can

be avoided, or the defective embryos can themselves be detected and discarded.[12]

I leave it to the reader to consider the ethical, social, legal, and political implications of Professor Glass's proposal, and to elaborate his own favorite objections. My point here is simply to show that even before Edwards opened Pandora's box, there were well-meaning, decent men already at work to find good uses for its contents.

A similar camel's-nose-under-the-tent argument was advanced by opponents of artificial insemination. Ironically, some of the same people who made light of these arguments in defending artificial insemination are now defending the camel's neck while again dismissing the camel's nose. It is true that the practice of artificial insemination has thus far been confined to the treatment of infertility, and has not, to my knowledge, been malevolently, despotically, or even, until recently, frivolously used—though one now must wonder about its growing uses for eugenic choice or surrogate gestation. But I am no longer talking about the problem of misuse or abuse of a given technique, but rather about the fact that one technical advance makes possible the next and in more than one respect. The first serves as a precedent for the second, the second for the third—not just technologically, but also in moral arguments. At least one good humanitarian ground can be found to justify each step. For these reasons, we must try to see more than a few feet in front of us before we set forth.

It was this kind of foresight that prompted James. D. Watson, co-discoverer of the structure of DNA, to bring before the public his concern over one technological prospect, the cloning of human beings, which Edward's work makes much more possible. In his very sober and careful testimony before the House panel on Science and Technology, Professor Watson concluded a discussion of the work on *in vitro* fertilization as follows:

> Some very hard decisions may soon be upon us. For it is not obvious that the vague potential of abhorrent misuse should weigh more strongly than the unhappiness which thousands of married couples feel when they are unable to have their own children. Different societies are likely to view the matter differently and it would be surprising if all come to the same conclusion. We must, therefore, assume that techniques for the *in vitro* manipulation of human eggs are likely to be general medical practice, capable of routine performance throughout the world within some ten to twenty years.
>
> The situation would then be ripe for extensive efforts, either legal or illegal, at human cloning. . . .
>
> Moreover, given the widespread development of the safe clinical procedures for handling human eggs, cloning experiments would not be prohibitively expensive. They need not be restricted to the super-powers—

medium sized, if not minor countries, all now possess the resources needed for eventual success. There furthermore need not exist the coercion of a totalitarian state to provide the surrogate mothers. There already are such widespread divergences as to the sacredness of the act of human reproduction that the boring meaninglessness of the lives of many women would be sufficient cause for their willingness to participate in such experimentation, be it legal or illegal. Thus, if the matter proceeds in its current nondirected fashion, a human being—born of clonal reproduction—most likely will appear on the earth within the next twenty to fifty years, and conceivably even sooner, if some nation actively promotes the venture.[13]

I now turn to consider this second new method for making babies, asexual reproduction, or cloning.

Cloning, or Asexual Reproduction—State of the Art

In genetic terms, asexual reproduction is distinguished from sexual reproduction (whether practiced in bed or in the test tube) by the following two characteristics: the new individuals are, first, derived from a single parent, and second, genetically identical to—are identical twins of—that parent. Asexual reproduction occurs widely in nature, and is the normal mode of reproduction of bacteria, many plants, and some lower animals. By means of a technique known as nuclear transplantation (also called nuclear transfer), experimental biologists have artificially achieved the asexual reproduction of organisms that naturally reproduce only sexually (so far, frogs, salamanders, and fruit flies). The procedure is conceptually simple. The nucleus of a mature but unfertilized egg is removed (by microsurgery or by irradiation) and replaced with a nucleus obtained from a specialized somatic cell of an adult organism (e.g., an intestinal cell or a skin cell). Since almost all the hereditary material (DNA) of a cell is contained within its nucleus, the renucleated egg and the individual into which it develops are genetically identical to the organism that was the source of the transferred nucleus. Thus, the origin of the new individual is not the chance union of egg and sperm, with the generation of a new and unique genetic arrangement or genotype, but rather the contrived perpetuation into another generation of an already existing genotype.

An unlimited number of identical individuals, all generated asexually from a single parent—that is, a clone—could be produced by nuclear transplantation. An adult organism comprises many millions of cells, all genetically identical, each a potential source of a nucleus for cloning. In addition, techniques for storage and subsequent laboratory culture of animal tissues permit the preservation and propagation of cells long after the deaths of the bodies from which they were removed. There would thus be the possibility of a virtually unlimited supply of genetically identical nuclei for cloning.

The extension of nuclear transplantation to mammals has not yet been achieved, although several people have been trying for a few years. The difficulties are technical; there is no theoretical reason to believe that clonal reproduction is not possible in mammals, including man. The technical problems when this work began included: (1) obtaining mature mammalian eggs; (2) removal of the egg nucleus; (3) insertion of the donor nucleus; and (4) transfer and implantation of the renucleated egg in the uterus of a female at the right stage in her menstrual cycle. As a result of the work on *in vitro* fertilization, the first and fourth problems have been solved. Recently, chemical methods have been perfected to remove the nucleus from mammalian cells in tissue culture, methods that can probably be used to enucleate egg cells. The only serious difficulty remaining is the introduction of the donor nucleus. And this difficulty may also be short-lived, since there are now very simple methods for fusing almost any two cells to produce a single cell containing the combined genetic material of both original cells. Fusion of an enucleated egg cell with a cell containing the donor nucleus might provide the method for getting the nucleus into the egg. In fact, Dr. Christopher Graham at Oxford has already succeeded in fusing mouse eggs with adult mouse cells. The fused egg divides several times but has thus far not gone on to form a blastocyst. Given the rate at which the other technical obstacles have fallen, and given the increasing number of competent people entering the field of experimental embryology, it is reasonable to expect the birth of the first cloned mammal sometime in the next few years. This will almost certainly be followed by a rush to develop cloning for other animals, especially livestock, in order to propagate in perpetuity the champion meat or milk producers.

With the human embryo culture and implantation technologies being perfected in parallel, the step to the first clonal man might require only a few additional years. Within our lifetime, possibly even as early as the year 2000, it may be technically feasible to clone a human being.

Among sensible men, the ability to clone a man would not be sufficient reason for doing so. Nevertheless, the apologists and the titillators have been at work, and the laundry list of possible applications keeps growing in anticipation of the perfected technology: (1) replication of individuals of great genius or great beauty to improve the species or to make life more pleasant; (2) replication of the healthy to bypass the risk of genetic disease contained in the lottery of sexual recombination; (3) provision of large sets of genetically identical hu-

mans for scientific studies on the relative importance of nature and
nurture for various aspects of human performance; (4) provision of a
child to an infertile couple; (5) provision of a child with a genotype of
one's own choosing—of someone famous, of a departed loved one, of
one's spouse or oneself; (6) control of the sex of future children (the
sex of a cloned offspring is the same as that of the adult from whom
the donor nucleus was taken); (7) production of sets of identical persons
to perform special occupations in peace and war (not excluding espionage); (8) production of embryonic replicas of each person, to be frozen away until needed as a source of organs for transplantation to
their genetically identical twin; and (9) to beat the Russians and the
Chinese, to prevent a "cloning gap."

Cloning—Some Ethical Questions

Some of the ethical and social questions raised in connection with *in
vitro* fertilization apply also to cloning: questions of experimenting
upon the unborn, discarding of embryos, problems of misuse and abuse
of power, questions concerning the camel and the tent. I will not repeat what has gone before, except to call special attention to the point
about the ethics of experimentation. A significant number of grossly
abnormal creatures have resulted from the frog experiments, and there
is no reason to be more optimistic about the early attempts in humans.
If the attempts to clone a man result in the production of a defective
"product," who will or should care for it, and what status and rights
will it have? If the offspring is subhuman, are we to consider it murder
to destroy it? The twin issues of the production and disposition of
defectives provide sufficient moral grounds for rebutting any first attempt to clone a man. Add to these the reasons, soon to be discussed,
for believing that one cannot presume a future cloned child's consent
to *be* a clone, even a healthy one, and we must support Paul Ramsey's
verdict: We cannot ethically get to know whether or not human cloning is feasible.

But there are other questions that apply to cloning and not to the
techniques discussed earlier. Among the most important are questions
concerning identity and individuality. One problem can be illustrated
by exploiting the ambiguity of the word "identity": The cloned person
may experience serious concerns about his identity (distinctiveness)
because his genotype, and hence his appearance, stand in a relationship of identity (sameness) to another human being.

The natural occurrence of identical twins in no way weakens the
argument against the artificial production of identical humans; there
are many things that occur accidentally that ought not to be done

deliberately. In fact, the problem of identity faced by identical twins should instruct us and enable us to recognize how much greater the problem might be for someone who was the "child" (or "father") of his twin. I cannot improve upon Paul Ramsey's reflections on this subject:

> Growing up as a twin is difficult enough anyway; one's struggle for self-hood and identity must be against the very human being for whom no doubt there is also the greatest sympathy. Who then would want to be the son or daughter of his twin? To mix the parental and the twin relation might well be psychologically disastrous for the young. Or to look at it from the point of view of parents, it is an awful enough responsibility to be the parent of a son or daughter as things now are. Our children begin with a unique genetic independence of us, analogous to the personal independence that sooner or later will have to be granted them or wrested from us. For us to choose to replicate ourselves in them, to make ourselves the foreknowers and creators of every one of their genetic predispositions, might well prove to be a psychologically and personally unendurable undertaking. For an elder to teach his "infant copy" is a repellant idea not because of the strangeness of it, but because we are altogether too familiar with the problems this would exponentially make more difficult.[14]

Perhaps this issue can be pressed even further, beyond such concerns for undesirable psychological consequences. Does it make sense to say that each person has a right not to be deliberately denied a unique genotype? Is one inherently injured by having been made the copy of another human being, independent of *which* human being? We should not be deterred by the strangeness of these questions, a strangeness due largely to the fact that the problem could not have arisen before.

Central to this matter is the idea of the dignity and worth of each human being. The question we must ask is this: Is individual dignity undermined by a lack of genetic distinctiveness? To which one might argue that indistinctiveness in appearance and capacity might produce a greater incentive to be distinct in deed and accomplishment. Certainly the latter are more germane to any measure of individual worth and self-esteem than the former. On the other hand, our personal appearance is, at the very least, symbolic of our individuality. Differences in personal appearance, genetically determined, reinforce (if not make possible) our sense of self, and hence lend support to the feelings of individual worth we seek in ourselves and from others.* Some put it more strongly and argue that a man not only *has* a body, but *is* his body.† By this argument, a man's distinctive countenance

*See Chapter 13, "Looking Good: Nature and Nobility."
†See Chapter 11, "Thinking About the Body."

not only makes possible his sense of self, but is in fact at one with
that self. Membership in a clone numbering five to ten would no doubt
threaten one's sense of self; membership in a clone of two might also.
To answer the question posed above, we may *not* be entitled, in principle, to a unique genotype, but we are entitled not to have deliberately
weakened the necessary supports for a worthy life. Genetic distinctiveness would seem to me to be one such support.

A second and related problem of identity and individuality is this:
The cloned individual is not simply denied genetic distinctiveness, he
is saddled with a genotype that has already lived. He will not be fully
a surprise to the world; people are likely always to compare his performance in life with that of his alter ego. He may also be burdened
by knowledge of his precursor's life history. Imagine living with the
knowledge that the person from whom you were cloned then developed
schizophrenia or suffered multiple heart attacks before the age of forty.
For these reasons, the cloned individual's belief in the openness of his
own future may be undermined, and with it, his freedom to be himself.
Ignorance of what lies ahead is a source of hope to the miserable, a
spur to the talented, a necessary support for a tolerable—let alone worthy—life for all.

But is the cloned individual's future really determinable or determined? After all, only his genotype has been determined; it is true that
his environment will exert considerable influence on who and what he
becomes. Yet isn't is likely that the "parents" will seek to manipulate
and control the environment as well, in an attempt to reproduce the
person who was copied? For example, if a couple decided to clone a
Rubinstein,* is there any doubt that early in life young Artur would
be deposited at the piano and encouraged to play? It would not matter
to the "parents" that the environment in which the true Rubinstein
blossomed can never be reproduced or even approximated. Nor would
it matter that no one knows what is responsible for the development
of genius, or even for the appearance of ordinary talents and traits for
which other people might have elected to clone. Such ignorance would
not deter the "parents." Why else did they clone young Artur in the
first place?

Thus, although the cloned individual's future is probably not determinable according to his "parents' " wishes, enough damage is done

---

*Frequently mentioned candidates for cloning are musicians and mathematicians, such
as Mozart, Newton, and Einstein, whose genius is presumed to have a large genetic
component. But all such suggestions ignore the wishes of these men. I suspect that
none of them would consent to having themselves replicated. Indeed, should we not
assert as a principle that any so-called great man who *did* consent to be cloned should
on that basis be disqualified, as possessing too high an opinion of himself and of his
genes? Can we stand an increase in arrogance?

by leading him to believe otherwise and by their believing otherwise. His own potential will in all likelihood be stunted and his outlook warped as he is forced into a mold he neither fits nor wants. True, some parents are already guilty of this same crime, but many more are restrained by their impotence in determining the raw material. The opportunity to clone would not only remove this restraint, but would openly invite and encourage more outrageous efforts to shape our children after our own desires.

Although these arguments would apply with even greater force to any large-scale efforts at human cloning, I find them sufficient to reject even the first attempts at human cloning. It cannot be repeated too often that these are human beings upon whom these eugenic or merely playful visions (shall I say hallucinations?) are to be worked.

Thus far, I have dealt separately with two technological prospects, one now upon us, the other on the horizon and fast approaching, in an effort to reason about and evaluate each piece of technology one at a time. I have been at pains to analyze the morally relevant features of each in order to show that real and important distinctions can and should be drawn among different technologies, that the practice of one should not *ipso facto* justify the introduction of another. I am far more concerned that this approach be found reasonable and useful than that any of my specific arguments be found convincing.

Yet despite its practical utility, this piece by piece approach has grave deficiencies. It ignores the great wave upon which each of these techniques is but a ripple. All of the technologies arise from and are part of the great project of modernity, the "conquest of nature for the relief of man's estate." They go beyond many earlier techniques in that they seek to relieve man's estate by directly changing man himself. We must therefore raise some broader questions concerning this project as these questions arise in connection with the technologies discussed above. Here, I am far more concerned that my arguments be found convincing than that they be useful.

Questions of Power

Though philosophically debatable, the Baconian principle "knowledge is power" is certainly correct when applied to that knowledge which has been sought under that principle. The knowledge of how to begin human life in these new ways is a human power to do so. But the power rests only metaphorically with humankind; it rests in fact with particular men—geneticists, embryologists, obstetricians. And these men will wield the power knowledge provides them over other men, and women, and children—especially children of unborn generations. Indeed, it is especially in the coming technologies of human reproduction and genetic engineering that one feels the force of C. S. Lewis's claim, as presented on page 28, that "each new power won *by* man is a power *over* man as well."

The new procedures for making babies all involve a new partner: the scientist-physician. The obstetrician is no longer just the midwife, but also the sower of seed. Even in the treatment of intramarital infertility, the scientist-physician who employs *in vitro* fertilization and laboratory culture of human embryos has acquired far greater power over human life than his colleague who simply repairs the obstructed oviduct. He presides over many creations in many patients. And once he goes beyond the bounds of marriage, he is not simply the Fertilizer General but the Matchmaker as well, as in the practice of artificial insemination (donor), where a small number of physicians have already arranged for the fathering of several hundred thousand children—many of them, nepotistically, by their professional offsprings, the medical students. I am not at present questioning this practice; my point is rather to illustrate how the new technologies lead to the concentration of power.

Both a cause and effect of the growing power of biomedical technologists is the growing complexity of scientific knowledge, and the related fragmentation of disciplines and extreme specialization of their practitioners. Science understands and explains the world in ideas and language that the layman cannot understand. I am not speaking now about the problem of jargon, but rather about a more fundamental matter. The phenomena of nature as they present themselves to us in ordinary experience are understood by the scientists in terms of abstract concepts such as molecules and genes, and ultimately in terms of mathematical formulae. This is to the layman a new cabala, but it is a cabala with a difference: It can create new life in test tubes and can send men to the moon. Small wonder that the scientists and technologists have become, for many people, the new priesthood.*

Because this new priesthood has promised its rewards here on

---

*I once attended a meeting at which scientists were summarizing for a group of educated laymen and potential patrons the current state of knowledge in human genetics and the promising fields for the future. At the conclusion of a summary of studies on mutation, a woman rose in the audience to ask about the meaning of one of the findings for the chance of her having an abnormal child. The answer came back that the matter was complicated, involving some function of "one over the square root of the mean." The woman seated herself with a look of bewilderment on her face, but shaking her head affirmatively, as if to say, "Amen."

There is of course a revolt in progress against this new priesthood, but only because it has been selling indulgences to the Pentagon or because its blessings are not biodegradable. Very few are questioning the intellectual foundations of the modern scientific conception of the world.

earth, it faces perhaps a heavier responsibility—especially when it fails to deliver. The public has acquired high expectations for technology, from which impatience and frustration are easily bred. This point has been surprisingly overlooked, but I think this is what is meant every time someone starts a complaint with "Well, if they can put a man on the moon, why can't they. . . ." Given this general disposition with regard to science and technology, is it not likely that the expectations of an infertile couple will be much higher for any baby given them by the rationalized, disinfected procedures of the laboratory, than for a baby born in the usual way and obtained via adoption? Even with adequate warnings, it will be hard to get the science-worshipping patient to face the reality of a possibly disastrous outcome. Imagine the heartache and then the outcry if the child conceived *in vitro* turns out to have hemophilia or is a mental retardate (even for reasons unrelated to the use of the procedure). On this ground alone, prudence dictates caution.

The problem of the specific abuses of specific powers cannot be overlooked. However, because it is more widely appreciated, I will spend little time on it. It is sufficient merely to mention the possibilities of coercive breeding, cloning tyrants, or producing whole cadres of gammas and deltas to handle the onerous tasks of an advanced civilization, or the more modest abuses of cloning quintuplets for the circus, for five complete sets of spare organs, or for partners in crime who can always have an alibi, to make people see how the technologies might be abused. Nevertheless, while events that have occurred in our lifetime should warn us not to dismiss these possibilities, I think we have greater reason to be concerned about the private, well-intentioned, and voluntary use of the new technologies. The major problem to be feared is not tyranny but voluntary dehumanization.

Questions of Dehumanization

Human procreation not only issues new human beings, it is itself a human activity (an activity of embodied men and women). The new forms of baby-making that I have discussed represent in themselves a radical change in human procreation as a human activity. As already noted, the new beginnings occur in a new locus, the laboratory, and involve a new partner, the scientist. Moreover, the techniques that at first serve merely to provide a child to a childless couple will soon be used to exert control over the quality of the child. A new image of human procreation has been conceived, and a new "scientific" obstetrics will usher it into existence. No more begetting or generating, procreating or even *re*producing: just plain *producing* or *making,* the attempt to supplant nature with rationality in the very mystery of life, all in the service of producing only wanted, willed, and flawless babies. The new reproduction shifts increasingly from home to laboratory, where it is transformed into manufacture. To repeat, increasing control over the product can only be purchased by increasing depersonalization of the process. In this continuum, artificial insemination represented the first step, genetic testing of embryos the second, *in vitro* fertilization the third, and so on and on.

These tendencies are not simply unwelcome. Perhaps for some techniques used for some purposes—artificial insemination (husband) to circumvent infertility—the benefits outweigh the costs. But we cannot say that there are *no* costs. The complete rationalization and depersonalization of procreation (i.e., as in ectogenesis), and its surrender to the demands of the calculating will, would be in itself thoroughly dehumanizing. Might not the initial steps in this direction also be dehumanizing, albeit only embryonically so?

How and why dehumanizing? Because human procreation is not simply an activity of our rational wills. Men and women are embodied as well as calculating creatures. It is for the gods to create in thought and by fiat ("Let the earth bring forth. . . ."). Some future race of demigods (or "demimen") may obtain its survivors from the local fertilization and decanting station, but *human* procreation is begetting. It is a more complete human activity precisely because it engages us bodily, erotically, and even spiritually, as well as rationally. Is there possibly some wisdom in the mystery of nature that joins the pleasure of sex, the inarticulate longing for union, the communication of love, and the deep and partly articulate desire for children in the very activity by which we continue the chain of human existence? Is biological parenthood a built-in "device" selected to promote the adequate caring for posterity? Before we embark on new modes of reproduction, we should consider the meaning of the union of sex, love, and procreation, and the meaning and consequences of its cleavage.

My point is almost certain to be misunderstood. I am not suggesting that one can be truly human only by engaging in procreation. I think there is a clear need for curtailing procreation, and I have no objections to the use of any and all contraceptive devices. I am not suggesting that there is something inhuman about adopting children instead of getting them through the pelvis, nor do I think that the most distinctively human activities center in the groin. My point is simply that there are more and less human ways of bringing a child into the world. I am arguing that the laboratory production of human beings is no longer *human* procreation, that making babies in laboratories—even "perfect" babies—means a degradation of parenthood.

There will be some who object to my calling the new technologies forms of manufacture—but I use the word "manufacture" in quite a literal sense: *handmade.* It matters not whether we are talking about small- or large-scale manufacture. With *in vitro* fertilization, the natural process of generating becomes the artificial process of making. In the case of cloning, the artistry is taken one step further. Not only is the process *in hand,* but the total genetic blueprint of the cloned offspring is selected and determined by the human artisan. To be sure, the subsequent development is still according to natural processes, and no so-called laws of nature have been or can be violated. What has been violated, even if only slightly, is the distinction between the natural and the artificial, and at its very root, the nature of man himself. For man is the watershed that divides the world of the familiar into those things that belong to nature and those things that are made by men. To lay one's hands on human generation is to take a major step toward making man himself simply another one of the man-made things. Thus, human nature becomes simply the last part of nature that is to succumb to the modern technological project, a project that has already turned all the rest of nature into raw material at human disposal, to be homogenized by our rationalized technique according to the artistic conventions of the day.

If the depersonalization of the process of reproduction, and its separation from human sexuality, dehumanizes the activity that brings new life, and if the manufacture of human life threatens its humanness, both together add up to yet another assault on the existence of marriage and the human family. The transfer of procreation to the laboratory, and the shuffling of paternity and maternity this shift makes possible, undermines the justification and support that biological parenthood gives to the monogamous marriage. Cloning adds an additional, more specific, and more fundamental threat: The technique renders males obsolete. All it requires are human eggs, nuclei, and (for the time being) uteri. All three can be supplied by women.

Curiously, both those who welcome and those who fear the new technologies for making babies agree that they will pose serious threats to marriage and the family. Indeed, it seems not unfair to say that one of the reasons, not always explicitly admitted, why the new technologies are endorsed in some quarters is precisely that they will help lay these institutions to rest. The congregation of deliberate family wreckers includes persons eager to remove all restraints from human sexuality or to render obsolete the biological differences between the sexes, others who see the destruction of marriage as a needed step in limiting population growth, and yet others who find the modern nuclear family a stifling and harmful institution for education and child-rearing. I will not deny that the modern nuclear family shows signs of cracking under various pressures. It may have intrinsic lim-

itations that make it seem, even at best, ill-fitted for a modern technological society, but perhaps this should be viewed as a problem of
this society rather than of the family. We really ought to be less frivolous and journalistic in discussing such matters, and should keep in
mind the essential question: Are we to accept as desirable the final
solution that eliminates biological kinship from the foundation of social organization? Yes, laboratory and government alternatives could
be devised for procreation and child-rearing. But at what cost? How
much stunting of our humanity would result from the totalitarian orientation that these alternatives require and foster?

I noted in the first chapter that we foolishly neglect some of the
family's perhaps irreplaceable functions and virtues: refuge in an increasingly impersonal and abrasive world; forge of connectedness to
past and future; nursery of affection and decent human sentiments;
teacher of concern and responsibility for other people and, especially,
for those who come after.* It would be a just irony if programs of
cloning or laboratory-controlled reproduction to improve the genetic
constitutions of future generations were to undermine the very institution that teaches us concern for the future. I repeat: There is every
reason to believe that the elimination of the family would weaken ties
to the past and future, and would throw us even more on the mercy
of an impersonal, lonely present. The burden of proof should fall to
those believing our humanness could survive even if the biological
family does not.

Finally, there may well be a dehumanizing effect on the scientist
himself, and through him, on all of us. The men who are at work on
new beginnings in life are out to subdue one of the most magnificent
mysteries, the mystery of birth and renewal. To some extent, the mystery has already been subdued. Those who do *in vitro* fertilization are
in the business of initiating new life. To the extent that they feel that
there is nothing unusual or awesome in what they are doing they have
already lost the appreciation of mystery, the sense of wonder. The
same can be said of the heart surgeon who sees the heart simply as a
pump, the brain surgeon who sees the brain simply as a computer, or
the pathologist who sees the corpse simply as a body containing demonstrable pathology. The sense of mystery and awe of which I am
speaking is demonstrated by most medical students on their first encounter with a cadaver in the gross anatomy laboratory.† Their uncomfortable feeling is more than squeamishness. It is a deep
recognition, no matter how inarticulate, that it is the mortal remains

---

*See Chapter 9, pp. 236–243, and Chapter 12, pp. 314–317, for further discussion of the
family and perpetuation.
†This topic becomes thematic in Chapter 11, "Thinking About the Body."

of a human being in which they are to be digging, and ultimately, it is a recognition of the mysterious phenomena of life and death. The loss of this sense of awe occurs in a matter of days or weeks; mastery drives out mystery in all but a very few.

There is, I admit, no reason in principle why the sense of mystery needs to be lost by the increase of knowledge or power. Indeed, in the case of the great men of science, knowledge served to increase rather than to decrease their sense of wonder and awe. Nevertheless, for most ordinary men of science and technology, and probably for most men in this technological age, once nature is seen as or transformed into material and given over for manipulation, the mystery and the appreciation are gone. Awed by nothing, freed from all so-called superstitions and so-called atavistic beliefs, they practice their power without even knowing what price they have paid.

Consider, in this connection, these excerpts from an editorial in *Nature* concerning adverse public reaction in Britain to the announcement (apparently erroneous) that Drs. Edwards and Steptoe were at that time about to do the first transfer of a laboratory-grown human embryo into a woman's uterus to circumvent her infertility.

What has all this to do with the test tube baby? In terms of scientific fact, almost nothing at all. The test tube baby, as this phrase is usually understood, refers to the growing of a human embryo to full term outside the body and the chief obstacle to this feat (not that anybody has proclaimed it as a goal) is the formidable problem of maintaining the embryo after the stage at which it would normally implant in the uterus. The Oldham procedure concerns only the pre-implantation embryo which, except in the trivial sense that fertilization is carried out *in vitro,* can hardly be equated with the test tube baby. Moreover, it is difficult to see that the wastage of embryos occasioned by the procedure raises moral problems any knottier than those to do with IUCD, a device which probably prevents the embryo from implanting in the womb. What then was all the fuss about? . . .

A curious feature of the public debate is that the letter-writing segment of the public, at least, seemed to believe that human life was about to be created from nothing in the test tube. For example, a correspondent in *The Times* voiced the fear that "The ability of scientists to develop the technique of creating life in a test tube is so serious that I feel human beings should be given the opportunity to express their views on whether or not this line of research should be pursued. . . . [sic] Personally I find the idea of creating life at man's will terrifying." These are indeed dark and atavistic fears which have been nurtured, perhaps, by the views of Dr. Edmund Leach that scientists have usurped the creative powers and should assume the moral responsibilities formerly attributed to gods. Whatever the merit of Dr. Leach's thesis, those who are engaged in research that is at all liable to be misinterpreted will doubtless take the present episode as a warning of the misunderstandings that can arise, particularly if the true facts are not readily available from authoritative sources. There is always the danger that lack of information or misinformation may convert legitimate public concern about new knowledge into a paranoia that impedes research.[15]

The moral is clear. Research, the supreme value, is to be protected from the dark "atavistic" fears of an ignorant public by giving out only the "true facts."

The first paragraph of the excerpt contains the hard, cold, technical facts, presented "scientifically" by the scientist-editor of *Nature*. That editor finds it "curious" that the public had a somewhat different view of what was done; he erroneously attributes this difference to the public's lack of the "true facts." The source of the difference is not the lack of information, but a difference in interpretation; the real reason for the difference is that the editor lacks, whereas the correspondent in *The Times* does not, the sense of mystery and awe concerning the initiation of new life. The editor is correct in distinguishing *in vitro* fertilization from full extracorporeal gestation, and partly correct in analogizing the question of disposal of embryos to the question concerning the IUCD (but not in thus disposing of the question). But by calling "trivial" the fact that fertilization has occurred *in vitro*, and that embryonic development has been initiated, he displays his own human impoverishment.* I do not insist that the embryo created is a human life worthy of protection, but it surely is alive, and it surely is potentially human. To look upon these embryos as anything less than potential human life in human hands is a misperception so gross as to be alarming.

We have paid some high prices for the technological conquest of nature, but none perhaps so high as the intellectual and spiritual costs of seeing nature as mere material for our manipulation, exploitation, and transformation. With the powers for biological engineering now gathering, there will be splendid new opportunities for a similar deg-

---

*Which account of the beginning of new life seems truer-to-*life*?:

The oocytes were suspended in droplets consisting of fluid from their own follicle (where available), and the medium being tested for fertilization. After incubation for 1–4 hours at 37°C the oocytes were washed through two changes of the medium under test before being placed in the suspensions of spermatozoa. . . . Ejaculated spermatozoa were supplied by the husband. The spermatozoa were washed twice by gentle centrifugation in the medium under test, and made up to a final concentration of between 8×10^5 and 2×10^6 per milliliter depending on the quality of the sample. The higher numbers were used with samples of poor quality containing many inactive spermatozoa, cellular inclusions, other debris, or viscous seminal fluid. The fertilization droplets were approximatley 0.05 milliliters.[16]

or

Now Adam knew Eve his wife, and she conceived and bore Cain, saying I have created a man (equally) with the Lord.[17]

radation in our view of man. Indeed, we are already witnessing the erosion of our idea of man as something splendid or divine, as a creature with freedom and dignity. And clearly, if we come to see ourselves as meat, then meat we shall become.

"Humanization"—Man as Self-Creator

Among those who would take strong exception to my remarks on dehumanization are those who argue that the new biomedical technologies, including those that make possible new methods for making babies, provide the means for human self-modification, and therefore improvement. They see man as imperfect, unfinished, but endowed with creative powers to complete and perfect himself. Included in this group are scientists, such as Robert Sinsheimer:

> For the first time in all time a living creature understands its origin and can undertake to design its future. Even in the ancient myths man was constrained by his essence. He could not rise above his nature to chart his destiny. Today we can envision that chance—and its dark companion of awesome choice and responsibility. . . .
> We are an historic innovation. We can be the agent of transition to a wholly new path of evolution. This is a cosmic event.[18]

and theologians, such as Karl Rahner:

> [F]reedom enables man to determine himself irrevocably, to be for all eternity what he himself has chosen to make himself.[19]

We note in passing that the theologians have done the scientists one better. Most scientists generally talk about what we are now able or free to do, about technique and its possible uses. At most, some take the fatalistic view that "what can be done, will be done." Those theologians-turned-technocrats sanctify the new freedoms: "What can be done, should be done."

The notion of man as a creature who is free to create himself, as a "freedom-event," to use one of Rahner's formulations, is problematic to say the least. It is an idea that is purely formal, not to say empty. It provides no boundaries that would indicate when what was subhuman became truly human, or when what was at first human became less than human. Moreover, the freedom to change one's nature includes the freedom to destroy (by genetic manipulation or brain modification) one's nature, and thereby the capacity and desire for freedom itself. It is, literally, a freedom that can end all freedom. Nor can it provide standards by which to measure whether the changes made are in fact improvements. Evolution simply means change—to measure progress requires a standard that this view cannot in principle supply.

The new technologies for human engineering may well be the "transition to a wholly new path of evolution." They may, therefore, mark the end of *human* life as we and all other humans have known it. It is possible that the nonhuman life that may take our place will be superior, but I think it most unlikely, and certainly not demonstrable. In either case, we are ourselves human beings; therefore, it is proper for us to have a proprietary interest in our survival, and in our survival as *human* beings. This is a difficult enough task without having to confront the prospect of a utopian, constant remaking of our biological nature with all-powerful means but no end in view.

A Matter of Wisdom

I had earlier raised the question of whether we have sufficient wisdom to embark upon new ways for making babies, on an individual scale as well as in the mass. By now it should be clear that I believe the answer must be a resounding no. To have developed to the point of introduction such massive powers with so little deliberation over the desirability of their use can hardly be regarded as evidence of wisdom. And to deny that questions of desirability, of better and worse, can be the subject of rational deliberation, to deny that rationality might dictate that there are some things that we can do that we must never do—in short, to deny the need for wisdom—can only be regarded as the height of folly. Let us simply look at what we have done in our conquest of nonhuman nature. We shall find there no grounds for optimism as we now consider offers to turn our technology loose on human nature. In the absence of standards to guide and restrain the use of this awesome power, we can only dehumanize man as we have despoiled our planet. The knowledge of these standards requires a wisdom we do not possess, and what is worse, we do not even seek.

But we have an alternative. In the absence of such wisdom, we can be wise enough to know that we are not wise enough. To repeat: When we lack sufficient wisdom to do, wisdom consists in not doing. Restraint, caution, abstention, delay are what this second-best (and maybe only) wisdom dictates with respect to baby manufacture, and with respect to various other forms of human engineering made possible by other new biomedical technologies. It remains for another time to discuss how to give practical effect to this conclusion: how to establish reasonable procedures for monitoring, reviewing, and regulating the new technologies; how to deal with the undesirable consequences of their proper use; how to forestall or prevent the introduction of the worst innovations; how to achieve effective international control so that one nation's folly does not lead the world into degradation.

Fortunately, there are no compelling reasons to proceed, certainly not rapidly, with these new methods for making babies. Though it saddens the life of many couples, infertility is hardly one of our major social problems. Moreover, there are other means of circumventing it that are free of the enormous moral and social problems discussed earlier. It is perhaps a questionable sentimentality that seeks to provide every couple with its own biological child rather than continue the practice of adoption. But it would certainly be a foolish sentimentality to unleash baby-making technologies for this purpose, especially before there are means to limit and control their use. The same arguments apply, with equal force, to the use of these technologies for the eradication of genetic disease. We probably do not know enough about the genetics of man, despite our well-meaning desires, to prevent genetic disease by the practice of eugenic abortion (e.g., following amniocentesis). We certainly don't know enough to escalate our tinkering by the eugenic use of the new baby-making techniques.

I am aware that mine is, at least on first glance, not the most compassionate view (although it may very well turn out to be so in the long run). I am aware that there are some who now suffer who will not get relief should my view prevail. Nevertheless, we must measure the cost—and I do not mean the financial cost—of seeking to eradicate that suffering by any and all means. In measuring the cost, we must, of course, evaluate each technological step in its own terms, but we can ill afford to ignore its place in the longer journey. For defensible step by defensible step, we can willingly walk to our own degradation. The road to Brave New World is paved with sentimentality—yes, even with love and charity. Have we enough sense to turn back?

CHAPTER THREE

Perfect Babies
Prenatal Diagnosis and the Equal Right to Life

We hold these truths to be self-evident, that all men are created equal, that they are endowed by their Creator with certain unalienable Rights, that among these are Life, Liberty, and the pursuit of Happiness.

—Declaration of Independence

All animals are equal, but some animals are more equal than others.

—George Orwell, *Animal Farm*

The chapter you are about to read might never have been written. The same, of course, could be said about *any* work of writing, for the usual and obvious reasons—not least, because the author might never have been born. But for the present author and the present readers of the present chapter, the accident of our births may now be seen to have been more than usually accidental. Reflect a moment, gentle reader, and take stock of yourself: I suppose that you, too, will discover how fortunate we are to be here. For we were conceived after the discovery of antibiotics yet before amniocentesis, late enough to have benefited from medicine's ability to prevent and control fatal infectious diseases, yet early enough to have escaped from medicine's ability to detect, and to prevent us from living to suffer, our genetic diseases. To be sure, my own genetic vices are, as far as I know them, rather modest, taken individually—myopia, asthma and other allergies, bilateral forefoot adduction, bowleggedness, loquacity, and pessimism, plus some four to eight as yet undiagnosed recessive lethal genes in the heterozygous condition—but, taken together, and if diagnosable prenatally, I might never have made it.

Over the past decade, thousands of our might-have-been friends and relations have, in fact, *not* made it. For but a single (albeit more serious) genetic fault, itself no fault of theirs, they were prematurely sentenced to death (or nonbirth, if you prefer), through amniocentesis and abortion. Their genetic offenses had become capital—somewhat arbitrarily, since others equally grave have not—only because scientists had developed tests that could detect them prenatally.

Genetic counseling had previously been limited to teaching couples at risk for having children with genetic abnormalities—usually, in families already known to have one or more such offspring—the probabilities of an afflicted child and helping them consider what to do with the knowledge. Only adoption or artificial insemination (with donor semen) stood as practical alternatives to taking the risk or abstaining from further procreation. Amniocentesis and the devising of sophisticated cytological and biochemical tests have changed the situation dramatically; today, though still selectively used, prenatal diagnosis is becoming an integral part of obstetrical care. Early in gestation, cells of the prospective child are obtained by withdrawing samples of the amniotic fluid in which it is suspended *in utero*. From studies on these fetal cells, a large and growing number of chromosomal and functional abnormalities (as well as biological gender) can be accurately diagnosed. Some day, probably in the distant future, genetic screening and prenatal diagnosis may be coupled to techniques of genetic engineering (see Chapter 5, "Patenting Life: Science, Politics, and the Limits of Mastering Nature") that would seek to correct these abnormalities, treating either the fetus *in utero* or, perhaps, the egg and sperm before fertilization (using technologies for reproduction discussed in the last chapter). Eventually, efforts to make perfect babies may go beyond treating known diseases to include also engineering desired improvements in biological capacities. But, for the present, abortion of the defective child-to-be is the only remedy medicine has to offer once the defect is prenatally diagnosed.*

Abortion and Genetic Abortion

Any discussion of the ethical issues of prenatal diagnosis will be unavoidably haunted by a ghost called the morality of abortion. This ghost I shall not vex. Neither shall I vex the reader by telling ghost stories. However, I would be neither surprised nor disappointed if my discussion of an admittedly related matter, the ethics of aborting the genetically defective, summons that hovering spirit to the reader's

---

*Relief of anxiety, if tests are negative, is, of course, a remedy for the apprehensive parents-to-be—though it must be said that the availability of genetic testing is itself probably responsible for much of our increased anxiety about genetic disease, especially where there has been no previous family history.

mind. For the morality of abortion is a matter not easily laid to rest, recent efforts to do so notwithstanding. A decision of the Supreme Court of the United States can indeed legitimate the disposal of fetuses, but not of the moral questions. The questions remain, and there is likely to be little new that can be said about them, and certainly not by me.

Yet before leaving the general question of abortion, let me pause to drop some anchors for the discussion that follows. Despite great differences of opinion regarding what to think and how to reason about abortion, nearly everyone agrees that abortion *is* a moral issue.* What does this mean? Formally, it means that a woman choosing or refusing an abortion (or a physician performing or refusing to perform an abortion) can expect to be asked (at least by herself) to justify her action. A moral choice begs for justification, whereas a mere preference, say for strawberry over vanilla, neither needs nor can get one. In other words, we expect that the woman (or physician) choosing abortion should be able to give reasons for her choice, beyond "I want it" or "I don't want it." Substantively, to say abortion is a moral issue means that, absent good reasons for termination, there is some presumption in favor of allowing the pregnancy to continue once it has begun. A common way, but by no means the only way, of expressing this presumption is to say that "a fetus has a right to continued life."† In this context, disagreement concerning the moral permissibility of abortion concerns what rights (or interests or needs), and whose, override (take precedence over, or outweigh) this fetal right. Even most of the opponents of abortion agree that the mother's right to live takes precedence, and that abortion to save her life is permissible, perhaps

---

*This strikes me as by far the most important inference to be drawn from the fact that human beings in different times and cultures have answered the abortion question differently. Seen in this light, the differing and changing answers themselves suggest that it is a question not easily put under, at least not for very long.

†Other ways include: one should not do violence to living or growing things; life is sacred; life is good; respect nature; respect life; fetal life has value; refrain from taking innocent life; protect and preserve life. It is evident that the terms chosen are of different weight, and would require reasons of different weight to tip the balance in favor of abortion. My choice of the "rights" terminology is not meant to beg the questions of whether such rights really exist, or of where they come from. The notion of a "fetal right to life" presents only a little more difficulty in this regard than does the notion of a "human right to life," since the former does not depend on a claim that the human fetus is already "human." In my sense of the terms "right" and "life," one might even say that a dog or a fetal dog has a "right to life," and that it would be cruel and immoral for a man to go around performing abortions even on dogs for no good reason.

While on the subject of terminology, I note that the choice of words to describe the intrauterine being may also be morally charged. Consider, for example, the differing pictures conjured by "embryo," "fetus," "potential child," "nascent life," "child-to-be," "unborn child," "child," or "baby," or by the pronoun "he" or "she" as opposed to "it." My usual choice will be "fetus"; my reason, a wish not to beg any questions nor to gain by naming what cannot be had by reasoned argument.

obligatory. Some believe that a woman's right to determine the number and spacing of her children takes precedence, while others argue that the need to curb population growth is, at least at this time, overriding.

This brief analysis of what it means to say that abortion is a moral issue should suffice to establish two points. First, the fetus is a living thing with some moral claim on us not to do it violence, and, therefore, second, justification must be given for destroying it.

Let us turn now from the ethical questions of abortion in general, to focus on the special ethical issues raised by the abortion of "defective" fetuses (so-called abortion for fetal indications or genetic abortion). I shall consider only the cleanest cases—those cases in which well-characterized genetic diseases are diagnosed with a high degree of certainty by means of amniocentesis—in order to sidestep the added moral dilemmas posed when the diagnosis is suspected or possible, but unconfirmed. However, much of the discussion will also apply to cases in which genetic analysis gives only a statistical prediction about the genotype of the fetus, and also to cases in which the defect has an infectious or chemical rather than a genetic cause (e.g., rubella, thalidomide).

My first and possibly most difficult task is to show that there is something left to discuss once we have agreed not to discuss the morality of abortion in general. There is a sense in which abortion for genetic defect is, after abortion to save the life of the mother, perhaps the most defensible kind of abortion. Certainly, it is a serious and not a frivolous reason for abortion, defended by its proponents in sober and rational speech—unlike justifications based upon the false notion that a fetus is a mere part of a woman's body, to be used and abused at her pleasure. Standing behind genetic abortion are serious and well-intentioned people, with reasonable ends in view: the prevention of genetic diseases; the elimination of suffering in families; the preservation of precious financial and medical resources; the protection of our genetic heritage. No profiteers, no profligates, no racists. No arguments about the connection of abortion with promiscuity and licentiousness, no perjured testimony about the mental health of the mother, no arguments about the seriousness of the population problem. In short, clear objective data, a worthy cause, decent men and women. If abortion, what better reason for it?

If we consider only the *fact* of abortion, that is, the emptying of the womb and the destruction of fetal life, genetic abortion would seem to raise no new questions. But if we attend to the *reason* for abortion, that is, the genetic defectiveness of the fetus, we confront an entirely new set of issues. Precisely because the quality of the fetus is central to the decision to abort, the practice of genetic abortion has implications beyond those raised by abortion in general. The new focus on quality challenges the old presumption of equality. At stake here is the belief in the radical moral equality of all human beings, the belief that all human beings possess equally and independent of merit certain fundamental rights, one among which is, of course, the right to life.

Equality and Equal Rights

The fate of belief in the equality of fundamental rights must be regarded as a weighty matter, and especially by Americans—regardless of their race, religion, or genotype of origin. For although one may claim that the principle of the sanctity of life, arguably violated in abortion as such, finally requires specific religious justification, no one can deny that the principle of human equality and equal rights has a secular and political-philosophical ground. Moreover this ground is the foundation of the American Republic. The Declaration of Independence states that the American people hold as a self-evident truth "that all men are created equal." Not equal in *all* respects, not created the *same*, but nonetheless equal in certain politically crucial ways. The politically relevant meaning of equality comes out in the next two clauses: "that they are endowed by their Creator with certain unalienable rights, that among these are Life, Liberty, and the pursuit of Happiness." Human beings are equal in the equal and equally unalienable possession of rights. These rights are said to belong to us by nature, by mere membership in the human species, without qualification according to differences in intelligence, virtue, wisdom, beauty, strength, health, or genetic endowment.

To be sure, the belief that fundamental human rights belong equally to all human beings has been but an ideal, never fully realized,* often ignored, sometimes shamelessly. Yet it has been perhaps the most powerful moral idea at work in the world for at least two centuries. It is this idea and ideal that animates most of the current political and social criticism around the globe, and that has inspired much progressive legislation and social change in our own country. It is ironic that we should acquire the power to detect and eliminate the genetically unequal at a time when we have finally succeeded in re-

---

*In the Gettysburg Address, Lincoln calls the principle of human equality a "proposition" to which our Fathers dedicated the nation, rather than, as the Declaration had it, a self-evident truth. Self-evident truths (axioms) neither require nor admit of proof; propositions require it. Lincoln appears to teach that the American Republic is founded to test or prove the truth of the proposition of equal rights, by realizing that ideal through practice dedicated to that end.

moving much of the stigma and disgrace previously attached to victims of congenital illness, in providing them with improved care and support, and in preventing, by means of education, feelings of guilt on the part of their parents. One might even wonder whether the development of amniocentesis and prenatal diagnosis may represent a backlash against these same humanitarian and egalitarian tendencies in the practice of medicine, which, by helping to sustain to the age of reproduction persons with genetic disease, has itself contributed to the increasing incidence of genetic disease, and with it, to increased pressures for genetic screening, genetic counseling, and genetic abortion.

No doubt our humanitarian and egalitarian principles and practices have caused us some new difficulties. I myself have argued in the preceding chapters that compassionate humanitarianism is an insufficient guide to wise practice, and that its logical conclusion would be the dehumanization of a Brave New World. Also, our current egalitarianism has wandered excessively far from that foundational equality of rights, and even foolishly threatens to sacrifice those rights (especially liberty) to achieve equality of condition. But these excesses ought not to lead us to reject the principles of human equality, equal rights, or generous humaneness. And, in any case, if we do mean to weaken or turn our backs on these beliefs and practices, we should do so consciously and thoughtfully. If, as I believe, the way in which genetic abortion is described, discussed, and justified is perhaps of even greater consequence than its practice for our notion of human rights and their equal possession by all human beings, we should pay special attention to questions of language and, in particular, to the question of justification. Before turning full attention to these matters, two points should be clarified.

First, moral questions surrounding genetic abortion, as well as the implications of this practice for the principle of human equality, are largely independent of the question, Who decides? The vast majority of genetic counselors endorse (and, for the most part, in practice adhere to) the principle that the decision to abort an abnormal fetus belongs solely to the woman (or couple).* When challenged about their practice, counselors and obstetricians have tended either to deny that they make any moral choices ("We just provide information; the decision is the woman's") or to assert that they serve the moral good of enhancing freedom ("We get the information *in order* to permit free choice"), thus transforming the substantive moral question (*"What* to do?"*) into a procedural one (*"Who* should *decide* what to do?"). Yet

---

*However, some physicians refuse to perform amniocentesis unless the woman first indicates that she will elect abortion if an abnormality is discovered.

the substantive questions—What decision, and why?—do not disappear simply because the decision is left in the hands of each pregnant woman (or couple); it remains *her* (their) moral questions. And because of the nature of the moral questions of genetic abortion, her decision has consequences that affect more than herself and her fetus. The moral and political health of the community, and, indirectly, of each of its members, is as likely to be affected by the aggregate of purely private and voluntary decisions on genetic abortion as by a uniform policy dictated by physicians or imposed by statute. Further, it seems especially disingenuous and even irresponsible for physicians and scientists to finesse the moral questions of genetic abortion and its implications and to take refuge behind the issue of who decides or behind the not-yet-widened skirts of their pregnant patients. For it is we who are responsible for choosing to develop the technology of prenatal diagnosis, for informing and promoting this technology among the public, and for the actual counseling of individual patients to avail themselves of this procedure.

Second, I wish to distinguish my discussion of what ought to be done from a descriptive account of what in fact is being done, and also from a consideration of what I myself might do if faced with the difficult decision. I cannot know with certainty what I would think, feel, do, or want done faced with the knowledge that my wife was carrying a child branded with Down's syndrome or Tay-Sachs disease. But an understanding of the issues is not advanced by personal anecdote or confession. We all know that what we and others actually do is often done out of weakness, rather than conviction. It is all too human to make an exception in one's own case (consider, e.g., busing our own child, doing national service, taking income tax deductions, claiming or paying a just wage, drinking before driving).

For what it is worth, I confess to feeling more than a little sympathy with would-be parents who choose abortions for severe genetic defects. Nevertheless, as I shall indicate later, in seeking for reasons to justify this practice, I can find none that are in themselves fully satisfactory and none that do not simultaneously justify the killing of "defective" infants, children, and adults. I am mindful that my arguments will fall far from the middle of the stream, yet I hope that the oarsmen of the flagship will pause and row more slowly while we all consider whither we are going.

Genetic Abortion and the Living Defective

The practice of abortion of the genetically defective will no doubt affect our view of and our behavior toward those abnormals who escape the net of detection and abortion. A child with Down's syndrome or hemophilia or muscular dystrophy born at a time when most of his (potential) fellow sufferers were destroyed prenatally is liable to be looked upon by the community as one unfit to be alive, as a second- (or even lower) class human type. He may be seen as a person who need not have been and who would not have been if only someone had gotten to him in time.

The parents of such children are also likely to treat them differently, especially if the mother would have wished but failed to get an amniocentesis because of ignorance, poverty, or distance from the testing station, or if the prenatal diagnosis was in error. In such cases, parents are especially likely to resent the child. They may be disinclined to give it the kind of care they might have before the advent of amniocentesis and genetic abortion, rationalizing that a second-class specimen is not entitled to first-class treatment. If pressed to do so, say by physicians, the parents might refuse, and the courts may become involved. This has already begun to happen, and with increasing frequency.

In one of the earliest cases, in the early 1960s in Maryland, parents of a child with Down's syndrome refused permission to have the child operated on for an intestinal obstruction present at birth. The physicians and the hospital sought an injunction to require the parents to allow surgery. The judge ruled in favor of the parents, despite what was then the weight of precedent to the contrary, on the grounds that the child was "Mongoloid"—i.e., had the child been "normal," the decision would have gone the other way. Although the decision was not appealed and hence not affirmed by a higher court, one could already see through the prism of this case the possibility that the new powers of human genetics would strip the blindfold from the lady of justice and would make official the dangerous doctrine that some men are more equal than others. A steady parade of such "Baby Jane Doe" cases and similar court decisions during the past decade seem to be establishing precisely this doctrine.

The abnormal child may also feel resentful. A child with Down's syndrome or Tay-Sachs disease will probably never know or care, but what about a child with hemophilia or with Turner's syndrome?* In the past two decades, with medical knowledge and power over the prenatal child increasing and with parental authority over the postnatal child decreasing, we have seen the appearance of a new type of legal action, suits for wrongful life. Children have brought suit against their parents (and others) seeking to recover damages for physical and social handicaps inextricably tied to their birth (e.g., congenital deformities, congenital syphilis, illegitimacy). Recently, genetic counselors

---

*Turner's syndrome, caused by a defect or absence of a second sex (X) chromosome (genotype XO, phenotype female), comprises short stature, undifferentiated gonads (and hence sterility), and other variably associated abnormalities.

and laboratories have been sued for not correctly diagnosing prenatally the disease now suffered by the child plaintiff, who, in the absence of such negligence, would have been prevented by abortion from
being born to suffer. Most American courts, though recognizing justice in the child's claim (that he suffers now through someone's negligence), have refused to award damages, due to policy considerations.
But even this precedent has been overturned. In the *Curlender* case
in California,[1] damages were awarded to a child with Tay-Sachs disease whose birth would have been prevented had accurate genetic information been provided her parents. With the spread of amniocentesis
and genetic abortion, we can only expect such cases to increase. And
here it will be the soft-hearted rather than the hard-hearted judges
who will establish the doctrine of second-class human beings because
of their compassion for the mutants who escaped the traps set out for
them.

It may be argued that I am dealing with a problem that, even if
it is real, will affect very few people. It may be suggested that very
few will escape the traps once we have set them properly and widely,
once people are informed about amniocentesis, once the power to detect prenatally grows to its full capacity, and once our allegedly "superstitious" opposition to abortion dies out or is extirpated. But in
order even to come close to this vision of success, amniocentesis or
some other technique for prenatal diagnosis* will have to become part
of every pregnancy—either by making it mandatory, like the test for
syphilis, or by making it routine medical practice, like the Pap smear.
Leaving aside the other problems with universal amniocentesis, we
must expect that the problem for the few who escape is likely to be
even worse precisely because they will be few.

The point, however, should be generalized. How will we come to
view and act toward the many abnormals that will remain among us—
the retarded, the crippled, the senile, the deformed, and the true mutants—once we embark on a program to root out genetic abnormality?
For it must be remembered that we shall always have abnormals—
some who escape detection or whose disease is undetectable *in utero,*
others whose defects are a result of new mutations, birth injuries, accidents, maltreatment, or disease—who will require our care and protection. The existence of "defectives" cannot be fully prevented, not

---

*Amniocentesis is not the only means of gathering information about the genetic state
of the fetus. New techniques that biopsy the chorionic villi (the fetal portion of the
placenta) are already in clinical use, and others that depend on finding fetal cells in the
maternal bloodstream are under investigation. Also, direct inspection of the fetus (fe-
toscopy) provides other information about the normality of development. The issues
discussed in this chapter are, however, independent of the technique used for prenatal
evaluation; they concern only the implications of and justifications for the practice of
aborting those deemed abnormal (by whatever means).

even by totalitarian breeding and weeding programs. Is it not likely that our principle with respect to these people will change from "we try harder" to "why accept second best?" The idea of "the unwanted because abnormal child" may become a self-fulfilling prophecy whose consequences may be worse than those of the abnormality itself.

Genetic and Other Defectives

The mention of other abnormals points to a second danger of the practice of genetic abortion. Genetic abortion may come to be seen not so much as the prevention of genetic diseases, but as the prevention of defective or abnormal children—and, in a way, understandably so. *For in the case of what other diseases does preventive medicine consist in the elimination of the patient at-risk?* Moreover, the very language used to discuss genetic disease leads us to the easy but wrong conclusion that the afflicted fetus or person *is* rather than *has* a disease. True, one is partly defined by his genotype, but only partly. A person is more than his disease. And yet we slide easily from the language of possession to the language of identity, from "he has hemophilia" to "he is a hemophiliac," from "she has diabetes" through "she is diabetic" to "she is a diabetic," from "the fetus has Down's syndrome" to "the fetus is a Down's." This way of speaking encourages the belief that it is defective persons (or potential persons) that are being eliminated, rather than diseases.

If this is so, then it becomes simply accidental that the defect has a genetic cause. Surely, it is only because of the high regard for medicine and science, and for the accuracy of genetic diagnosis, that genotypic defectives are likely to be the first to go. But once the principle, "defectives should not be born," is established, grounds other than cytological and biochemical may very well be sought. Even ignoring racialists and others equally misguided—of course, they cannot be ignored—we should know that there are social scientists, for example, who believe that one can predict with a high degree of accuracy how a child will turn out from a careful, systematic study of the socioeconomic and psychodynamic environment into which he is born and in which he grows up. They might press for the prevention of sociopsychological "disease," even of criminality, by means of prenatal *environmental* diagnosis and abortion. A crude unscientific form of eliminating potential, "phenotypic defectives" may already be operative in some cities in the sense that submission to abortion is allegedly being made a condition for the receipt of welfare payments. " 'Defectives' should not be born" is a principle without limits. We can ill-afford to have it established.

Up to this point, I have been discussing the possible implications of the practice of genetic abortion for our belief in and adherence to the idea that, at least in fundamental human matters such as life and liberty, all men are to be considered as equals, and that, for these matters, we should ignore as irrelevant the real qualitative differences among men, however important these differences may be for other purposes. Those who are concerned about abortion in general fear that the permissible time of eliminating the unwanted will be moved forward along the time continuum, against newborns, infants, and children. Analogously, I suggest that we should be concerned lest the attack on gross genetic inequality in fetuses be advanced both along the continuum of quality and into the later stages of life.

I am not engaged in predicting the future; the point is not that amniocentesis and genetic abortion *will* lead down the road to Nazi Germany. Rather, by examining the logic of justification, we discover that the principles underlying genetic abortion simultaneously justify many further steps down that road. The point was very well made by Abraham Lincoln:

> If A. can prove, however conclusively, that he may, of right, enslave B.— why may not B. snatch the same argument, and prove equally, that he may enslave A?—
>
> You say A. is white, and B. is black. It is *color*, then; the lighter, having the right to enslave the darker? Take care. By this rule, you are to be slave to the first man you meet, with a fairer skin than your own.
>
> You do not mean *color* exactly?—You mean the whites are *intellectually* the superiors of the blacks, and, therefore have the right to enslave them? Take care again. By this rule, you are to be slave to the first man you meet, with an intellect superior to your own.
>
> But, say you, it is a question of *interest;* and, if you can make it your *interest,* you have the right to enslave another. Very well. And if he can make it his interest, he has the right to enslave you.[2]

Perhaps I have exaggerated the dangers; perhaps we will not abandon our inexplicable preference for generous humanitarianism and equal treatment over consistency. But we should indeed be cautious and move slowly as we give serious consideration to the question: What price the perfect baby?[3] In particular, we should attend carefully to the principles and standards that might guide and justify our practice.

Standards for Justifying Genetic Abortion

What would constitute an adequate justification of the decision to abort a genetically defective fetus? Let me suggest the following formal characteristics, each of which still leaves open many questions.

(1) The reasons given should be logically consistent and should lead to relatively unambiguous guidelines—note that I do not say rules—for action in most cases. (2) The justification should make evident to a reasonable person that the interest or need or right being served by abortion is sufficient to override the otherwise presumptive claim on us to protect and preserve the life of the fetus. (3) The justification ought to be such as to help provide intellectual support for drawing distinctions between acceptable and unacceptable kinds of genetic abortion, and between genetic abortion itself and the further practices we would all find abhorrent. (4) The justification ought to be generalizable to cover all persons in identical circumstances. (5) The justifications should not lead to different actions from month to month or from year to year. (6) The justification should be grounded on standards that can, both in principle and in fact, sustain and support our actions in the case of genetic abortion without contradicting or subverting our notions of fundamental and equal human rights.

The reader would do well to consider all these criteria, but I shall focus here primarily on the last. According to what standards can and should we judge a fetus with genetic abnormalities unfit to live (i.e., abortable)? It seems to me that there are at least three dominant standards to which we are likely to repair.

The Social Standard

The first is social or public good. The needs and interests of society are often invoked to justify the practices of prenatal diagnosis and abortion of the genetically abnormal. The argument, full blown, runs something like this. Society has an interest in the genetic fitness of its members. It is foolish for society to squander its precious resources ministering to and caring for the unfit, especially for those who will never become "productive," or who will never in any way benefit society. Therefore, the interests of society are best served by the elimination of the genetically defective prior to their births.

The social standard is all too often reduced to its lowest common denominator: money. Thus, one physician, claiming that he has "made a cost-benefit analysis of Tay-Sachs disease," notes that "the total cost of carrier detection, prenatal diagnosis, and termination of at-risk pregnancies for all Jewish individuals in the United States under thirty who will marry is $5,730,281. If the program is set up to screen only one married partner, the cost is $3,122,695. The hospital costs for the 990 cases of Tay-Sachs disease these individuals would produce over a thirty-year period in the United States is $34,650,000." Another physician, apparently less interested or able to make such a precise audit, has written: "Cost benefit analyses have been made for the total prospective detection and monitoring of Tay-Sachs disease, cystic fibrosis (when prenatal detection becomes available for cystic fibrosis), and other disorders, and in most cases, the expenditures for hospitalization and medical care far exceed the cost of prenatal detection in properly selected risk populations, followed by selective abortion." Yet a third physician has calculated that the costs to the state of caring for children with Down's syndrome is more than three times that of detecting and aborting them. (These authors all acknowledge the additional nonsocial [i.e., private] costs of personal suffering, but insofar as they consider *society,* the costs are purely economic.)

Many questions can be raised about this approach. First, how accurate are the calculations? Not all the costs have been reckoned. The aborted "defective" child will in most cases be "replaced" by a "normal" child. In keeping the ledger, the costs to society of his care and maintenance cannot be ignored—costs of educating him, or removing his wastes and pollutions, not to mention the costs in nonreplaceable natural resources that he consumes. Who is a greater drain on society's precious resources, the average inmate of a home for the retarded or the average graduate of Berkeley? I doubt that we know or can even find out. Then there are the costs of training the physicians and genetic counselors, equipping their laboratories, supporting their research, and sending them and us to conferences to worry about what they are doing. An accurate economic analysis seems to me to be impossible, even in principle. (And even were it possible, the economic argument must still face or silence the challenge posed by that ordinary language philosopher, Andy Capp, who, when his wife said that she was getting really worried about the cost of living, replied: "Sweet'eart, name me one person who wants t'stop livin' on account of the cost.")

A second, more serious, defect of the economic analysis: There are matters of social importance that are not reducible to financial costs, and others that may not be quantifiable at all. How does one quantify the costs of real and potential social conflict, either between children and parents, or between the community and the "deviants" who refuse amniocentesis and continue to bear abnormal children? Can one measure the effect on racial tensions of government-supported programs to screen for and prevent the birth of children homozygous (or heterozygous) for sickle cell anemia? What numbers does one attach to any decreased willingness or ability to take care of the less fortunate, or to cope with difficult problems? And what about the costs of rising expectations? Will we become increasingly dissatisfied with anything short of the "optimum baby"? How does one quantify anxiety? Humiliation? Guilt? Might not the medical profession pay an unmeasurable price if genetic abortion and other revolutionary activities bring about changes in medical ethics and medical practice that lead to the further erosion of trust in the physician? Finally, who is able accurately to compute the costs of weakening our dedication to the proposition of equal rights?

An appeal to social worthiness or usefulness is a less vulgar form of the standard of social good. It is true that great social contributions are unlikely to be forthcoming from persons who suffer from most serious genetic diseases, especially since many of them die in childhood. Yet consider the following remarks by Pearl Buck on the subject of being a mother of a child retarded from phenylketonuria (PKU):*

> My child's life has not been meaningless. She has indeed brought comfort and practical help to many people who are parents of retarded children or are themselves handicapped. True, she has done it through me, yet without her I would not have had the means of learning how to accept the inevitable sorrow, and how to make that acceptance useful to others. Would I be so heartless as to say that it has been worthwhile for my child to be born retarded? Certainly not, but I am saying that even though gravely retarded it has been worthwhile for her to have lived.
>
> It can be summed up, perhaps, by saying that in this world, where cruelty prevails in so many aspects of our life, I would not add the weight of choice to kill rather than to let live. A retarded child, a handicapped person, brings its own gift to life, even to the life of normal human beings. That gift is comprehended in the lessons of patience, understanding, and mercy, lessons which we all need to receive and to practice with one another, whatever we are.[4]

The standard of potential social worthiness is little better in deciding about abortion in particular cases than is the standard of economic cost. To drive home the point, each one of us might consider retrospectively whether he would have been willing, when he was a fetus, to stand trial for his life, pleading only his worth to society as he now can evaluate it. How many of us are not socially "defective" and with none of the excuses possible for a child with PKU? If there is to be human life at all, potential social worthiness cannot be its entitlement.

Finally, we should take note of the ambiguities in the very notion of social good. Some use the term "society" to mean their own particular political community, others to mean the whole human race, and still others speak as if they mean both simultaneously, following that all-too-human belief that what is good for me and mine is good for mankind. Who knows what is genetically best for mankind, even with

---

*PKU is a genetic disorder (autosomal recessive in inheritance) due to an inborn absence of an enzyme that catalyzes the conversion of the amino acid phenylalanine to tyrosine. The children are fair-skinned and suffer from mental retardation, other neurological symptoms (e.g., epilepsy), and eczema, unless they are treated by a diet low in phenylalanine.

respect to Down's syndrome? And even if we knew, we are unlikely to secure it through prenatal diagnosis and abortion. For the genetic heritage of the human species is largely in the care of persons who do not live, and may never live, along the amniocentesis frontier. Moreover, for those who live in the industrialized West, it is certainly a mistake to regard genetic abortion as our first duty to the human gene pool. If we are truly serious about the genetic future of the entire species, we should concentrate our attack on mutagenesis, and especially on *our* large contribution to the pool of environmental mutagens.

But even the more narrow definition of society is ambiguous. Do we mean our society as it is today? Or do we mean our society as it ought to be? If the former, our standards will be ephemeral, for ours is a faddish society. (By far the most worrisome feature of the changing attitudes on abortion is the suddenness with which they changed.) Any such socially determined standards are likely to provide too precarious a foundation for decisions about genetic abortion, let alone for our notions of human rights. If we mean the latter, then we have transcended the social standard, since the good society is not to be found in society itself, nor is it likely to be discovered by taking a vote. In sum, social or public good as a standard for justifying genetic abortion seems to be unsatisfactory. It is hard to define in general, difficult to apply clearly to particular cases, susceptible to overreaching and abuse (hence, very dangerous), and not sufficient unto itself if considerations of the *good* community are held to be automatically implied.

The Familial Standard

A second major alternative is the standard of parental or familial good. Here the argument of justification might run as follows: Parents have a right to determine, according to their own wishes, and based upon their own notions of what is good for them, the qualitative as well as the quantitative character of their families. If they believe that the birth of a seriously deformed child will be the cause of great sorrow and suffering to themselves and to their other children and a drain on their time and resources, then they may ethically decide to prevent the birth of such a child, even by abortion.

This argument, I expect, is more attractive to most people than the argument appealing to the good of society. For one thing, we are more likely to trust a person's conception of what is good for him than his notion of what is good for society. Also, for each decision, the number of persons involved is small, making it seem less impossible to weigh all the relevant factors in determining the good of the family. Most powerfully, one can see and appreciate the possible harm done to healthy children if the parents are obliged to devote most of their energies to caring for the afflicted child.

Yet there are ambiguities and difficulties perhaps as great as with the standard of social good. In the first place, it is not entirely clear what *would* be good for the other children. In a strong family, the experience with a suffering and dying child might help the healthy siblings learn to face and cope with adversity. Some have even speculated that the lack of experience with death and serious illness in our affluent young people is responsible for their immaturity and lack of gravity and their inability to respond patiently and steadily to the serious problems they encounter, in private or community life. Doubtless many American parents have unwittingly fostered childishness by their well-meaning efforts to spare their children any confrontation with harsh reality. Still, I suspect that one can not generalize. In some children, and in some families, experience with suffering may be strengthening, and in others, disabling. My point here is that the matter is uncertain, and that parents deciding *on this basis* are as likely as not to be mistaken.

The familial or parental standard, like the social standard, is unavoidably elastic because suffering does not come in discontinuous units, and because parental wishes and desires know no limits. Both are utterly subjective, relative, and notoriously subject to change. Some parents claim that they could not tolerate raising a child of the undesired sex; I know of one case where, in the delivery room, the mother, on being informed that her child was a son, told the physician that she did not even wish to see it and that he should get rid of it. We may judge her attitude to be pathological, but even pathological suffering is suffering. Would such suffering justify aborting her normal male fetus?

Or take the converse case of two parents, who for their own very peculiar reasons, wish to have an abnormal child, say a child who will suffer from the same disease as grandpa or a child whose arrested development would preclude the threat of adolescent rebellion and separation, or a dwarf who could work with them in the circus. Are these acceptable grounds for the abortion of "normals"?

Granted, such cases will be rare. But they serve to show the dangers inherent in talking about a parental right to determine, according to their wishes, the quality of their children. Indeed, the whole idea of parental rights with respect to children strikes me as problematic. It suggests that children are like property, that they exist *for* the parents. One need only look around to see some of the results of this notion of parenthood. The language of duties to children—including, at minimum, the duty not to violate their unalienable rights as human beings—would be more in keeping with the heavy responsibility we bear in affirming the continuity of life with life, and in trying to transmit what wisdom we have acquired to the next generation. Our children are not *our* children. Reflection on these matters could lead to a greater appreciation of why it is people do and should have children. No better consequence can be hoped for from the advent of amniocentesis and other technologies for controlling human reproduction.

If one speaks of familial good in terms of parental duty, one could still argue that parents have an obligation to do what they can to insure that all their children are born healthy and sound. But this formulation transcends the limitation of parental wishes and desires. As in the case of the good society, the idea of healthy and sound requires a nonarbitrary or so-called objective standard, a standard in the nature of things. Hard as it may be to uncover it, this is what we are seeking.

The Natural Standard

Nature as a standard is thus the third alternative. The justification according to the natural standard might run like this: As a result of our knowledge of genetic diseases, we know that persons afflicted with certain diseases will never be capable of living the full life of a human being. Just as a no-necked giraffe could never live a giraffe's life, or a needleless porcupine would not attain true "porcupine-hood," so a child or fetus with Tay-Sachs disease or Down's syndrome, for example, will never be truly human. He will never be able to care for himself, nor have even the potential for developing, to any significant extent, the distinctively human capacities for thought or self-consciousness. Nature herself has aborted many similar cases and has provided for the early death of many who happen to get born. There is no reason to keep them alive; instead, we should prevent their birth by contraception or sterilization if possible, and abortion if necessary.

The advantages of this approach are clear. The standards are "objective" and in the fetus itself, thus avoiding the relativity and ambiguity we observed in the notions of social and parental good. The standard can be easily generalized to cover all such cases and will be resistant to the shifting sands of public opinion.

This standard, I would suggest, is the one that most physicians and genetic counselors appeal to in their heart of hearts, no matter what they say or do about letting the parents choose. Why else have they developed genetic counseling and amniocentesis? Indeed, the notions of disease, of abnormal, of defective, make no sense at all in the absence of a natural norm of health. This norm is the foundation of the art of the physician and of the inquiry of the health scientist. Yet the standard is elusive. Ironically, we are gaining increasing power to manipulate and control our own nature at a time in which we are increasingly confused about what is normal, healthy, and fit.*

Although possibly acceptable in principle, the natural standard runs into problems in application, when attempts are made to fix the boundary between potentially human and potentially not human. Professor Jerome Lejeune has demonstrated the difficulty, if not the impossibility, of setting clear molecular, cytological, or developmental signposts for this boundary.[5] Attempts to obtain such signposts by induction, say by considering the phenotypes of the worst cases, are equally difficult. Which features would we take to be the most relevant in, say, Tay-Sachs disease, Lesch-Nyhan syndrome, Cri du chat,† Down's syndrome? Certainly severe mental retardation. But how severe is severe? As Abraham Lincoln and I argued earlier, mental retardation admits of degree. It, too, is highly variable and relative. Moreover, it is not clear that certain other defects and deformities might not equally foreclose the possibility of a truly or fully human life. What about blindness or deafness? Quadriplegia? Aphasia? Several of these in combination? Not only does each kind of defect admit a continuous scale of severity, but it also merges with other defects on a continuous scale of defectiveness. Where on this scale is the line to be drawn: after mental retardation? Blindness? Muscular dystrophy? Cystic fibrosis? Hemophilia? Diabetes? Galactosemia? Turner's syndrome? XYY? Clubfoot? Asthma? Moreover, the identical two continuous scales—of kind and severity—are found also among the living. In fact, it is this natural standard that most threatens the notion of human equality. For it leads most directly to the idea that there are second-class human beings and subhuman human beings, not equally entitled to the rights of life or the pursuit of their own happiness.

But the story is not complete. The very idea of "nature" is ambiguous. According to one view, the one I have been using, nature points to or implies a peak, a fullness, a perfection. According to this view, human rights depend upon attaining the status of full humanness. The fetus is only potential; it has no rights, according to this view. But all kinds of people also fall short of the norm: children, idiots, "defective" adults. It is this understanding of nature that has been used to justify not only abortion and infanticide, but also slavery.

---

*For a discussion of the nature of health, see Chapter 6, "The End of Medicine and the Pursuit of Health."

†Lesch-Nyhan syndrome is an inherited (sex-linked) severe neurological disease (of males), characterized by mental retardation, involuntary writhing motions, and compulsive self-mutilation through biting of lips and fingertips; it is due to an absence of an enzyme in purine metabolism. Cri du chat is a hereditary condition characterized by severe mental deficiency, skull abnormalities, and a plaintive cat-like cry; it is due to a deletion of a specific chromosome.

There is another notion of nature, less splendid, more humane, and, though less able to sustain a notion of health, more acceptable to the findings of modern science: nature not as the norm or perfection, but nature as the innate, the given, the inborn. Our animal nature is characterized by impulses toward self-preservation and by the capacity to feel pleasure and to suffer pain. On this understanding of nature, man is fundamentally like the other animals, differing only in his conscious awareness of these inborn impulses, capacities, and concerns. The right to life rests on the fact that we are self-preserving and suffering creatures.* Yet on this understanding of nature, the fetus—even a "defective" fetus, not to speak of a child, "defective" or no—is not potential, but actual. The right to life belongs to him. For this reason, this understanding of nature does not provide and seems even to deny us what we are seeking, namely, a justification for genetic abortion, adequate unto itself, which does not simultaneously justify infanticide, homicide, and enslavement of the genetically abnormal.

There is a third understanding of nature, nature as sacrosanct, nature as created by a Creator. Indeed, to speak about this reminds us that there is a fourth possible standard for judgments about genetic abortion: the religious standard. I shall leave the discussion of this standard to those who are able to speak of it in better faith. I suspect, however, that it, too, will not answer our justificatory needs.

Now that I am at the end, the reader can better share my sense of frustration. I have failed to provide myself with a satisfactory intellectual and moral justification for the practice of genetic abortion. Perhaps others more able than I can supply one. Perhaps the pragmatists can persuade me that we should abandon the search for principled justification, that if we just trust people's situational decisions or their gut reactions, everything will turn out fine. Maybe they are right. But we should not forget the sage observation of Bertrand Russell: "Pragmatism is like a warm bath that heats up so imperceptibly that you don't know when to scream." Before we submerge ourselves irrevocably in amniotic fluid, we should note its connection to our own baths, into which we have started the hot water running.

---

*The whole tradition of natural rights, in which the American Founders firmly stand, goes back to Thomas Hobbes, the first to speak of natural right in this now commonplace sense. "The right of nature, which writers commonly call *jus naturale,* is the liberty each man hath, to use his own power, as he will himself, for the preservation of his own nature; that is to say, of his own life; and consequently, of doing any thing, which in his own judgment, and reason, he shall conceive to be the aptest means thereunto" (*Leviathan,* Chapter XIV). As the example of Hobbes plainly shows, the doctrine of natural and equal rights is connected with a nonteleological view of nature, that is, with a view of nature that knows not natural perfection or fullness.

CHAPTER FOUR

The Meaning of Life— in the Laboratory

People will not look forward to posterity who never look backward to their ancestors.

—Edmund Burke, *Reflections on the Revolution in France*

What's a nice embryo like you doing in a place like this?

—Traditional

The readers of Aldous Huxley's novel, like the inhabitants of the society it depicts, enter into the Brave New World through "a squat gray building . . . the Central London Hatchery and Conditioning Centre," beginning, in fact, in the Fertilizing Room. There, three hundred fertilizers sit bent over their instruments, inspecting eggs, immersing them "in a warm bouillon containing free-swimming spermatozoa," and incubating the successfully fertilized eggs until they are ripe for bottling (or Bokanovskification).[1] Here, most emphatically, life begins with fertilization—in the laboratory. Life in the laboratory is the gateway to the Brave New World.

We stand today fully on the threshold of that gateway. How far and how fast we travel through this entrance is not a matter of chance or necessity but rather a matter of human decision—*our* human decision. Indeed, it seems to be reserved to the people of this country and this century, by our conduct and example, to decide also this important question.

Should we allow or encourage the initiation and growth of human life in the laboratory? This question, in one form or another, has been an issue for public policy for nearly a decade, even before the birth of the first test-tube baby in the summer of 1978. Back in 1975, after prolonged deliberations, the National Commission for the Protection of Human Subjects of Biomedical and Behavioral Research issued its report and recommendations for research on the human fetus. The secretary of health, education, and welfare then published regulations re-

garding research, development, and related activities involving
fetuses, pregnant women, and *in vitro* fertilization.[2] These provided
that no Federal monies should be used for *in vitro* fertilization of human eggs until a special Ethics Advisory Board reviewed the special
ethical issues and offered advice about whether government should
support any such proposed research. Perhaps for the first time in the
modern era of biomedical research, public deliberation and debate
about ethical matters led to an effective moratorium on Federal support for experimentation—in this case, for research on human *in vitro*
fertilization.

A few years later the whole matter once again became the subject
of intense policy debate, when an Ethics Advisory Board was established to consider whether the United States government should fi-
nance research on human life in the laboratory. The question had been
placed on the policy table by a research proposal submitted to the
National Institute of Child Health and Human Development, by Dr.
Pierre Soupart of Vanderbilt University. Dr. Soupart requested
$465,000 for a study to define in part the genetic risk involved in ob-
taining early human embryos by tissue culture methods. He proposed
to fertilize about 450 human ova, obtained from donors undergoing
gynecological surgery (i.e., not from women whom the research could
by expected to help), with donor sperm, to observe their development
for five to six days, and to examine them microscopically for chromosomal and other abnormalities before discarding them. In addition,
he proposed to study whether such laboratory-grown embryos can be
frozen and stored without introducing abnormalities; for it was
thought that temporary cold storage of human embryos might improve the success rate in the subsequent embryo transfer procedure
used to produce a child. Though Dr. Soupart did not then propose to
do embryo transfers to women seeking to become pregnant, his research was intended to serve that goal: He thought to reassure us that
baby-making with the help of *in vitro* fertilization is safe, and he sought
to perfect the techniques of laboratory growth of human embryos introduced by Drs. Edwards and Steptoe in England.

Dr. Soupart's application was approved for funding by the National Institutes of Health review process in October 1977, but because of the administrative regulations, it could not be funded without
review by an Ethics Advisory Board. The then secretary of HEW,
Joseph Califano, constituted such a board, and charged it not only
with a decision on the Soupart proposal, but with an inquiry into all
the scientific, ethical, and legal issues involved, urging it "to provide
recommendations on broad principles to guide the Department in fu-

ture decision-making." After six months of public hearings all over the United States and another six months of private deliberation, the board issued its report in 1979, recommending that research funding be permitted for some *in vitro* experimentation—including the sort proposed by Dr. Soupart. But no secretary of health and human services—then or thereafter—has been willing to act on that recommendation. In fact, Dr. Soupart died in 1981 without having received a clear answer from the government. Thus, we still have no clear policy regarding our question: Should we allow or encourage the initiation and growth of human life in the laboratory?

The Meaning of the Question: The Question of Meaning

How should one think about such ethical questions, here and in general? There are many possible ways, and it is not altogether clear which way is best. For some people, ethical issues are immediately matters of right and wrong, of purity and sin, of good and evil. For others, the critical terms are benefits and harms, risks and promises, gains and costs. Some will focus on so-called rights of individuals or groups (e.g., a right to life or childbirth); still others will emphasize so-called goods for society and its members, such as the advancement of knowledge and the prevention and cure of disease. My own orientation here is somewhat different. I wish to suggest that before deciding what to do, one should try to understand the implications of doing or not doing. The first task, it seems to me, is not to ask "moral or immoral?" or "right or wrong?" but to try to understand fully the meaning and significance of the proposed actions.

This concern with significance leads me to take a broad view of the matter. For we are concerned here not only with some limited research project of the sort proposed by Dr. Soupart, and the narrow issues of safety and informed consent it immediately raises; we are concerned also with a whole range of implications, including many that are tied to definitely foreseeable consequences of this research and its predictable extensions—and touching even our common conception of our own humanity. As most of us are at least tacitly aware, more is at stake than in ordinary biomedical research or in experimenting with human subjects at risk of bodily harm. At stake is the *idea* of the *humanness* of our human life and the meaning of our embodiment, our sexual being, and our relation to ancestors and descendants. In thinking about necessarily particular and immediate decisions, say, for example, regarding Dr. Soupart's research, we must be mindful of the larger picture and must avoid the great danger of trivializing the matter for the sake of rendering it manageable.

The Status of Extracorporeal Life

The meaning of "life in the laboratory" turns in part on the nature and meaning of the human embryo, isolated in the laboratory and separate from the confines of a woman's body. What is the status of a fertilized human egg (i.e., a human zygote) and the embryo that develops from it? How are we to regard its being? How are we to regard it morally (i.e., how are we to behave toward it)? These are, alas, all too familiar questions. At least analogous, if not identical, questions are central to the abortion controversy and are also crucial in considering whether and what sort of experimentation is properly conducted on living but aborted fetuses. Would that it were possible to say that the matter is simple and obvious, and that it has been resolved to everyone's satisfaction!

But the controversy about the morality of abortion continues to rage and divide our nation. Moreover, many who favor or who do not oppose abortion do so despite the fact that they regard the previable fetus as a living human organism, even if less worthy of protection than a woman's desire not to give it birth. Almost everyone senses the importance of this matter for the decision about laboratory culture of and experimentation with human embryos. Thus, we are obliged to take up the question of the status of the embryo in our search for the outlines of some common ground on which many of us can stand. To the best of my knowledge, the discussion that follows is not informed by any particular sectarian or religious teaching, though it may perhaps reveal that I am a person not devoid of reverence and the capacity for awe and wonder, said by some to be the core of the religious sentiment.

I begin by noting that the circumstances of laboratory-grown blastocysts (i.e., three-to-six-day-old embryos) and embryos are not identical with those of the analogous cases of (1) living fetuses facing abortion and (2) living aborted fetuses used in research. First, the fetuses whose fates are at issue in abortion are unwanted, usually the result of so-called accidental conception. Here, the embryos are wanted, and deliberately created, despite certain knowledge that many of them will be destroyed or discarded. Moreover, the fate of these embryos is not in conflict with the wishes, interests, or alleged rights of the pregnant women. Second, though the Federal guidelines governing fetal research permit studies conducted on the not-at-all viable aborted fetus, such research merely takes advantage of available "products" of abortions not themselves undertaken for the sake of the research. No one has proposed and no one would sanction the deliberate production of life fetuses to be aborted for the sake of research, even very beneficial research.* In contrast, we are here considering the deliberate production of embryos for the express purpose of experimentation.

The cases may also differ in other ways. Given the present state of the art, the largest embryo under discussion is the blastocyst, a spherical, relatively undifferentiated mass of cells, barely visible to the naked eye. In appearance it does not look human; indeed, only the most careful scrutiny by the most experienced scientist might distinguish it from similar blastocysts of other mammals. If the human zygote and blastocyst are more like the animal zygote and blastocyst than they are like the twelve-week-old human fetus (which already has a humanoid appearance, differentiated organs, and electrical activity of the brain), then there would be a much diminished ethical dilemma regarding their deliberate creation and experimental use. Needless to say, there are articulate and passionate defenders of all points of view. Let us try, however, to consider the matter afresh.

First of all, the zygote and early embryonic stages are clearly alive. They metabolize, respire, and respond to changes in the environment; they grow and divide. Second, though not yet organized into distinctive parts or organs, the blastocyst is an organic whole, self-developing, genetically unique and distinct from the egg and sperm whose union marked the beginning of its career as a discrete, unfolding being. While the egg and sperm are alive as cells, something new and alive *in a different sense* comes into being with fertilization. The truth of this is unaffected by the fact that fertilization takes time and is not an instantaneous event. For after fertilization is *complete,* there exists a new individual, with its unique genetic identity, fully potent for the self-initiated development into a mature human being, if circumstances are cooperative. Though there is some sense in which the lives of egg and sperm are continuous with the life of the new organism (or, in human terms, that the parents live on in the child-to-be [or child]), in the decisive sense there is a discontinuity, a new beginning, with fertilization. *After* fertilization, there is continuity of subsequent development, even if the locus of the new living being alters with implantation (or birth). Any honest biologist must be impressed by these facts, and must be inclined, at least on first glance, to the view that

---

*Though perhaps a justifiable exception would be a universal plague that fatally attacked all fetuses *in utero.* To find a cure for the end of the species may entail deliberately "producing" (and aborting) live fetuses for research.

a human life begins at fertilization.* Even Dr. Robert Edwards had apparently stumbled over this truth, perhaps inadvertently, in his remark about Louise Brown, his first successful test-tube baby: "The last time I saw *her, she* was just eight cells in a test-tube. *She* was beautiful *then,* and she's still beautiful *now!*"[3]

Granting that a human life begins at fertilization, and comes to be via a continuous process thereafter, surely—one might say—the blastocyst itself can hardly be considered a human being. I myself would agree that a blastocyst is not, in a *full* sense, a human being— or what the current fashion calls, rather arbitrarily and without clear definition, a person. It does not look like a human being nor can it do very much of what human beings do. Yet, at the same time, I must acknowledge that the human blastocyst is (1) human in origin and (2) *potentially* a mature human being, if all goes well. This, too, is beyond dispute; indeed it is precisely because of its peculiarly human potentialities that people propose to study *it* rather than the embryos of other mammals. The human blastocyst, even the human blastocyst *in vitro,* is not humanly nothing; it possesses a power to become what everyone will agree is a human being.

Here it may be objected that the blastocyst *in vitro* has today no such power, because there is now no *in vitro* way to bring the blastocyst to that much later fetal stage in which it might survive on its own. There are no published reports of culture of human embryos past the blastocyst stage (though this has been reported for mice). The *in vitro* blastocyst, like the twelve-week-old aborted fetus, is *in this sense* not viable (i.e., it is at a stage of maturation before the stage of possible independent existence). But if we distinguish, among the *not*-viable embryos, between the *pre*viable and the *not-at-all* viable—on the basis that the former, though not yet viable is capable of *becoming* or *being made* viable[4]—we note a crucial difference between the blastocyst and the twelve-week-old abortus. Unlike an aborted fetus, the blastocyst is possibly salvageable, and hence potentially viable, *if it is transferred to a woman for implantation.* It is not strictly true that the *in vitro* blastocyst is *necessarily* not-viable. Until proven otherwise, by embryo transfer and attempted implantation, we are right to consider the human blastocyst *in vitro* as potentially a human being and, in this respect, not fundamentally different from a blastocyst *in utero.* To put the matter more forcefully, the blastocyst *in vitro* is

---

*The truth of this is not decisively affected by the fact that the early embryo may soon divide and give rise to identical twins or by the fact that scientists may disaggregate and reassemble the cells of the early embryos, even mixing in cells from different embryos in the reaggregation. These unusual and artificial cases do not affect the natural norm, or the truth that *a* human life begins with fertilization—and does so always, if nothing abnormal occurs.

more viable, in the sense of more salvageble, than aborted fetuses at most later stages, up to say twenty weeks.

This is not to say that such a blastocyst is therefore endowed with a so-called right to life, that failure to implant it is negligent homicide, or that experimental touchings of such blastocysts constitute assault and battery. (I myself tend to reject such claims, and indeed think that the ethical questions are not best posed in terms of rights.) But the blastocyst is not nothing; it is *at least* potential humanity, and as such it elicits, or ought to elicit, our feelings of awe and respect. In the blastocyst, even in the zygote, we face a mysterious and awesome power, a power governed by an immanent plan that may produce an indisputably and fully human being. It deserves our respect not because it has rights or claims or sentience (whch it does not have at this stage), but because of what it is, now *and* prospectively.

Let us test this provisional conclusion by considering intuitively our response to two possible fates of such zygotes, blastocysts, and early embryos. First, should such an embryo die, will we be inclined to mourn its passing? When a woman we know miscarries, we are sad—largely for *her* loss and disappointment, but perhaps also at the premature death of a life that might have been. But we do not mourn the departed fetus, nor do we seek ritually to dispose of the remains. In this respect, we do not treat even the fetus as fully one of us.

On the other hand, we would, I suppose, recoil even from the thought, let alone the practice—I apologize for forcing it upon the reader—of eating such embryos, should someone discover that they would provide a great delicacy, a "human caviar." The human blastocyst would be protected by our taboo against cannibalism, which insists on the humanness of human flesh and does not permit us to treat even the flesh of the dead as if it were mere meat. *The human embryo is not mere meat; it is not just stuff; it is not a "thing."** Because of its origin and because of its capacity, it commands a higher respect.

How much more respect? As much as for a fully developed human being? My own inclination is to say probably not, but who can be certain? Indeed, there might be prudential and reasonable grounds for an affirmative answer, partly because the presumption of ignorance

---

*Some people have suggested that the embryo be regarded in the same manner as a vital organ, salvaged from a newly dead corpse, usable for transplantation or research, and that its donation by egg and sperm donors be governed by the Uniform Anatomical Gift Act, which legitimates premortem consent for organ donation upon death. But though this acknowledges that embryos are not "things," it is a mistake to treat embryos as mere organs, thereby overlooking that they are early stages of a complete, whole human being. The Uniform Anatomical Gift Act does not apply to, nor should it be stretched to cover, donation of gonads, gametes (male sperm or female eggs) or—especially—zygotes and embryos.

ought to err in the direction of never underestimating the basis for
respect of human life (not least, for our own self-respect), partly because so many people feel very strongly that even the blastocyst is
protectably human. As a first approximation, I would analogize the
early embryo *in vitro* to the early embryo *in utero* (because both are
potentially viable and human). On this ground alone, the most sensible
policy is to *treat the early embryo as a previable fetus, with constraints imposed on early embryo research at least as great as those
on fetal research.*

To some this may seem excessively scrupulous. They will argue
for the importance of the absence of distinctive humanoid appearance
or the absence of sentience. To be sure, we would feel more restraint
in invasive procedures conducted on a five-month-old or even a twelve-
week-old living fetus than on a blastocyst. But this added restraint
on inflicting suffering on a look-alike, feeling creature in no way denies the propriety of a prior restraint, grounded in respect for individuated, living, potential humanity. Before I would be persuaded to
treat early embryos differently from later ones, I would insist on the
establishment of a reasonably clear, naturally grounded boundary that
would separate "early" and "late," and provide the basis for respecting the "early" less than the "late." This burden must be accepted by
proponents of experimentation with human embryos *in vitro* if a decision to permit the creation of embryos for such experimentation is
to be treated as ethically responsible.

The Treatment of Extracorporeal Embryos

Where does the above analysis lead in thinking about treatment of
human embryos in the laboratory? I indicate, very briefly, the lines
toward a possible policy, though that is not my major intent.

The *in vitro* fertilized embryo has four possible fates: (1) implantation, in the hope of producing from it a child; (2) death, by active
killing or disaggregation, or by a "natural" demise; (3) use in manipulative experimentation—embryological, genetic, etc.; and (4) use in
attempts at perpetuation *in vitro,* beyond the blastocyst stage, ultimately, perhaps, to viability. Let us consider each in turn.

On the strength of my analysis of the status of the embryo, and
the respect due it, no objection would be raised to implantation. *In
vitro* fertilization and embryo transfer to treat infertility, as in the
case of Mr. and Mrs. Brown, is perfectly compatible with a respect
and reverence for human life, including potential human life. Moreover, no disrespect is intended or practiced by the mere fact that several eggs are removed to increase the chance of success. Were it

possible to guarantee successful fertilization and normal growth with a single egg, no more would need to be obtained. Assuming nothing further is done with the unimplanted embryos, there is nothing disrespectful going on. The demise of the unimplanted embryos would be analogous to the loss of numerous embryos wasted in the normal *in vivo* attempts to generate a child. It is estimated that over 50 percent of eggs successfully fertilized during unprotected sexual intercourse fail to implant, or do not remain implanted, in the uterine wall, and are shed soon thereafter, before a diagnosis of pregnancy could be made. Any couple attempting to conceive a child tacitly accepts such embryonic wastage as the perfectly acceptable price to be paid for the birth of a (usually) healthy child. Current procedures to initiate pregnancy with laboratory fertilization thus differ from the natural process in that what would normally be spread over four or five months *in vivo* is compressed into a single effort, using all at once a four or five months' supply of eggs.*

Parenthetically, we should note that the natural occurrence of embryo and fetal loss and wastage does not necessarily or automatically justify all deliberate, humanly caused destruction of fetal life. For example, the natural loss of embryos in early pregnancy cannot in itself be a warrant for deliberately aborting them or for invasively experimenting on them *in vitro,* any more than stillbirths could be a justification for newborn infanticide. There are many things that happen naturally that we ought not do deliberately. It is curious how the same people who deny the relevance of nature as a guide for evaluating human interventions into human generation, and who deny that the term "unnatural" carries any ethical weight, will themselves appeal to "nature's way" when it suits their purposes.† Still, in this present matter, the closeness to natural procreation—the goal is the same, the embryonic loss is unavoidable and not desired, and the amount of loss is

---

*There is a good chance that the problem of surplus embryos may be avoidable, for purely technical reasons. Some researchers believe that the uterine receptivity to the transferred embryo might be reduced during the particular menstrual cycle in which the ova are obtained because of the effects of the hormones given to induce superovulation. They propose that the harvested eggs be frozen and then defrosted one at a time each month for fertilization, culture, and transfer, until pregnancy is achieved. By refusing to fertilize all the eggs at once—not placing all one's eggs in one uterine cycle—there will not be surplus embryos, but at most only surplus eggs. This change in the procedure would make the demise of unimplanted embryos exactly analogous to the "natural" embryonic loss in ordinary reproduction.

†The literature on intervention in reproduction is both confused and confusing on the crucial matter of the meanings of "nature" or "the natural," and their significance for the ethical issues. It may be as much a mistake to claim that the natural has no moral force as to suggest that the natural way is best, because natural. Though shallow and slippery thought about nature, and its relation to "good," is a likely source of these confusions, the nature of nature may itself be elusive, making it difficult for even careful thought to capture what is natural. (See Part III.)

similar—leads me to believe that we do no more intentional or unjustified harm in one case than in the other, and practice no disrespect.

But must we allow the unimplanted *in vitro* embryos to die? Why should they not be either transferred for adoption into another infertile woman, or else used for investigative purposes, to seek new knowledge, say about gene action? The first option raises questions about lineage and the nature of parenthood to which I will return. But even on first glance, it would seem likely to raise a large objection from the original couple who were seeking a child of their own, and not the dissemination of their biological children for prenatal adoption.

But what about experimentation on such blastocysts and early embryos? Is that compatible with the respect they deserve? This is the hard question. On balance, I would think not. Invasive and manipulative experiments involving such embryos very likely presume that they are things or mere stuff, and deny the fact of their possible viability. Certain observational and noninvasive experiments might be different. But on the whole, I would think that the respect for human embryos for which I have argued—I repeat, not their so-called right to life—would lead one to oppose most potentially interesting and useful experimentation. This is a dilemma, but one which cannot be ducked or defined away. Either we accept certain great restrictions on the permissible uses of human embryos or we deliberately decide to override—though I hope not deny—the respect due to the embryos.

I am aware that I have pointed toward a seemingly paradoxical conclusion about the treatment of the unimplanted embryos: leave them alone, and do not create embryos for experimentation only. To let them die naturally would be the most respectful course, grounded on a reverence, generically, for their potential humanity, and a respect, individually, for their being the seed and offspring of a particular couple, who were themselves seeking only to have a child of their own. An analysis that stressed a right to life, rather than respect, would, of course, lead to different conclusions. Only an analysis of the status of the embryo that denies both its so-called rights or its worthiness of all respect would have no trouble sanctioning its use in investigative research, donation to other couples, commercial transactions, and other activities of these sorts.

I have to this point ignored the fourth and future fate of life in the laboratory, perpetuation in the bottle beyond the blastocyst stage, ultimately, perhaps, to viability. As a practical matter, this repugnant Huxleyan prospect probably need not concern us much for the time being. But as a thought experiment, it permits us to test further our intuitions about the meaning of life in the laboratory, and to discover thereby the limitations of the previous analysis. For these unimplanted and cultivated embryos raise even more profound difficulties. Bad as it may now be to discard or experiment upon them in these primordial stages, it will be far worse once we learn how to perpetuate them to later stages in their laboratory existence—especially when the technology arrives that can bring them to viability *in vitro*. For how long and up to what stage of development will they be considered fit material for experimentation? When ought they to be released from the machinery and admitted into the human fraternity, or, at least, into the premature nursery? The need for a respectable boundary defining protectable human life cannot be overstated. The current boundaries, gerrymandered for the sake of abortion—namely, birth or viability—may now satisfy both women's liberation and the United States Supreme Court, and may someday satisfy even a future pope, but they will not survive the coming of more sophisticated technologies for growing life in the laboratory.*

But what if perpetuation in the laboratory were to be sought not for the sake of experimentation but in order to produce a healthy living child—say, one with all the benefits of a scientifically based gestational nourishment and care? Would such treatment of a laboratory-grown embryo be compatible with the respect it is owed? If we consider only what is owed to its vitality and potential humanity *as an individuated human being,* then the laboratory growth of an embryo into a viable full-term baby (i.e., ectogenesis) would be perfectly compatible with the requisite respect. (Indeed, for these reasons one would guess that the right to life people, who object even to the destruction of blastocysts, would find infinitely preferable any form of their preservation and perpetuation to term, in the bottle if necessary.) But the practice of ectogenesis would be incompatible with the *further* respect owed to our humanity on account of *the bonds of lineage, kinship, and descent.* To be human means not only to have human form and powers; it means also to have a human context and to be humanly connected. The navel, no less than speech and the upright posture, is a mark of our being.† It is for these sorts of reasons that we find the Brave New World's Hatcheries dehumanizing.

Assisted by these reflections on the futuristic prospect of ecto-

---

*In *Roe* v. *Wade,*[5] the U.S. Supreme Court ruled that state action regarding abortion was unconstitutional in the first trimester of pregnancy, permissible after the first trimester in order to promote the health of the "mother," and permissible in order to protect "potential life" only at viability (about 24 weeks), prior to which time the state's interest in fetal life was deemed not "compelling." This rather careless and arbitrary placement of boundaries is already something of an embarrassment, thanks to growing knowledge about fetal development and, especially, sophisticated procedures for performing surgery on the intrauterine fetus—even in the second trimester. Also, because viability is, in part, a matter of available outside support, technical advances—such as an artifical placenta or even less spectacular improvements in sustaining premature infants—will reveal that viability is a movable boundary and that development is a continuum without clear natural discontinuities.

†See Chapter 11, "Thinking About the Body," for additional treatment of these bodily marks and the meaning of our (peculiar) embodiment.

genesis, we return for a closer look at the present practices of implantation. For just as the laboratory is not a fitting home for nascent human life, so, too, some human homes are more appropriate than others.

Questions of Lineage and Parenthood, Embodiment and Gender

Many people rejoiced at the birth of Louise Brown. Some were pleased by the technical accomplishment, many were pleased that she was born apparently in good health. But most of us shared the joy of her parents, who after a long, frustrating, and fruitless period, at last had the pleasure and blessing of a child of their own. (Curiously, the perspective of the child was largely ignored. It will thus be easier to come at the matter of lineage by looking at it first from the side of the progenitors rather than the descendants.) The desire to have a child of one's own is acknowledged to be a powerful and deep-seated human desire—some have called it instinctive—and the satisfaction of this desire, by the relief of infertility, is said to be one major goal of continuing work with *in vitro* fertilization and embryo transfer. That this is a worthy goal few, if any, would deny.

Yet let us explore what is meant by "to have a child of one's own." First, what is meant by "to have"? Is the crucial meaning that of gestating and bearing? Or is it to have as a possession? Or is it to nourish and to rear, the child being the embodiment of one's activity as teacher and guide? Or is it rather to provide someone who descends and comes after, someone who will replace oneself in the family line or preserve the family tree by new sproutings and branchings, someone who will renew and perpetuate the vitality and aspiration of human life? (See Chapter 12, "Mortality and Morality: The Virtues of Finitude," pages 314–317.)

More significantly, what is meant by "one's own"? What sense of one's own is important? A scientist might define one's own in terms of carrying one's own genes. Though in some sense correct, this cannot be humanly decisive. For Mr. Brown or for most of us, it would not be a matter of indifference if the sperm used to fertilize the egg were provided by an identical twin brother—whose genes would be, of course, the same as his. Rather, the humanly crucial sense of one's own, the sense that leads most people to choose their own, rather than to adopt, is captured in such phrases as "my seed," "flesh of my flesh," "sprung from my loins." More accurately, since one's own is not the own of one but of two, the desire to have a child of one's own is a *couple's* desire to embody, out of the conjugal union of their separate bodies, a child who is flesh of their separate flesh made one. This archaic language may sound quaint, but I would argue that this is precisely what is being celebrated by most people who rejoice at the birth of Louise Brown, whether they would articulate it this way or not. Mr. and Mrs. Brown, by the birth of their daughter, embody themselves in another, and thus fulfill this aspect of their separate sexual natures and of their married life together. They also acquire descendants and a new branch of their joined family tree. Correlatively, the child, Louise, is given solid and unambiguous roots from which she has sprung and by which she will be nourished.

If this were to be the only use made of embryo transfer, and if providing *in this sense* "a child of one's own" were indeed the sole reason for the clinical use of the techniques, there could be no objection. Here indeed is the natural and proper home for the human embryo. Here indeed is an affirmation of transmission and the importance of lineage and connectedness. Yet there will almost certainly be—in fact, there already are—other uses, involving third parties, to satisfy the desire to have a child of one's own in different senses of "to have" and "one's own." I am not merely speculating about future possibilities. With the technology to effect human *in vitro* fertilization and embryo transfer comes the immediate possibility of egg donation (egg from donor, sperm from husband), embryo donation (egg and sperm from outside of the marriage), and foster pregnancy (host surrogate for gestation). Clearly, the need and demand for extramarital embryo transfers are real and probably large, probably eventually even greater than the intramarital ones.

Nearly everyone agrees that these circumstances are morally and perhaps psychologically more complicated than the intramarital ones. The reasons touch the central core of gestation. Here the meaning of one's own is no longer so unambiguous; neither is the meaning of motherhood and the status of pregnancy. Indeed, one of the clearest meanings of having life in the laboratory is the rupture of the normally necessary umbilical connection between mother and child. This technical capacity to disrupt the connection has in fact been welcomed, curiously, for contradictory reasons. On the one hand, it is argued that embryo donation, or prenatal adoption, would be superior to present adoption, precisely because the woman would have the experience of pregnancy and the child would be born of the adopting mother, rendering the maternal tie that much closer. On the other hand, the mother-child bond rooted in pregnancy and delivery is held to be of little consequence by those who would endorse the use of surrogate gestational mothers, say for a woman whose infertility is due to uterine disease rather than ovarian disease or oviduct obstruction. But in both cases, the new techniques will serve not to ensure and preserve lineage, but rather to confound and complicate it. The principle truly at work in bringing life into the laboratory is not to provide married couples with a child of their own—or to provide a home of their own for children—but to provide anyone who wants one with a child, by whatever possible or convenient means.

So what? it will be said. First of all, we already practice and encourage adoption. Second, we have permitted artificial insemination—though we have, after roughly fifty years of this practice, yet to resolve questions of legitimacy. Third, what with the high rate of divorce and remarriage, identification of mother, father, and child are already complicated. Fourth, there is a growing rate of illegitimacy and husbandless parentages. Fifth, the use of surrogate mothers for foster pregnancy is becoming widespread with the aid of artificial insemination.[6] Finally, our age in its enlightenment is no longer so certain about the virtues of family, lineage, and heterosexuality, or even about the taboos against adultery and incest. Against this background it will be asked, Why all the fuss about some little embryos that stray from their nest?

It is not an easy question to answer. Yet consider. We practice adoption because there are abandoned children who need good homes. We do not, and would not, encourage people deliberately to generate children for others to adopt; partly we wish to avoid baby markets, partly we think it unfair to deliberately deprive the child of his natural ties. Recent years have seen a rise in our concern with roots, against the rootless and increasingly homogeneous background of contemporary American life. Adopted children, in particular, are pressing for information regarding their biological parents, and some states now require that information be made available (on that typically modern rationale of freedom of information, rather than because of the profound importance of lineage for self-identity). Even the importance of children's ties to grandparents are being reasserted, as courts are granting visitation privileges to grandparents, over the objections of divorced-and-remarried former daughters- or sons-in-law. The practice of artificial insemination has yet to be evaluated, the secrecy in which it is practiced being an apparent concession to the dangers of publicity.* Indeed, most physicians who practice artificial insemination (donor) routinely mix in some semen from the husband, to preserve some doubt about paternity—again, a concession to the importance of lin-

---

*There are today numerous suits pending, throughout the United States, because of artificial insemination with donor semen (AID). Following divorce, the ex-husbands are refusing child support for AID children, claiming, minimally, no paternity, or maximally that the child was the fruit of an adulterous "union." In fact, a few states still treat AID as adultery. The importance of anonymity is revealed in the following bizarre case: A woman wanted to have a child, but abhorred the thought of marriage or of sexual relations with men. She learned a do-it-yourself technique of artificial insemination, and persuaded a male acquaintance to donate his semen. Now some ten years after this virgin birth, the case has gone to court. The semen donor is suing for visitation privileges, to see his son.

eage and legitimacy. Finally, what about the changing mores of marriage, divorce, single-parent families, and sexual behavior? Do we applaud these changes? Do we want to contribute further to this confusion of thought, identity, and practice?*

Our society is dangerously close to losing its grip on the meaning of some fundamental aspects of human existence. In reviewing the problem of the disrespect shown to embryonic and fetal life in our efforts to master them, we noted a tendency—we shall meet it again shortly—to reduce certain aspects of human being to mere body, a tendency opposed most decisively in the nearly universal prohibition of cannibalism. Here, in noticing our growing casualness about marriage, legitimacy, kinship, and lineage, we discover how our individualistic and willful projects lead us to ignore the truths defended by the equally widespread prohibition of incest (especially parent-child incest). Properly understood, the largely universal taboo against incest, and also the prohibitions against adultery, defend the integrity of marriage, kinship, and especially the lines of origin and descent. These time-honored restraints implicitly teach that clarity about who your parents are, clarity in the lines of generation, clarity about who is whose, are the indispensable foundations of a sound family life, itself the sound foundation of civilized community. Clarity about your origins is crucial for self-identity, itself important for self-respect. It would be, in my view, deplorable public policy to erode further such fundamental beliefs, values, institutions, and practices. This means, concretely, no encouragement of embryo adoption or especially of surrogate pregnancy. While it would perhaps be foolish to try to proscribe or outlaw such practices,† it would not be wise to support or foster them.

The existence of human life in the laboratory, outside the confines of the generating bodies from which it sprang, also challenges the meaning of our embodiment. People like Mr. and Mrs. Brown, who seek a child derived from their flesh, celebrate in so doing their self-identity with their own bodies, and acknowledge the meaning of the living human body by following its pointings to its own perpetuation. For them, their bodies contain the seeds of their own self-transcendence and enable them to strike a blow for the enduring goodness of the life in which they participate. Affirming the gift of their embodied life, they show their gratitude by passing on that gift to their children.

---

*To those who point out that the bond between sexuality and procreation has already been effectively and permanently cleaved by the pill, and that this is therefore an idle worry in the case of *in vitro* fertilization, it must be said that the pill—like earlier forms of contraception—provides only sex without babies. Babies without sex is the truly unprecedented and radical departure.

†But see first footnote on p. 61.

Only the body's failure to serve the transmission of embodiment has led them—and only temporarily—to generate beyond its confines. But life in the laboratory also allows other people—including those who would donate or sell sperm, eggs, or embryos; or those who would bear another's child in surrogate pregnancy; or even those who will prefer to have their children rationally manufactured entirely in the laboratory—to declare themselves independent of their bodies, in this ultimate liberation. For them the body is a mere tool, ideally an instrument of the conscious will, the sole repository of human dignity. Yet this blind assertion of will against our bodily nature—in contradiction of the meaning of the human generation it seeks to control—can only lead to self-degradation and dehumanization.

In this connection, the case of surrogate wombs bears a further comment. While expressing no objection to the practice of foster pregnancy itself, some people object that it will be done for pay, largely because of their fear that poor women will be exploited by such a practice. But if there were nothing wrong with foster pregnancy, what would be wrong with making a living at it? Clearly, this objection harbors a tacit understanding that to bear another's child for pay is in some sense a degradation of oneself—in the same sense that prostitution is a degradation primarily because it entails the loveless surrender of one's body to serve another's lust, and only derivatively because the woman is paid. It is to deny the meaning and worth of one's body to treat it as a mere incubator, divested of its human meaning. It is also to deny the meaning of the bond among sexuality, love, and procreation. The buying and selling of human flesh and the dehumanized uses of the human body ought not to be encouraged. To be sure, the practice of womb donation could be engaged in for love not money, as it apparently has been in some cases, including the original case in Michigan. A woman could bear her sister's child out of sisterly love. But to the degree that she escapes in this way from the degradation and difficulties of the sale of human flesh and bodily services and the treating of the body as undignified stuff (the problem of cannibalism), once again she approaches instead the difficulties of incest and near incest.

To this point we have been examining the meaning of the presence of human life in the laboratory, but we have neglected the meaning of putting it there in the first place, that is, the meaning of extracorporeal fertilization *as such*. What is the significance of divorcing human generation from human sexuality, precisely for the meaning of our bodily natures as male and female, as both gendered and engendering? To be male or to be female derives its deepest meaning only in relation to the other, and therewith in the gender-mated prospects for generation through union. Our separated embodiment prevents us as lovers from attaining that complete fusion of souls that we as lovers seek; but the complementarity of gender provides a bodily means for transcending separateness through the children born of sexual union. As the navel is our bodily mark of lineage, pointing back to our ancestors, so our genitalia are the bodily mark of linkage, pointing ultimately forward to our descendants. Can these aspects of our being be fulfilled through the rationalized techniques of laboratory sexuality and fertilization? Does not the scientist-partner produce a triangle that somehow subverts the meaning of "two"? Even in the best of cases, do we not pay in coin of our humanity for electing to generate sexlessly?

Future Prospects

Before proceeding to look at some questions of public policy, we need first to consider the likely future developments regarding human life in the laboratory. In my view, we must consider these prospects in reaching our decisions about present policy. For, clearly, part of the meaning of what we are now doing consists in the things it will enable us sooner or later to do hereafter.

What can we expect for life in the laboratory, as an outgrowth of present studies? To be sure, prediction is difficult. One can never know with certainty what will happen, much less how soon. Yet uncertainty is not the same as simple ignorance. Some things, indeed, seem likely. They seem likely because (1) they are thought necessary or desirable, at least by some researchers and their sponsors; (2) they are probably biologically possible and technically feasible; and (3) they will be difficult to prevent or control (especially if no one anticipates their development or sees a need to worry about them). Wise policymakers will want to face up to reasonable projections of future accomplishments, consider whether they are cause for social concern, and see whether or not the principles now enunciated and the practices now established are adequate to deal with any such concerns.

I project at least the following:

1. The growth of human embryos in the laboratory will be extended beyond the blastocyst stage. Such growth must be deemed desirable under all the arguments advanced for developmental research up to the blastocyst stage; research on gene action, chromosome segregation, cellular and organic differentiation, fetus-environment interaction, implantation, etc., cannot answer all its questions with the blastocyst. Such *in vitro* postblastocyst differentiation has apparently been achieved in the mouse, in culture; the use of other mammals as temporary hosts for human embryos is also a possibility. How far such

embryos will eventually be perpetuated is anybody's guess, but full-term ectogenesis cannot be excluded. Neither can the existence of laboratories filled with many living human embryos, growing at various stages of development.

2. Experiments will be undertaken to alter the cellular and genetic composition of these embryos, at first without subsequent transfer to a woman for gestation, perhaps later as a prelude to reproductive efforts. Again, scientific reasons now justifying research like Dr. Soupart's already justify further embryonic manipulations, including formations of hybrids or chimeras (within species and between species); gene, chromosome, and plasmid insertion, excision, or alteration; nuclear transplantation or cloning; etc. The techniques of DNA recombination, coupled with the new skills of handling embryos, make prospects for some precise genetic manipulation much nearer than anyone would have guessed ten years ago. And embryological and cellular research in mammals is making astounding progress. Not long ago the cover of *Science* featured a picture of a hexaparental mouse, born after reaggregation of an early embryo with cells disaggregated from three separate embryos. (Note: That sober journal called this a "Handmade mouse"—literally a *manu-factured* mouse—and went on to say that it was "manufactured by genetic engineering techniques.")[7]

3. Storage and banking of living human embryos (and ova) will be undertaken, perhaps commercially. After all, commercial sperm banks are already well established and prospering.

I can here do no more than identify a few kinds of questions that must be considered in relation to such possible coming control over human heredity and reproduction: questions about the wisdom required to engage in such practices; questions about the goals and standards that will guide our interventions; questions about changes in the concepts of being human, including embodiment, gender, love, lineage, identity, parenthood, and sexuality; questions about the responsibility of power over future generations; questions about awe, respect, humility; questions about the kind of society we will have if we follow along our present course.*

Though I cannot discuss these questions now, I can and must face

---

*Some of these questions were addressed, albeit only briefly, in the latter part of Chapter 2. It has been pointed out to me by an astute colleague that the tone of the present chapter is less passionate and more accommodating than the first (which was written earlier), change he regards as an ironic demonstration of the inexorable way in which we get used to, and accept, our technological nightmares. I myself share his concern. I cannot decide whether the decline of my passion is to be welcomed; that is, whether it is due to greater understanding bred of more thought and experience, or to greater callousness and the contempt of familiarity bred from too much thought and experience. Adaptiveness is our glory and our curse: as Raskolnikov put it, "Man gets used to everything, the beast!"

a serious objection to considering them at all. Most people would agree that the projected possibilities raise far more serious questions than do simple fertilization of a few embryos, their growth *in vitro* to the blastocyst stage, and their subsequent use in experimentation or possible transfer to women for gestation. Why burden present policy with these possibilities? Future abuses, it is often said, do not disqualify present uses (though these same people also often say that "future benefits justify present practices, even questionable ones"). Moreover, there can be no certainty that *A* will lead to *B*. This thin-edge-of-the-wedge argument has been open to criticism.

But such criticism misses the point for two reasons. First, critics often misunderstand the wedge argument, which is not primarily an argument of prediction, that *A* will lead to *B*, say on the strength of the empirical analysis of precedent and an appraisal of the likely direction of present research. It is primarily an argument about the logic of justification. Do not the principles of justification *now* used to justify the current research proposal already justify *in advance* the further developments? Consider some of these principles:

1. It is desirable to learn as much as possible about the processes of fertilization, growth, implantation, and differentiation of human embryos and about human gene expression and its control.
2. It would be desirable to acquire improved techniques for enhancing conception and implantation, for preventing conception and implantation, for the treatment of genetic and chromosomal abnormalities, etc.
3. In the end, only research using *human* embryos can answer these questions and provide these techniques.
4. There should be no censorship or limitation of scientific inquiry or research.

This logic knows no boundary at the blastocyst stage, or, for that matter, at any later stage. For these principles *not* to justify future extensions of current work, some independent additional principles (e.g., a principle limiting such justification to particular stages of development) would have to be found. (Here, the task is to find such a biologically defensible distinction that could be respected as reasonable and not arbitrary, a difficult—perhaps impossible—task, given the continuity of development after fertilization.) Perhaps even more important than any present decision to encourage bringing human life into the laboratory will be the reasons given to support that decision. We will want to know *precisely* what grounds our policymakers will give for endorsing such research, and whether their principles have not already sanctioned future developments. If they do give such wedge-opening justifications, let them do so deliberately, candidly, and intentionally.

A better case to illustrate the wedge logic is the principle offered for the embryo transfer procedure as treatment for infertility. Will we support the use of *in vitro* fertilization and embryo transfer because it provides a child of one's own, in a strict sense of "one's own," to a married couple? Or will we support the transfer because it is treatment of involuntary infertility, which deserves treatment in or out of marriage, hence endorsing the use of any available technical means (that would produce a healthy and normal child), including surrogate wombs, or even ectogenesis?

Second, logic aside, the opponents of the wedge argument do not counsel well. It would be simply foolish to ignore what might come next, and to fail to make the best possible assessment of the implications of present action (or inaction). Let me put the matter very bluntly: The decisions we must now make may very well help to determine whether human beings will eventually be produced in laboratories. I say this not to shock—and I do not mean to beg the question of whether that would be desirable or not. I say this to make sure that we and our policymakers face squarely the full import and magnitude of this decision. Once the genies let the babies into the bottle, it may be impossible to get them out again.

What Should We Do? The Question of Federal Funding

So much, then, for the meanings of initiating, housing, and manipulating human embryos in the laboratory. We are now better prepared to consider the original practical question: Should we allow or encourage these activities? The foregoing reflections still make me doubt the wisdom of proceeding with these practices, both in research and in their clinical application, notwithstanding that valuable knowledge might be had by continuing the research and identifiable suffering might be alleviated by using it to circumvent infertility. To doubt the wisdom of going ahead makes one at least a fellow traveler of the opponents of such research, but it does not, either logically or practically, require that one join them in trying to prevent it, say by legal prohibition. Not every folly can or should be legislated against. Attempts at prohibition here would seem to be both ineffective and dangerous, ineffective because impossible to enforce, dangerous because the costs of such precedent-setting interference with scientific research might be greater than the harm it prevents. To be sure, we already have legal restrictions on experimentation with human subjects, restrictions that are manifestly not incompatible with the progress of medical science. Neither is it true that science cannot survive if it must take some direction from the law. Nor is it the case that all research, because it is research, is or should be absolutely protected. But it does not seem to me that *in vitro* fertilization and growth of human embryos or embryo transfer deserve, at least at present, to be treated as sufficiently dangerous for legislative interference.

But if to doubt the wisdom does not oblige one to seek to outlaw the folly, neither does a decision to permit require a decision to encourage or support. A researcher's freedom to do *in vitro* fertilization, or a woman's right to have a child with laboratory assistance, in no way implies a public (or even a private) obligation to pay for such research or treatment. A right *against* interference is not an entitlement *for* assistance. The question before the Department of Health and Human Services is not whether such research should be permitted or outlawed, but only whether the Federal government should fund it. This is the policy question that needs to be discussed.

I propose to discuss it here, and at some length, not because it is itself timely or relatively important—it is neither—but because it is exemplary. Policy questions regarding controversial new biomedical technologies and practices—as well as other morally and politically charged matters on the border between private and public life (e.g., abortion, racial discrimination, developing the artificial heart, or affirmative action)—frequently take the form of arguments over Federal support. Social control and direction of new developments is often given not in terms of yes or no, but rather, how much, how fast, or how soon? Thus, much of the present analysis—the only one such in the book—can be generalized and made applicable to other specific developments in the field and to the field as a whole.

The arguments in favor of Federal support are well known. First, the research is seen as continuous with, if not quite an ordinary instance of, the biomedical research that the Federal government supports handsomely; roughly two-thirds of the money spent on biomedical research in the United States comes from Uncle Sam. Why is this research different from all other research? Its scientific merit has been attested to by the normal peer review process of NIH. For some, that is a sufficient reason to support it.

Second, there are specific practical fruits expected from the anticipated successes of this new line of research. Besides relief for many cases of infertility, the research promises new birth-control measures based upon improved understanding of the mechanisms of fertilization and implantation, which in turn could lead to techniques for blocking these processes. Also, studies on early embryonic development hold forth the promise of learning how to prevent some congenital malformations and certain highly malignant tumors (e.g., hydatidiform mole) that derive from aberrant fetal tissue.

Third, as he who pays the piper calls the tune, Federal support would make easy Federal regulation and supervision of this research. For the government to abstain, so the argument runs, is to leave the control of research and clinical application in the hands of profit hungry, adventurous, insensitive, reckless, or power-hungry private physicians, scientists, or drug companies, or, on the other hand, at the mercy of the vindictive, mindless, and superstitious civic groups that will interfere with this research through state and local legislation. Only through Federal regulation—which, it is said, can only follow with Federal funding—can we have reasonable, enforceable, and uniform guidelines.

Fourth is the chauvinistic argument that the United States should lead the way in this brave new research, especially as it will apparently be going forward in other nations. Indeed, one witness testifying before the Ethics Advisory Board deplored the fact that the first test-tube baby was British and not American, and complained, in effect, that the existing moratorium on Federal support has already created what one might call an "*in vitro* fertilization gap." The preeminence of American science and technology, so the argument implies, is the center of our preeminence among the nations, a position that will be jeopardized if we hang back out of fear.

Let me respond to these arguments, in reverse order. Conceding—even embracing—the premise of the importance of American science for American strength and prestige, it is far from clear that failure to support *this* research would jeopardize American science. Certainly the use of embryo transfer to overcome infertility, though a vital matter for the couples involved, is hardly a matter of vital national interest—at least not unless and until the majority of American women are similarly infertile. The demands of international competition, admittedly often a necessary evil, should be invoked only for things that really matter; a missile gap and an embryo-transfer gap are chasms apart. In areas not crucial to our own survival, there will be many things we should allow other nations to develop, if that is their wish, without feeling obliged to join them. Moreover, one should not rush into potential folly to avoid being the last to commit it.

The argument about governmental regulations has much to recommend it. But it fails to consider that there are other safeguards against recklessness, at least in the clinical applications, known to the high-minded as the canons of medical ethics and to the cynical as liability for malpractice. Also, Federal regulations attached to Federal funding will not in any case regulate research done with private monies, say by the drug companies. Moreover, there are enough concerned practitioners of these new arts who would have a compelling interest in regulating their own practice, if only to escape the wrath and interference of hostile citizen's groups in response to unsavory goings-on. The available evidence does not convince me that a sensible practice of *in vitro* experimentation requires regulation by the Federal government.

In turning to the argument about anticipated technological powers, we face difficult calculations of unpredictable and more-or-less likely costs and benefits, and the all-important questions of priorities in the allocation of scarce resources. Here it seems useful to consider separately the techniques for generating children and the anticipated techniques for birth control or for preventing developmental anomalies and malignancies.

First, accepting that providing a child of their own to infertile couples is a worthy goal—and it is both insensitive and illogical to cite the population problem as an argument for ignoring the problem of infertility—one can nevertheless question its rank relative to other goals of medical research. One can even wonder whether it is indeed a *medical* goal, or a worthy goal for *medicine*, that is, whether alleviating infertility, especially in this way, is part of the art of healing.[8] Just as abortion for genetic defect is a peculiar innovation in medicine (or in preventive medicine) in which a disease is treated by eliminating the patient (or, if you prefer, a disease is prevented by "preventing" the patient), so laboratory fertilization is a peculiar treatment for oviduct obstruction in that it requires the creation of a new life to "heal" an existing one. All this simply emphasizes the uniqueness of the reproductive organs in that their proper function involves other people, and calls attention to the fact that infertility is not a disease, like heart disease or stroke, even though obstruction of a normally patent tube or vessel is the proximate cause of each.

However this may be, there is a more important objection to this approach to the problem. It represents yet another instance of our thoughtless preference for expensive, high-technology, therapy-oriented approaches to disease and dysfunctions. What about spending this money on discovering the causes of infertility? What about the prevention of tubal obstruction? We complain about rising medical costs, but we insist on the most spectacular and the most technological—and thereby often the most costly—remedies.

The truth is that we do know a little about the causes of tubal obstruction, though much less than we should or could. For instance, it is estimated that at least one-third of such cases are the aftermath of pelvic inflammatory disease, caused by that uninvited venereal guest, gonococcus. Leaving aside any question about whether it makes sense for a Federally funded baby to be the wage of aphrodisiac indiscretion, one can only look with wonder at a society that will have "petri-dish babies"* before it has found a vaccine against gonorrhea.

True, there are other causes of blocked oviducts, and blocked oviducts are not the only cause of female infertility. True, it is not logically necessary to choose between prevention and cure. But *practically* speaking, with money for research as limited as it is, research funds targeted for the relief of infertility should certainly go first to epidemiological and preventive measures—especially where the costs of success in the high-technology cure are likely to be great.

What about these costs? I have already explored some of the nonfinancial costs, in discussing the meaning of this research for our images of humanness. Let us, for now, consider only the financial costs. How expensive was Louise Brown? We do not know, partly because Drs. Edwards and Steptoe did not tell us how many failures preceded their success, how many procedures for egg removal and for fetal monitoring were performed on Mrs. Brown, and so on. To the costs of laparoscopy, fertilization and growth *in vitro,* and transfer, one must add the costs of monitoring the baby's development to check on her "normality" and, should it come, the costs of governmental regulation. A conservative estimate might place the cost of a successful pregnancy of this kind to be between $5,000 and $10,000. If we use the conservative figure of 500,000 for estimating the number of infertile women with blocked oviducts in the United States whose *only* hope of having children lies *in vitro* fertilization,† we reach a conservative

---

*There has been much objection, largely from the scientific community, to the phrase "test-tube baby." More than one commentator has deplored the exploitation of its "flesh-creeping" connotations. They point out that a flat petri-dish is used, not a test tube— as if that mattered—and that the embryo spends only a few days in the dish. But they don't ask why the term "test-tube baby" remains the popular designation, and whether it does not embody more of the deeper truth than a more accurate, laboratory appellation. If the decisive difference is between "in the womb" or "in the lab," the popular designation conveys it (see Afterword," page 125). And it is right on target, and puts us on notice, if the justification for the present laboratory procedures tacitly also justifies future extensions, including full ectogenesis, say, if that were the only way a wombless woman could have a child of her own, without renting a human womb from a surrogate bearer.

†This figure is calculated from estimates that between 10 and 15 percent of all couples are involuntarily infertile, and that in more than half of these cases the cause is in the female. Blocked oviducts account for perhaps 20 percent of the causes of female infertility. Perhaps 50 percent of these women might be helped to have a child by means of reconstructive surgery on the oviducts; the remainder could conceive *only* with the aid of laboratory fertilization and embryo transfer. These estimates do not include additional candidates with uterine disease (who could "conceive" only by embryo transfer to surrogate-gestators), nor those with ovarian dysfunction who would need egg donation as well, nor that growing population of women who have had tubal ligations and who could later turn to *in vitro* fertilization. It is also worth noting that not all the infertile couples are childless; indeed, a surprising number are seeking to enlarge an existing family.

estimated cost of $2.5 to $5 billion. Is it really even fiscally wise for the Federal government to start down this road?

Clearly not, if it is also understood that the costs of providing the service, rendered possible by a successful technology, will also be borne by the taxpayers. Nearly everyone now agrees that the kidney machine legislation, obliging the Federal government to pay an average of $25,000 to $30,000 per patient per year for kidney dialysis for anyone in need (cost to the taxpayers in 1983 was over $1 billion), is an impossible precedent—notwithstanding that individual lives have been prolonged as a result. But once the technique of *in vitro* fertilization and embryo transfer is developed and available, how should the baby-making be paid for? Should it be covered under medical insurance? If a National Health Insurance program is enacted, will and should these services be included? (Those who argue that they are part of medicine will have a hard time saying no.) Failure to do so will make this procedure available only to the well-to-do, on a fee-for-service basis. Would that be a fair alternative? Perhaps, but it is unlikely to be tolerated. Indeed, the principle of equality—equal access to equal levels of medical care—is the leading principle in the press for medical reform. One can be certain that efforts will be forthcoming to make this procedure available equally to all, independent of ability to pay, under Medicaid or National Health Insurance or in some other way. (Only a few years ago, an egalitarian Boston-based group concerned with infertility managed to obtain private funding to pay for artificial insemination for women on welfare!)

Much as I sympathize with the plight of infertile couples, I do not believe that they are entitled to the provision of a child at the public expense, especially now, especially at this cost, especially by a procedure that also involves so many moral difficulties. Given the many vexing dilemmas that will surely be spawned by laboratory-assisted reproduction, the Federal government should not be misled by compassion to embark on this imprudent course.

In considering the Federal funding of such research for its other anticipated technological benefits, independent of its clinical use in baby-making, we face a more difficult matter. In brief, as is the case with all basic research, one simply cannot predict what kinds of techniques and uses it will yield. But here, also, I think good sense would at present say that before one undertakes *human in vitro* fertilization to seek new methods of birth control (e.g., by developing antibodies to the human egg that would physically interfere with its fertilization) one should make adequate attempts to do this in animals. One simply can't get large-enough numbers of human eggs to do this pioneering research well—at least not without subjecting countless women to additional risks not for their immediate benefit. Why not test this conceit first in the mouse or rabbit? Only if the results were very promising—and judged also to be relatively safe in practice—should one consider trying such things in humans. Likewise, the developmental research can and should be first carried out in animals, especially in primates. Though *in vitro* fertilization has yet to be achieved in monkeys, embryo transfer of *in vivo* fertilized eggs has been accomplished, thus permitting the relevant research to proceed. Purely on scientific grounds, the Federal government ought not *now* to be investing its funds in this research for its promised technological benefits—benefits that, in the absence of pilot studies in animals, must be regarded as mere wishful thoughts in the imaginings of scientists.

There does remain, however, the first justification: research for the sake of knowledge itself—knowledge about cell cleavage, cell-cell and cell-environment interactions, and cell differentiation; knowledge of gene action and gene regulation; knowledge of the effects and mechanisms of action of various chemical and physical agents on growth and development; knowledge of the basic processes of fertilization and implantation. This is all knowledge worth having, and though much can be learned using animal sources—and these sources have barely begun to be sufficiently exploited—the investigation of these matters in *man* would, sooner or later, require the use of human-embryonic material. Here, again, there are questions of research priority about which there is room for diagreement, among scientists and laymen alike. But there is also a more fundamental matter.

Is such research consistent with the ethical standards of our community? The question turns in large part on the status of the early human embryo. If, as I have argued, the early embryo is deserving of respect because of what it is, now and potentially, it is difficult to justify submitting it to invasive experiments, and especially difficult to justify *creating* it solely for the purpose of experimentation. The reader should test this conclusion against his or her reaction to imagining the Fertilizing Room of the Central London Hatchery or, more modestly, to encountering an incubator or refrigerator full of living human embryos.

But even if this argument fails to sway our policymakers, another one should. For their decision, I remind you, is not whether *in vitro* fertilization should be permitted in the United States, but whether our tax dollars should encourage and foster it. One cannot, therefore, ignore the deeply held convictions of a sizeable portion of our population—it may even be a majority on this issue—that regards the human embryo as protectable humanity, not to be experimented upon except for its own benefit. Never mind if these beliefs have a religious foundation—as if that should ever be a reason for dismissing them! The presence, sincerity, and depth of these beliefs, and the grave importance of their subject, is what must concern us. The holders of these beliefs have been very much alienated by the numerous court decisions and legislative enactments regarding abortion and research on fetuses. Many who by-and-large share their opinions about the humanity of prenatal life have with heavy heart gone along with the liberalization of abortion, out of deference to the wishes, desires, interests, or putative rights of pregnant women. But will they go along here with what they can only regard as gratuitous and willful assaults on human life, or at least on potential and salvageable human life, and on human dignity? We can ill afford to alienate them further, and it would be unstatesmanlike, to say the least, to do so, especially in a matter of so little importance to the national health and one so full of potential dangers.

Technological progress can be but one measure of our national health. Far more important is the affection and esteem in which our citizenry holds its laws and institutions. No amount of relieved infertility is worth the further disaffection and civil contention that the lifting of the moratorium on Federal funding is likely to produce. People opposed to abortion and people grudgingly willing to permit women to obtain elective abortion but at their own expense, will not tolerate having their tax money spent on scientific research requiring what they regard as at best cruelty, at worst murder. A prudent and wise secretary of health and human services should take this matter most seriously, and continue to refuse to lift the moratorium—at least until he is persuaded that public opinion will overwhelmingly support him. Imprudence in this matter may be the worst sin of all.

An Afterword

This has been for me a long and difficult exposition. Many of the arguments are hard to make. It is hard to get confident people to face unpleasant and future prospects. It is hard to get people to take seriously such "soft" matters as lineage, identity, respect, and self-respect when they are in tension with such "hard" matters as a cure for infertility or new methods of contraception. It is hard to claim respect for human life in the laboratory from a society that does not respect human life in the womb. It is hard to talk about the meaning of sexuality and embodiment in a culture that treats sex increasingly as sport and has trivialized the significance of gender, marriage, and procreation. It is hard to oppose Federal funding of baby-making in a society that increasingly expects the Federal government to satisfy all demands, and that—contrary to so much evidence of waste, incompetence, and corruption—continues to believe that only Uncle Sam can do it. And, finally, it is hard to speak about restraint in a culture that seems to venerate very little above man's own attempt to master all. Here, I am afraid, is the biggest question and one which we perhaps can no longer ask or face: the question about the reasonableness of the desire to become masters and possessors of nature, human nature included.

Here we approach the deepest meaning of *in vitro* fertilization. Those who have likened it to artificial insemination are only partly correct. With *in vitro* fertilization, the human embryo emerges for the first time from the natural darkness and privacy of its own mother's womb, where it is hidden away in mystery, into the bright light and utter publicity of the scientist's laboratory, where it will be treated with unswerving rationality, before the clever and shameless eye of the mind and beneath the obedient and equally clever touch of the hand. What does it mean to hold the beginning of human life before your eyes, in your hands—even for five days (for the meaning does not depend on duration)? Perhaps the meaning is contained in the following story.

Long ago there was a man of great intellect and great courage. He was a remarkable man, a giant, able to answer questions that no other human being could answer, willing boldly to face any challenge or problem. He was a confident man, a masterful man. He saved his city from disaster and ruled it as a father rules his children, revered by all. But something was wrong in his city. A plague had fallen on generation; infertility afflicted plants, animals, and human beings. The man confidently promised to uncover the cause of the plague and to cure the infertility. Resolutely, confidently, he put his sharp mind to work to solve the problem, to bring the dark things to light. No secrets, no reticences, a full public inquiry. He raged against the representatives of caution, moderation, prudence, and piety, who urged him to curtail his inquiry; he accused them of trying to usurp his rightfully earned power, to replace human and masterful control with submissive reverence. The story ends in tragedy: He solved the problem but, in making visible and public the dark and intimate details of his origins, he ruined his life and that of his family. In the end, too late, he learns about the price of presumption, of overconfidence, of the overweening desire to master and control one's fate. In symbolic rejection of his desire to look into everything, he punishes his eyes with self-inflicted blindness.

Sophocles seems to suggest that such a man is always in principle—albeit unwittingly—a patricide, a regicide, and a practitioner of incest. These are the crimes of the tyrant, that misguided and vain seeker of self-sufficiency and full autonomy, who loathes being reminded of his dependence and neediness and who crushes all opposition to the assertion of his will, and whose incest is symbolic of his desire to be even the godlike source of his own being. His character is his destiny.

We men of modern science may have something to learn from our philosophical forebear Oedipus. It appears that Oedipus, being the kind of man an Oedipus is (the chorus calls him a paradigm of man), had no choice but to learn through suffering. Is it really true that we, too, have no other choice?

CHAPTER FIVE

Patenting Life
Science, Politics, and the Limits of Mastering Nature

We have consultations, which of the inventions and experiences which we have discovered shall be published, and which not: and take all an oath of secrecy for the concealing of those which we think fit to keep secret; though some of those we do reveal sometimes to the state, and some not.

— "Father of Salomon's House," in Francis Bacon's *New Atlantis*

We reduce things to mere Nature in order that we may "conquer" them. We are always conquering Nature, because "Nature" is the name for what we have, to some extent, conquered. The price of conquest is to treat a thing as mere Nature. Every conquest over Nature increases her domain.

—C. S. Lewis, *The Abolition of Man*

Every once in a while, we come upon an event of seemingly minor import which, on reflection, turns out to betoken deep and problematic truths about our culture. The "Patenting of Life" decision is such a significant event.

On June 16, 1980, the Supreme Court of the United States ruled that a living microorganism was patentable matter, under the provision of patent laws enacted by Congress in 1952. In 1972, Ananda Chakrabarty, a microbiologist at the University of Illinois, had filed a patent application, assigned to the General Electric Company, asserting multiple claims related to a novel bacterial strain that he had obtained with the aid of techniques of genetic engineering, a strain capable of degrading many components of crude oil and thus potentially useful in the biological control of oil spills. In addition to readily granted process claims for the method of producing the bacterium, and claims relating to the mode of carrying such bacteria to waterborne oil spills, Chakrabarty claimed patent rights to the bacteria themselves. This last claim, at first rejected by the patent examiner and then by the Patent Office Board of Appeals, was finally granted on appeal by the United States Court of Customs and Patent Appeals in 1979, in the decision affirmed by a narrow five–four vote of the Supreme Court a year later.[1]

The case attracted considerable attention, but the Court's decision fell short of the momentous ruling some had anticipated. For one thing, the Court was divided. For another, both sides agreed that the question before them was simply "a narrow one of statutory interpretation," requiring the Court to construe the language of that section of the patent law which defined patentable matter. The Court's opinion, and the dissent, were largely technical. Thus, readers of the opinion who looked for large philosophical dicta about man's art and living nature or about genetic engineering came away disappointed. Alas, it looked as if the Court was, for a change, being simply judicious, doing no more than its proper work.

Yet the decision was not inconsequential. Indeed, it has already contributed to numerous recent practices. Patent claims are now pending for other living microorganisms, as well as for animal cell lines propagated in tissue culture, allegedly valuable for uses ranging from a cheaper means to make penicillin to novel treatments for specific cancers. Genetic engineering firms are springing up all around. Academic molecular biologists are being courted by industry, with astounding financial incentives. Major grants for genetic engineering research to universities have been given by industries in exchange for patent rights to any resulting useful and profitable discoveries. Under such an agreement, Hoechst, the German chemical company, in 1981 gave Harvard University $50 million for a new genetic-engineering institute—the university, due to considerable faculty opposition, had only just before abandoned plans to form its own genetic-engineering company. Many industries are tooling up in anticipation of the flood of new organisms and cell lines to be brought into being with the aid of human ingenuity, spurred on by our ingenious stroke to encourage genius, the patent laws. True, the art of genetic engineering was born and would grow without the *Chakrabarty* decision. But there is no question that it will now grow much, much faster.

The *Chakrabarty* decision is useful in still another, perhaps more fundamental, respect. It is useful for thought, for reflection on the relation between modern science and politics, and between science and the American polity in particular, especially as that relation is embodied in and exemplified by the patent laws. Indeed, the *Chakrabarty* case provides a wonderful mirror in which we can see fundamental features of the American polity, and therewith of modernity itself, and discern some of its deeper tensions: the relation of private interests or rights and the common good; the purposes of science and thought in their relation to practice and to the public interest; and, finally, the prevailing view of man's place in and attitude toward the natural world. Before looking into that mirror, we need to describe some contours of the broader background.

Science in the Public Interest

Science in the public interest is a guiding intention of modern science and has been since its origins in the seventeenth century. Though we hear much about the distinction between pure and applied science— and I, too, shall distinguish them later—it is important to grasp the essentially practical and social intention of modern science as such. Ancient science had sought knowledge of *what* things are, to be contemplated as an end in itself satisfying to the knower. In contrast, modern science seeks knowledge of *how* things work, to be used as a means for the relief and comfort of all humanity, knowers and non-knowers alike. Francis Bacon, the man who first enunciated (1620) the revolutionary assertion that "human knowledge and human power meet in one," complained of the useless and vain speculations of all previous philosophers and urged "that knowledge be used not as a courtesan, for pleasure and vanity only . . . but as a spouse for generation, fruit, and comfort," all in the service of "the relief of man's estate." And standing on the threshold of the new science of mathematical physics, René Descartes (1637) appeals, over the heads of Church and the Schoolmen, to the public and its rulers, for popular support of his researches by announcing the good news of knowledge "very useful in life":

> So soon as I had acquired some general notions concerning Physics . . . I believed that I could not keep them concealed *without greatly sinning against the law which obliges us to procure, as much as in us lies, the general good of all mankind.* For they caused me to see that it is possible to attain knowledge which is very useful in life, and that, instead of that speculative philosophy which is taught in the Schools, we may find a practical philosophy by means of which, knowing the force and the action of fire, water, air, the stars, heaven and all the other bodies that environ us, as distinctly as we know the different crafts of our artisans, we can in the same way employ them in all those uses to which they are adapted, and *thus render ourselves the masters and possessors of nature.* This is not merely to be desired with a view to the invention of an infinity of arts and crafts which enable us to enjoy without any trouble the fruits of the earth and all the good things which are to be found there, but also principally because it brings about the preservation of health, which is without doubt the chief blessing and the foundation of all other blessings in this life. For the mind depends so much on the temperament and dispo-

sition of the bodily organs that, if it is possible to find a means of rendering men wiser and cleverer than they have hitherto been, I believe that it is in medicine that it must be sought.[2]

The announced goal of the new science is the mastery and possession of nature, and the purposes of mastery are humanitarian: the conquest of external necessity; the promotion of bodily health and longevity; the provision of psychic peace or a new kind of practical wisdom.

Even the notions and ways of science manifest a conception of knowledge for the sake of power: nature is conceived mechanistically, and explanation is in terms of efficient or moving causes; hidden truths are gained by acting on nature (i.e., through experiment); inquiry is made "methodical," through the imposition of order and schemes of measurement "made" by the intellect; knowledge, embodied in *laws* rather than theorems, becomes "systematic," under the rules of a new mathematics expressly *invented* for this purpose. Modern science rejects, as meaningless or useless, questions that cannot be answered by application of the method. In all these fundamental ways, modern science has a practical cast. This remains true of the science practiced even by those great scientists who are driven by curiosity and the desire for truth and who have themselves no interest in that mastery and possession of nature for which science is largely esteemed by the rest of us.

Though essentially linked to practice, modern science is, in certain important respects, morally neutral. It does not itself seek knowledge of the good. Indeed, it looks upon nature, its object, as neutral and indifferent to the good or the beautiful. Moreover, the technical power it yields can be used for good or ill. Nevertheless, modern science is guided overall by this moral and political intention: a lifting up of downtrodden humanity; a reversal of the curses laid upon Adam and Eve; and, ultimately, a restoration of the tree of life, by means of the tree of knowledge. Never mind the question how a science invincibly ignorant of and in principle skeptical about standards of better and worse can *know* how to do *good* for mankind. The new humanitarians simply point to the seemingly self-evident truth that life becomes better as it becomes less poor, nasty, brutish, and short.

Gradually, and increasingly, as it began to make good its promise of technological fruit from the tree of useful knowledge, science was welcomed into partnership with the political community. Yet thoughtful men disagreed sharply about *how* science and the useful arts would and should relate to morals and politics. Close to one extreme was the view that popular enlightenment—and particularly the teachings of modern science—undermined ruling opinions and beliefs, especially religious beliefs, necessary for a good regime, and that unbridled progress would lead to luxury, the liberation and inflation of vain and

foolish desires, and the debasement of morals and taste. Though they
appreciated science's contributions to health and plenty, these thinkers argued the need for settled laws, customs, and mores to restrain
the turbulent and licentious souls of men. In the absence of such restraints, the conquest of nature without could enslave us to unruly
nature within.

Even Francis Bacon, perhaps the greatest proponent of the marriage of science and politics, understood that the novelty sought by
the former was not always congenial to the stability required by the
latter:

> But surely there is a great distinction between matters of state and the
> arts; for the danger from new motion and from new light is not the same.
> In matters of state a change even for the better is distrusted, because it
> unsettles what is established; these things resting on authority, consent,
> fame, and opinion, not on demonstration. But arts and sciences should
> be like mines, where the noise of new works and further advances is heard
> on every side.[3]

Bacon's image of the best community, presented in his *New Atlantis*,
does indeed award a central place to Baconian science: The jewel and
lantern of the kingdom is a prodigious, state-supported scientific research foundation called Salomon's House or the College of the Six
Days Works (which, by the way, artfully creates new species through
genetic manipulation). But the community is not enlightened. The
populace has little access to the scientific goings-on, and the scientists
practice self-censorship to avoid publicizing dangerous knowledge. A
benevolent state, with the help of or perhaps under the direction of
the scientists, apparently closely regulates the lives of its inhabitants
by means of austere rituals, state-supported (albeit tolerant) religion,
and, one suspects, perhaps even some scientifically based means of
behavior modification. According to Bacon, the mixture of science and
politics, though desirable and even urgent, was potentially explosive
and needed delicate handling.

In contrast, some of the enlightenment thinkers of the eighteenth
century and their descendants were much more sanguine about the
easy compatibility of science and society. The most optimistic ones
prophesied an unlimited and *coupled* progress of science and morality:
The progress of science and technology would conquer necessity and
alleviate human misery, and man thus emancipated from nature's
harsh and cruel necessities would flower morally into the good creature only his neediness prevents him from being. Once man was liberated and enlightened, the external restraints imposed on him by law,
mores, and religion would, eventually, become unnecessary. In the end,
the state would wither away; politics, the rule over men, would be re-

placed by administration, the management of things. Our various species of Marxism are the lineal descendants of this messianic view of human perfectability based on progress in the arts and sciences.

To summarize, whereas premodern political thinkers and statesmen placed their trust in law and morals, and doubted the ethical and social benefits of inquiry, modern science, devoted to the public good, found a political home and able defenders in modern, liberal regimes. Nevertheless, the proper balance and relation between science-technology and law or morals, between change and stability, remained an open question.

The American Contract

Many Americans today—including, curiously, the critics no less than the friends of technology—share, explicitly or tacitly, this enlightenment dream of human perfectibility. But the Founders of the American Republic, though influenced by optimistic enlightenment thought, were hardly utopians; they pursued a middle course. They knew human nature well enough not to underestimate the crucial importance of good laws, education, and also religion for the preservation of decency and public spiritedness. But they also appreciated fully the promise of science. The American Republic is, to my knowledge, the first regime explicitly to embrace scientific and technical progress and officially to claim its importance for the public good. The United States Constitution, which is silent on education and morality, speaks up about scientific progress. It does so in the course of defining the powers of Congress (Article I, Sec. 8):

> The Congress shall have Power . . . To promote the Progress of Science and useful Arts, by securing for limited Times to Authors and Inventors the exclusive Right to their respective Writings and Discoveries.

"To promote the Progress of Science and the useful Arts." It is curious that this provision has come to be known as the Copyright and Patent Provision rather than the Provision on Progress in the Sciences and Useful Arts, for such progress is the explicit goal and purpose of the congressional power to enact the copyright and patent laws. These statutes, which we think of largely as protecting so-called intellectual property, were in the first instance thought of as useful to scientific and technical progress.

But progress was not itself the final end. Congress was given power to promote the *useful* arts, not the useless ones (e.g., the liberal arts or the fine arts). In the *Federalist* (No. 43), Madison speaks of the unquestionable utility of this congressional power to promote progress, and the context suggests that by its utility he means its usefulness to the public good.* From this we infer that the useful arts and sciences were meant to be subordinated to, and in the service of, the well-being of the nation. Not progress for progress's sake, but progress that might serve the enduring and unchanging goals set forth in the Preamble to the Constitution, among them, to provide for the common defense and promote the general Welfare, and thus, indirectly, to establish Justice, to insure domestic Tranquility, and to secure the Blessings of Liberty to ourselves and our Posterity. The American Republic embraces change, but in the service of duration; science, but in the service of liberty and justice, defined by law. In this respect, the Copyright and Patent Provision is perhaps only the most obvious example of the American way. For the entire Constitution is a deliberate embodiment of balanced tensions between science and law and between stability and novelty, inasmuch as the Founders self-consciously sought to *institutionalize* the improvements of the new "science of politics," and in such a way that would stably perpetuate openness to future change.

How best to promote the arts and sciences? How to induce talented men to behave for the common good? The Constitution once again makes a clear and measured choice: private enterprise, governed and protected by law. Other possibilities were considered by the Convention. Madison had proposed, among the powers of Congress, "To establish a university" and "To encourage by premiums and provisions, the advancement of useful knowledge and discoveries." Yet the Convention rejected the establishment of a national university and the Federal support of science through prizes and provisions, and adopted, apparently without debate, the provision that encourages progress by adding the fuel of interest to the fire of genius.

This reliance on self-interest and the motive of gain might be attributable to the Founders' hardheaded appraisal of the selfish tendencies of most human beings; and cynics have sometimes attributed such motives to the Founders themselves. But a careful look at the

---

*The *Federalist*'s explanation and defense of this provision, to which we shall refer again, comprises the following brief paragraph:[4]

> The utility of this power will scarcely be questioned. The copyright of authors has been solemnly adjudged in Great Britain to be a right of common law. The right to useful inventions seems with equal reason to belong to the inventors. The public good fully coincides in both cases with the claims of individuals. The States cannot separately make effectual provision for either of the cases, and most of them have anticipated the decision of this point by laws passed at the instance of Congress.

The fourth sentence seems to be a conclusion from the first three. Since the second and third deal with "claims of individuals," we infer that the first considers "the public good." It is worth noting that the extension of the common law teaching on copyright to cover "the right to useful inventions" is treated here as an American innovation, albeit one that can be adjudged "with equal reason."

constitutional text indicates that the patent provision is a matter not only of calculation but also of justice. Congress is empowered to secure, that is, to make safe and protect, a *right* of authors and inventors to the fruits of their genius and energy, a right that, by implication, antedates the Constitution. Indeed, this "Right of Authors and Inventors to their respective Writings and Discoveries" is the first and *only* right mentioned in the body of the original Constitution of 1787 (that is, before the Amendments and Bill of Rights). To quote Madison: "The copyright of authors has been solemnly adjudged in Great Britain to be a right of common law. The right to useful inventions seems with equal reason to belong to the inventors."[5]

There is justice, then, in the claims of copyright and patent. To be sure, doing justice will be complicated if the patent prize is awarded only for finishing first in a race in which the winner ran only the last leg of a long relay, tens or hundreds having assisted him. Nevertheless, everyone sees the at least *prima facie* claim that justice requires protecting the labors of the imaginative and industrious against theft by the sly and lazy. If theft of property is wrong, the right of patent is right, at least to some extent. The foundation of the patent law is not only utilitarian, but also ethical.

Indeed, it is ethical also in its consequences for character. The law not only protects individual rights and prevents injustice; it also rewards and encourages the energetic cultivation of the mind and the intellectual virtues of inventiveness, order, and precision, and promotes, in publicly beneficial ways, the moral virtues of ambition and industry. These likely consequences were in fact very important to many of the Founders, and their decision to fuel private enterprise was partly based on these hoped-for improvements in character and mind. To be sure, the mind has other and higher objects than inventions, and ambition and industry do not exhaust the moral virtues. Still a respect for the human mind and an appreciation of efforts to realize its potential are built into our constituting law. One errs to see here only greed and base calculation.

Patent laws serve the public interest at the same time as they protect private rights. The community gains publication, likely development of inventions, a share in the resulting prosperity, and, should it desire it, some legislative hand on the throttle of progress. The patent laws of 1790, enacted by the First Congress, thus established what can rightly be called an ethical-social contract of science in the public interest. In order to secure their rights, authors and inventors had to disclose, that is, make gift to the public of their findings: No protection without publication. (In choosing to promote the widest possible publication, the Founders showed less concern than Bacon for the problem of dangerous knowledge, a matter to which we shall return.)

Moreover, the exclusive right was obtained for only a limited period, to encourage prompt development and production of new inventions; thus, society might reap the benefits of innovation more quickly than if the right were of unlimited duration. All in all, the Copyright and Patent Provision and the patent law are most ingenious, public-spirited, and just inventions—themselves worthy of patent protection. Madison praised the former, saying, "The public good fully coincides with the claims of individuals."[6] Abraham Lincoln (in his lecture on "Discoveries and Inventions," February 11, 1859) listed the latter, along with the arts of writing and printing and the discovery of America, among those few inventions and discoveries in the history of the world, most valuable "on account of their great efficiency in facilitating all other inventions and discoveries."[7]

Time has vindicated these judgments. We are showered on all sides by countless benefits of this farsighted invention of the American mind, which harnessed science and artful intelligence to the carriage of state and which kept it moving by means of the carrot of self-interest. It would seem hypocritical and, what is worse, ungrateful to question this arrangement, all the more so in the light of the marvelous contributions to our health and prosperity that we can now obviously expect from the industrial exploitation of learning how to get microorganisms to do our manufacturing.

And yet, honesty compels us to acknowledge that we are not as sanguine about the future for our children as our forefathers were for ours. Under our present circumstances, the entanglements of science and politics present clear and present dangers to human freedom and dignity, above and beyond the threat of nuclear catastrophe. Honesty also compels us to point out certain peculiarities in the peculiarly American arrangement between science and politics, peculiarities that might eventually give rise to serious difficulties, not only for the union of science and the American polity, but also for each of the partners taken separately. First, it should be observed that the contract formed by the patent law brings together, in stressful if fertile union, certain contradictory, or at least inhospitable, partners and principles: self-interest and common good; monopoly and liberty; the ownership of ideas and the shareability or publicity of speech and thought. The patent law seeks to promote the common good by licensing private interest, thus running the risk of fostering a crass selfishness that in any particular instance might sacrifice public interest to private gain, and that eventually renders men generally indifferent, or even hostile, to the common good. It seeks indirectly, by means of progress and prosperity, to safeguard political liberty, but it does so by legitimating monopoly—albeit of limited duration—which is the antithesis of liberty. It rewards publication and, therefore, presupposes the shareability of thoughts and ideas, yet it does so by licensing the private ownership of these works of mind.

Second, there is the already noted built-in tension between progress and stability. Indeed, the very idea of a patent law is something of an oxymoron: It is a hybrid of two opposing principles, change and order, that live always in tension with each other. Law as law stands for order and stability. It not only sets limits and restrains undesirable conduct, it also embodies our opinions, albeit our variable opinions, about what is just and good. Though subject to change, law as such points to what is permanent. A law to encourage progress is thus, at bottom, a paradoxical law. In a way, though it promotes change, as an expression of *legitimacy* the patent law still, at least formally, accords primacy to order. Absent such a law, innovation would lack legal protection and even legitimacy. Thus, the supremely ingenious invention of the patent law could not itself be patentable, due to an absence of law.

In principle, the Constitution goes further than this formal subordination. The constitutional Patent Provision, we have suggested, maintains a balance by subordinating progress to the unchanging, substantive goals of justice and liberty. But in practice, the patent law threatens to tip the scale in favor of runaway change. Increasingly encouraged, the horses of technological progress break into full gallop, seemingly out of anyone's control, and the community is left with the difficult task of adjusting after the fact to the paths traveled and the changes wrought. Sometimes, when progress comes before the Bar—as in the present case—even learned men judicially charged with upholding the law choose instead injudiciously to redefine it, in order to keep pace with novelty.

Finally, there are potential strains in the American polity's contract with science, insofar as the polity accepts without reservations the methods, principles, and purposes of modern natural science. For example, the practice of experimentation, when extended to human subjects, often places science on a collision course with the rights of individuals. Worse yet, our fundamental political principles, the *natural* rights enunciated in the Declaration of Independence, acquire no support from the "nature" described by the laws of physics and chemistry. The nature of the physicists, to say the least, offers no ground for rights, let alone for the belief that we have these rights as endowments from our Creator. Further, in biology, the teachings of evolution seem to deny to human beings any special place in the whole. And when, encouraged by these teachings, the project to relieve man's estate through mastery and possession of nature approaches making fundamental alterations in human nature itself, Americans—everyone—must begin to wonder whether the goals and presuppositions of the entire venture are sound and even whether modern science's notions of knowledge and nature are simply and unqualifiedly true.

Curiously, the recent Supreme Court decision in the *Chakrabarty* case points up all these difficulties, notwithstanding the narrow question it decided and the limited character of its holding. Various commentators have raised broader questions about the meaning of the *Chakrabarty* decision and its consequences: questions about the desirability of genetic engineering; about the dangers of the further commercialization of science; and about the propriety of owning an entire living species. By examining each of these questions, we shall be led to discover some of the limitations of the contract between modern science and the American polity, as it is embodied in our patent law.

Questions and Difficulties

Private Interest and Public Good: Genetic Engineering, A Dangerous Technology?

First: Does the protection of private rights and interests in new discoveries and inventions always serve the public good? Is the awarding of patents always in the public interest? The answer to these questions necessarily turns, in any given case, on the nature of the particular discovery and invention. More generally, it turns on the question, Is progress or technical innovation always in the public interest? If the innovations are simply or largely beneficial, their encouragement through award of patents would still reflect harmony between private interest and common good. But what about dangerous discoveries and inventions? Does the community serve its best interests when it stimulates their development through patent grants? Might not even the publication of the existence of the dangerous invention prove harmful to the public interest? What is the American polity's remedy for this problem of dangerous innovation (the problem which, you will recall, Bacon dealt with through the voluntary prevention of publication, self-imposed by the innovating scientists themselves)?

Genetic engineering is regarded by many as just such a dangerous technology, and one posing no ordinary dangers. For in human genetic engineering, the previous beneficiary of the power to alter nature becomes himself subject to that power and those alterations. The power to engineer the engineer sharply raises questions about the meaning and limits of progress.

It was, I am sure, concern about the dangers of genetic engineering, especially human genetic engineering, that gave the *Chakrabarty* case such wide interest. In argument before the Supreme Court, grave risks allegedly associated with genetic manipulation were cited as a reason why the patent should be denied. The majority opinion states:

> We are told that genetic research and related technological developments may spread pollution and disease, that it may result in a loss of genetic diversity, and that its practice may tend to depreciate the value of human life.

These opinions were advanced and are held by reputable scientists, among others, whose concerns range from fears about new biohazards to doubts about our possessing the wisdom requisite to redesign human genes or to interfere designedly in the course of evolution. There is, to be sure, much disagreement about the degree to which these fears and doubts are warranted, but there is no doubt that the matters at stake are serious. The Court indeed acknowledged the seriousness of such considerations, but held them nonetheless irrelevant to its decision, partly on the ground that its negative decision would not prevent such research and partly on the ground that it lacked the competence or the constitutional authority to decide how much and what kind of genetic research our society should foster.

The Court's judgment seems to me to be sound. Under our Constitution, it is for the legislature to decide such questions, and the Courts ought not to rewrite the rules. Further, denial of individual patent applications seems a poor way for society to decide questions about allegedly dangerous research and technology. Yet this very fact calls attention to a defect in the relation between science and society, insofar as that relation is largely defined by or exemplified in the contract of the patent laws. The patent laws assume that innovations proposed by inventors are, because innovative and useful to some, simply good for the community at large. Instituted well before many people recognized the communal price everyone pays for certain kinds of technological change, they reflect a once little-questioned faith in progress. Thus, as they are instruments for encouraging innovation, they are poorly designed for regulating or controlling it. It is no surprise that the mechanism for making the individual horses run turns out to be incapable of slowing them down, should one later discover that, as a team, they are in danger of running away with the rider.

And yet one wonders. The Court says, "Whether respondent's claims are patentable may determine whether research efforts are accelerated by the hope of reward or slowed by the want of incentives, *but that is all*" (emphasis added). But that "all" is not nothing. True, something unpatentable could still be legal-and-profitable; one cannot assume that lack of a patent will prevent development. Nevertheless, the awarding of patents *is* a communal hand on the throttle, a gentle hand to be sure, but by no means ineffective. Further, it is, as it happens, a hand less threatening to science than the legislative power to prohibit and make illegal. Moreover, one might argue that, in the statutory criterion of utility, the Patent Office has been given the power— indeed, the duty—to judge the social merits of a given invention in deciding whether to encourage its development. According to the patent laws, only *useful* inventions may be patented, and rightly so, if some usefulness to the public good is society's share of the patenting contract. Though it is generally sound to believe that fueling private incentives serves the public good, allowing the market to decide "usefulness," this is notoriously not always the case (especially if by "public good" we mean more than economic growth).

How does the Patent Office understand "the useful"? In general, its presumption being to favor development, any definable use is sufficient. But is this always sound? How should it judge the usefulness of a manufacture that has obvious and likely misuses and abuses, along with some clear and well-defined use? For example, how should it judge the usefulness of a perfected pleasure drug, admittedly beneficial in the treatment of depression, but almost certainly subject to widespread social or political abuse? What about improved devices for subliminal advertising? Or new and improved miniature recording and photographic devices that would no doubt increase snooping and invasions of privacy? Should the inventor of selective spermicides and his financial backers be able to decide by patenting that our society should be able to practice sex-selection of offspring?

One would think that a well-developed and nuanced doctrine of utility might already be embodied in court decisions involving patent claims. But a brief survey of the legal literature shows otherwise. True, precedent denies patentable utility to inventions whose contemplated use is for purposes deemed illegal or immoral (bogus coin detectors for slot machines, for example) or which always cause bodily harm to the user when used in its intended manner (for example, a drug effective against depression but toxic to the point of lethality). "A composition *unsafe* for use by reason of *extreme* toxicity *to point of immediate death* under *all* conditions of its *sole* contemplated use in treating disease of human organisms would not be 'useful' within the meaning of patent laws" (emphasis added), is the limited, almost grudging, concession to such considerations made by the United States Court of Customs and Patent Appeals, in a well-known case, *Application of Anthony.* In that case (1969), the Court in fact argued that, short of such uniform catastrophe, safety as an ingredient of utility is a relative matter, and it overruled the U.S. Patent Office, which had denied a patent for an antidepressant drug, Monase, a drug voluntarily taken off the commercial market by its manufacturers because of a dozen fatalities reported among its many users. Commenting in a footnote on the more general question of social harm from inventions capable of affecting public morals, health, and order, the Court in *Anthony* endorses a turn of the century U.S. Circuit Court opinion (*Fuller* v. *Berger et al.*): An invention is "useful within the meaning of the law, if it is used (or is designed and adapted to be used) to accomplish a good result, though in fact it is oftener used (or is as well or even better adapted to be used) to accomplish a bad one."*

During the 1960s, this doctrine—that likely abuse does not negate use—caused some embarrassment to the Patent Office, indeed, as a consequence of its function as publicist. A patent had been earlier awarded for LSD, shortly before its hallucinogenic properties were known. When the drug found its way into street use, the Patent Office helped a whole generation learn how to manufacture it, being obliged to divulge the details of its chemical synthesis to anyone who requested them. The Patent Office did so until the supply of printed matter about LSD was exhausted.

Perhaps such precedents reflect our long-standing and naive belief in the beneficence, or at least the innocence, of all innovation. Would a similar court today allow a patent for Monase, for the Colt revolver, or for LSD? Perhaps the future might bring us a more complex and refined doctrine of utility, one that was willing to make balancing judgments in protecting the public's side of the contract. But, at least for now, it seems that any licit and nonlethal use suffices for the stat-

---

*The court opinion quotes at length from the still authoritative doctrine, formulated by Albert Henry Walker in his 1880s textbook of patent law.[8] The court's own additions to his text are noted by parenthesis:

> An important decision, relevant to utility in this aspect, may hereafter arise and call for judicial decision. It is perhaps true, for example, that the invention of Colt's revolver was injurious to the morals, and injurious to the health, and injurious to the good order of society. . . . On the other hand, the revolver, by furnishing a ready means of self-defense, may sometimes have promoted morals and health and good order. By what test is utility to be determined in such cases? Is it to be done by balancing the good functions with the evil functions? *Or is everything useful within the meaning of the law, if it is used (or is designed and adapted to be used) to accomplish a good result, though in fact it is oftener used (or is as well, or even better, adapted to be used) to accomplish a bad one?* Or is utility negatived by the mere fact that the thing in question is sometimes injurious to morals, or to health, or to good order? The third hypothesis cannot stand, because if it could, it would be fatal to patents for steam engines, dynamos, electric railroads, and indeed many of the noblest inventions of the nineteenth century. The first hypothesis cannot stand, because if it could, it would make the validity of the patents to depend on a question of fact to which it would *often* be impossible to give a reliable answer. The second hypothesis is the only one which is consistent with the reason of the case, and with the practical construction which the courts have given to the statutory requirement of utility.

Does the doctrine of utility enunciated in the italicized passage truly serve the public interest? Are Walker's three options exhaustive? And is not even the first hypothesis, as stated, a plausible principle for at least those cases in which it would *not* be impossible to give a reliable answer to the balance between benefits and harms to the general welfare?

utory test of utility, all likely abuses notwithstanding. Under these circumstances, our second thoughts confirm our first: The Court in *Chakrabarty* was right in not allowing concerns about the possible dangers of genetic engineering to influence its decision.

If patent decisions do not and cannot consider these broad questions of use and possible abuse, if restriction of patents is an inappropriate mechanism for setting the pace in the realm of potentially dangerous technologies, the contract between science and society needs additional clauses. To be sure, many regulations already exist (e.g., Regulations of the Food and Drug Administration, Guidelines for the Use of Human Subjects in Research, etc), yet most of them deal only with questions of health and safety. We have few means of assessing and regulating with regard to the massive consequences of new technologies to our mores, institutions, and ways of life. With the vast powers, now being accumulated, that would bring the mastery of nature to bear on human nature itself, some have begun to wonder whether the simply permissive contract between innovators and society needs to be renegotiated.

Such a response seems to me excessive. We have, and will continue to have, a commitment to scientific and technological progress. We have reason to expect that the social and political results of such progress will continue to be largely beneficial, and that the union of science and politics cemented by the patent laws will continue to serve us well. It would be foolish to dismantle our instruments of progress just because they require some additional devices and mechanisms. It would be foolish to shackle our accelerator just because it does not function as a brake. The difficult question—one that we have only begun to face—is what kinds of political arrangements and institutions are best suited to reviewing the direction and pace of certain "dangerous" developments and to applying the brakes, if necessary. One thing seems clear: The responsibility lies with the legislature. Courts may raise questions about the need for brakes, but it must be Congress that applies them. How to do so is, of course, the difficult question. The task of inventing suitable braking mechanisms will require even more ingenuity than the invention of the patent laws. We are all aware of the serious risks and costs of governmental regulation. Yet unless some means of control are found for those technologies reasonably regarded as *potentially dangerous to the public interest*—and, for the long run, who can be certain about genetic engineering—the motives of gain, when added to ingenuity and stimulated by patent protection, are likely to subvert the common good. With big money fanning the flames—consider the difficulties in regulating the tobacco or automobile industries—the fire of innovation could be out of control before anyone gets warm enough to worry.

We have argued that the job of brakeman does not belong to the Courts. But it does not follow that the Courts should be free to remove or revise brakes applied by the legislature. The Court should be neither partisan nor opponent of progress; it is, instead, the guardian of law and, implicitly, a teacher of law-abidingness. The Court in *Chakrabarty* rightly resisted encroaching upon the legislature's domain when it refused to become society's arbiter regarding genetic engineering, but how well did it discharge its own task of guarding the law? An examination of the decision reveals that the Court showed itself partial to progress, with the so-called conservative members leading the way.

The Court was asked to decide not whether living organisms *ought* to be, but only whether they *are* patentable matter, as this is defined by statute. The revelant portion of the patent law (35 United States Code §101) provides:

> Whoever invents or discovers any new and useful process, machine, manufacture, or composition of matter, or any new and useful improvement thereof, may obtain a patent therefor, subject to the condition and requirements of this title.

To decide affirmatively, the Court majority had to construe both the novel microorganism and the operative clause in the statute, which defines patentable matter as "any new and useful process, machine, manufacture, or composition of matter," such that a living bacterium could be understood to be either a "manufacture" or a "composition of matter" or both; the Court minority argued that "manufacture" or "composition of matter" were not intended by Congress to encompass living organisms. Though the majority opinion does not directly *argue* that the microorganism in question, is, say, a composition (i.e., a putting together) of matter, it treats its aliveness as irrelevant to its patentability. It ignores altogether the nature of the object, arguing: "In choosing such expansive terms as 'manufacture' and 'composition of matter,' modified by the comprehensive 'any,' Congress plainly contemplated that the patent laws would be given wide scope." Finally, the Court argues that novelty, utility, and the fact that Chakrabarty's discovery "is not nature's handiwork but his own" renders the bacterium patentable subject matter and Chakrabarty and General Electric the proud owners of "his" new species.

I happen to think the Court opinion mistaken on its reading of statute. The terms "manufacture" and "composition of matter" go back to Jefferson's 1793 Patent Law, and Congress has retained then without change in all subsequent revisions. (In another category of patentable matter, when Congress became dissatisfied with Jefferson's concepts, it replaced his term; the present "process" is a replacement for Jefferson's "art.") Did Jefferson regard a living organism as a mere "composition of matter"? Certainly, in the ordinary sense of these terms, no one should. The majority goes too far in extrapolating from its correct belief that Congress "contemplated that the patent laws would be given wide scope." It sustains the opinion that Congress intends statutory subject matter to "include anything under the sun made by man." But if so, why did Congress *in fact* make and preserve categorical distinctions among the *kinds* of patentable man-made things—"processes, machines, manufactures, and compositions of matter"—distinctions that would be unnecessary if "anything under the sun," so long as of artificial origin, were the sufficient mark of patentable matter—of course, along with novelty, utility, and nonobviousness? And why, the minority rightly asks, would Congress enact separate plant patent laws (in 1930 and again in 1970), to permit patenting of new plant varieties, if Congress understood "manufacture" and "composition of matter" as broadly as the Court majority now claims? Indeed, as the minority again points out, in the 1970 Plant Patent Act, Congress had specifically *excluded* bacteria from patentability under the Act: "Congress has included bacteria within the focus of its legislative concern, but not within the scope of patent protection. . . . The fact is that Congress, assuming that animate objects as to which it had not specifically legislated could not be patented, excluded bacteria from the set of patentable organisms."

The Court majority ignored these specific facts about the written statutes. It took its stand instead on what it calls the "broad general language" of the patent laws and on its own construction of the legislative intent: "The subject matter provisions of the patent law have been cast in broad terms to fulfill the constitutional and statutory goal of promoting 'the Progress of Science and useful Arts' with all that means for the social and economic benefits envisioned by Jefferson." It is an insult to Jefferson to suggest that his friendship for progress made him imprecise and vague as a legislator. He said what he meant and he meant what he said, always careful about his choice of words. Courts would do less mischief if they treated all law and legislatures as if they meant what they in fact explicitly said. The present Court's love of innovation extends to its reading of law. We must wonder whether such "progressive" jurisprudence is not too high a price for progress.

Is Practice Good for Theory?: Science, Profits, and the University

If patenting and the patent laws do not always serve the public or political good, are they simply good for science? As we shall see, this is best understood as a special case of the question, Is practice always good for theory? It also ultimately invites us to reconsider the purposes of science and thought more generally. But nearer at hand are questions about science and money.

The *Chakrabarty* decision has prompted discussion about possible corrupting tendencies of the profit motive, not so much for the society at large, but, curiously, for the present practice of biological science, especially in universities. For roughly a quarter century, biomedical research has flourished, largely funded by the Federal government and philanthropic foundations, much of it done in universities. Though ultimately interested in the practical benefits, the government, albeit not without frequent prodding by basic scientists, has wisely and patiently supported many outstanding minds in so-called basic research, largely without regard to its *immediate* utility. Progress over the past three decades is simply staggering. Though competition is keen, and there are well-known cases of secretive and even unscrupulous behavior, on the whole the field has thrived on free and cooperative exchange of information and materials, including strains of microorganisms.

Now that new discoveries and techniques in cell-biology and molecular genetics have brought these fields fully into the industrial area, many are worried that the profit motive will distort, not to say corrupt, scientific practices. Concerns are expressed for the effects on the behavior of scientists, on the balance in fields of research, and on universities. Warnings are heard about an impending restriction of the free flow of information and a rise of secretiveness, deception, and other unsavory conduct, not excluding espionage. Others are concerned that profits will dictate the direction of scientific research, deflecting the scientific mind from going where it will or should. With several universities, under threat of rising costs and dwindling financial support, already established entrepreneurs in genetic technology, and many fine scientists entering industry in a variety of capacities, often retaining their academic tenure, there is argument that such goings-on in principle violate the spirit and will, in practice, threaten the purpose of the university.

These are serious and complicated questions that cannot be addressed adequately here. On some matters the concerns seem exaggerated. The rise of industrial chemistry and applied physics has not in itself, it seems to me, corrupted basic research in universities, nor led to undue secrecy or unsavory practices. And in any case, such problems as appear are due more to the large amounts of money involved rather than to patenting (though the two are not unrelated). For the need to protect profitable discoveries through patent should lead not to secrecy, but to publication (though the anticipation of future patent application does lead many into temporary secrecy, and reports are now increasing of biologists who, looking to protect future patents, have become silent and stingy with new information and materials). Once a patent is granted, for a payment of royalties new information, materials, and techniques are potentially more widely shareable. Moreover, the disdain of many academic biologists for the practical applications of their work can only be regarded as hypocritical, especially considering the hopes of, and their promises to, their public patrons. Academic scientists have for years played upon the public's utilitarian concerns and always promised and even emphasized the probable long-run practical benefits when seeking congressional support to satisfy their own private curiosity. Science, even university science, is, to some extent, a kept woman, and the question sometimes seems to be only who shall keep her and what is her price. Her virtue and her fruitfulness may not suffer further from wedding herself to industry.

But this is no matter for levity; the stakes are very high. There is reason to be concerned about the growth of the academic-industrial complex, but not because industry is corrupt or corrupting or because there is something reprehensible about utility or even money-making. Rather, one is concerned because one knows that universities exist not only to generate useful discoveries and because one suspects that knowledge for the sake of power and utility is not the whole truth about knowledge, that thought at its best—including scientific thought—seeks truth for its own sake. For these reasons, we can ill afford to be indifferent to the fate and character of university science and to the climate for free and fundamental thought. The remarkable record of American scientists in *basic* discovery in biology is a credit to public support and especially to the university setting, with its great freedom of inquiry and its relative immunity to demands for prompt success or useful results. Here, fundamental thought is frequently stimulated by the collegiality of scholars in diverse areas of inquiry, scholars who are also teachers, somehow still heirs to a great tradition that often gave more than lip service to the disinterested pursuit of the truth. Professors are often pushed to fundamentals also by their undergraduate students, who are not yet sufficiently "educated" to know that there are some questions one should avoid asking. One wonders how theory will fare if the universities are increasingly drawn to practice. One wonders whether the search for the truth will flourish, should the universities and their scientists try to be increasingly relevant and useful.

Though largely unanticipated by the American Founders, the rise of universities and of science within them has added a new dimension to the original relation between science and politics, a dimension that acquisitive, democratic, and egalitarian regimes very much require. The point was made brilliantly by Tocqueville, in his *Democracy in America,* in the chapter, "Why the Americans Are More Concerned with the Applications than with the Theory of Science":

> The higher sciences or the higher parts of all sciences require meditation above everything else. But nothing is less conducive to meditation than the setup of democratic society. . . . Everyone is on the move, some in quest of power, others of gain. In the midst of this universal tumult, this incessant conflict of jarring interests, this endless chase for wealth, where is one to find the calm for the profound researches of the intellect? How can the mind dwell on any single subject when all around is on the move and when one is himself swept and buffeted along by the whirling current which carries all before it?
>
> Not only is the meditation difficult for men in democracies, but they naturally attach little importance to it. . . . In democratic centuries when almost everyone is engaged in active life, the darting speed of a quick, superficial mind is at a premium, while slow, deep thought is excessively undervalued. . . .
>
> Most of the people in these [democratic] nations are extremely eager in the pursuit of immediate material pleasures and are always discontented with the position they occupy and always free to leave it. They think about nothing but ways of changing their lot and bettering it. For people in this frame of mind every new way of getting wealth more quickly, every machine which lessens work, every means of diminishing the costs of production, every invention which makes pleasures easier or greater, seems the most magnificent accomplishment of the human mind. It is chiefly from this line of approach that democratic peoples come to study sciences, to understand them, and to value them. In aristocratic ages the chief function of science is to give pleasure to the mind, but in democratic ages to the body.
>
> It may be assumed that the more democratic, enlightened, and free a people is, the greater will be the number of these selfish admirers of scientific genius, and the more profit will be made out of discoveries immediately applicable to industry, bringing reknown and even power to their inventors. . . . It is easy to see how, in a society organized on these lines, men's minds are unconsciously led to neglect theory and devote an unparalleled amount of energy to the applications of science. . . .

On the strength of this analysis, Tocqueville gives this advice:

> If those who are called on to direct the affairs of nations in our time can clearly and in good time understand these new tendencies which will soon be irresistible, they will see that, granted enlightenment and liberty, people living in a democratic age are quite certain to bring the industrial side of science to perfection anyhow and that henceforth the whole energy of organized society should be directed to the support of higher studies and the fostering of a passion for pure science.

> Nowadays the need is to keep men interested in theory. They will look after the practical side of things for themselves. So, instead of perpetually concentrating attention on the minute examination of secondary effects, it is good to distract it therefrom sometimes and lift it to the contemplation of first causes. . . .
>
> We therefore should not console ourselves by thinking that the barbarians are still a long way off. Some people may let the torch be snatched from their hands, but others stamp it out themselves.[9]

I do not wish to exaggerate the dangers to pure science or to universities from the new privilege to patent microorganisms, hybridomas, and products of genetic engineering. Nor, unfortunately, are universities or academic scientists today the embodiment of thoughtfulness and disinterested inquiry that Tocqueville rightly argues we so urgently need. But the climate is not being helped by the eruption among scientists and administrators of what must frankly be called greed, nor is it likely to be improved by the continuing growth of the academic-industrial complex. When the president of Harvard University devotes his entire annual address to his Board of Overseers to the theme of "technology transfer"—the translation of scientific knowledge into useful products and processes—and argues that it must become a central task of the university, one has reason to believe that big winds may soon blow the academy off its present course.

American universities are, for all their faults, precious and precarious institutions. In fact, the present balance within them between the busy and the deliberate, the clever and the wise, the useful and the true, is already tipped so far toward the former that we must be cautious about all further changes that tend to diminish the latter. It should now be evident that my concern for universities and for theory and fundamental thought goes beyond my concern for so-called pure science. The earlier discussion should have made clear the importance of careful and thorough thinking about the relationship between science and the American polity and about the implications of our new forays into genetic engineering. Indeed, especially now, when the goal and direction of the scientific project for the mastery of nature seems less clear than ever, and when, despite this confusion about the end, the means are being amassed to directly and deliberately affect all forms of life on the planet, we stand in urgent need of the far-seeking and high-minded reflection about science, ethics, and society that the patent laws, industry, and even such fine institutions as the National Institutes of Health cannot encourage or foster.

But theory is urgent not only because basic research pays dividends in applications, nor even because we need theory to think about whither we are tending. Theory is urgent also because it is in itself elevating and liberating. Thoughtfulness, speculation, genuine inqui-

ry beyond mere problem-solving, philosophical reflection on our condition and our place in the world, in short, *liberal* learning and *liberal*
education—and not only the advancement of Baconian learning—are
necessary for a truly free people. Liberty, secured by the progress of
science and useful arts, would be little blessed if our minds become
enslaved in and to the process of serving our bodies.

Once again, the task is to restore the balance, to give weight to
the weaker side. And once again, it is difficult to see how and by whom
the countervailing forces for liberal learning and philosophic reflection
are to be generated and supported, especially now when economic
troubles aggravate the natural tendency of modern thought to serve
utility. The task is beyond the competence both of our science—not
least because of its antispeculative self-definition—and of our law. No
one would say that the practice or encouragement of philosophical reflection is the business of our courts. But, at the same time, it is sad
when the Supreme Court, the closest approximation in the American
polity to the rule of thoughtful reason, promulgates ill-considered
opinions about weighty matters. For in justifying its decisions, the
Court functions also as a teacher, helping to form what become *our*
ruling opinions. Indeed, the opinions of the Court are often more important for what they teach than for what they decide. We take one
last look at the *Chakrabarty* case, with a view to the Court as teacher.

Owning Life?: The Limits of Mastering Nature

What has the *Chakrabarty* decision accomplished? A rather modest
gain for Chakrabarty, a rather sizeable boost for the burgeoning hy-
bridoma and genetic technology industry, but—by means of negative
example—a most important lesson, if only we can learn it, about how
close we have come in our thinking, if not yet in our practice, to over-
stepping the sensible limits of the project for mastery and possession
of nature. This project makes sense only if we fully understand and
accept the limited meanings of mastery and possession and only if we
appreciate the nature of living nature and our place within it. On these
deep matters, the Court was here a teacher of shallowness.

Consider first the implicit teaching of our wise men, that a living
organism is no more than a composition of matter, no different from
the latest perfume or insecticide. What about other living organisms—
goldfish, bald eagles, horses? What about human beings? Just compositions of matter? Here arise deep philosophical questions to which
the Court has given little thought; but in its eagerness to serve innovation, it has, perhaps unwittingly, become the teacher of philosophical materialism—the view that all forms are but accidents of

underlying matter, that matter is what truly *is*—and therewith, the teacher also of the homogeneity of the given world, and, at least in principle, of the absence of any special dignity in all of living nature, our own included.

A similar teaching is also implicit in the enlargement of the sphere of what may be owned and possessed. By the arguments of the Court, it now seems that anything under the sun made of tangible stuff falls under "composition of matter," and is therefore patentable, so long as its origin is in human art. Nothing in the Court's opinion would permit one to argue that the "inventor" of the mule, were the mule to be a new invention, could not claim patentability. If the Chinese succeed in their present attempts with artificial insemination to cross-breed a human being with a chimpanzee, producing the novel and useful "humanzee," it would be arguably patentable matter—if the Court sticks to its interpretation and Congress does not act. These examples may be farfetched but they serve to illustrate the point: There is something obviously and immediately disquieting about the human ownership of an entire living species, even one brought into being wth the partial aid of art.

This bizarre new prospect, that one man could own—albeit for a limited time—an entire species, does indeed invite us to rethink the reasons why we permit ownership of any animals. There is a sense in which the former is but the logical extension of the latter, both instances of the possession and exploitation of living nature for human needs and wants, and this logical extension as a limiting case might, in fact, illuminate problematic aspects of our age-old and familiar practice of domesticating plants and animals. Still, there are significant differences which, though they do not fully explain our repugnance to the notion of owning a species, suggest that our disquiet is not due just to the novelty and audacity of the idea.

If usefulness justifies ownership, it also defines its justifiable limits. Ownership of animals, even of large herds, presupposes the usefulness of each animal to the owner. Even when animals are kept for their beauty, or companionship, possession is reasonable only on a human scale—that is, on a scale that permits individual appreciation or relation. We do not endorse possession for the sake of possession: The thought of a man buying up and collecting all the world's camels or giraffes or horses is repulsive, though nothing in the law prevents it. To own more of living nature than what one needs for one's own life and livelihood is hard to justify. It is even harder to justify such monopoly when the sole purpose is to exclude others from similar benefits.

Ownership also carries with it responsibility, not only for the living beings but also to other human beings for what animals inadvertently do. Indeed, living things, unlike true artifacts, have a life of their own and ways that we cannot simply predict or control.

> And if one man's ox hurt another's, so that it dieth; then they shall sell the live ox, and divide the price of it; and the dead also they shall divide. Or if it be known that the ox was wont to gore in time past, and its owner hath not kept it in; he shall surely pay ox for ox, and the dead beast shall be his own.[10]

Can one exercise responsibility for an entire species, especially species that reproduce prodigiously and are hard to confine? If one of Chakrabarty's bacteria escaped from his laboratory, can he be held responsible for the mischief it causes? If Chakrabarty's bacteria find their way into an oil well or oil storage tank, shall he pay drop for drop? For they were wont to gore in time past, and the owner hath not kept them in. And (while thinking about fugitive bacteria) if one of Chakrabarty's technicans going on vacation inadvertently carries—on his skin or clothing or in his digestive tract—one of the microbes from its laboratory confinement in Illinois to freedom in Missouri where it becomes fruitful and multiplies, must all the billions of progeny be returned to Illinois? Will the Supreme Court, in upholding Chakrabarty's patent claims of ownership, write a new Dred Scott decision?

Be this as it may, the implicit teaching about ownership of life in the present Supreme Court decision is indeed problematic. It is one thing to own *a* mule; it is another to own *mule.** Admittedly, bacteria are far away from mules. But the principles invoked, the reasoning, and the stance toward nature go all the way to mules, and beyond.

What is the principled limit to this beginning extension of the domain of private ownership and dominion over living nature? Is it not clear, if life is a continuum, that there are no visible or clear limits once we admit living species under the principle of ownership? The principle used in *Chakrabarty* says that there is nothing *in the nature of a being,* no, not even in the human patenter himself, that makes him immune to being patented: not what he is, but only the "accident" of his non-man-made origin renders man himself a nonpatentable organism. If a genetically engineered organism may be owned *because* it was genetically engineered, what would we conclude about a genetically altered or engineered human being? To be sure, in general it makes sense to allow people to own what they made, because they have artfully made it. But to respect art without respect for life is finally self-contradictory. For human art depends on the human artificer, whose inventive mind depends on his living body, not only to sustain it that he might practice its cleverness, but also because the

---

*This argument should cause us to reconsider the wisdom of permitting ownership even of plant species, which was made possible by the plant patent laws of 1930 and 1970.

ends of his artfulness emerge from the inner needs and aspirations of his embodied life.

Finally, the exalted and mastering status of human art claims too much and too little for itself. It claims too much because it ignores that art can only put together or alter what natural powers beyond human control will allow. In the present case, our inventor even had nature's active assistance; for it is not strictly true, as the Court claims, that "his discovery is not nature's handiwork, but his own." Chakrabarty did not himself *create* the new bacterium. Rather, he played the matchmaker for a shotgun wedding and the selector of its progeny, while the living organisms did the work. He mixed together plasmids (carrying genes for metabolizing hydrocarbons) *produced* by and isolated from certain oil-degrading bacterial species and incubated them with the hardier Pseudomonas species, which bacteria, *all by themselves,* incorporated the plasmids. By selecting conditions that would support growth only of the plasmid-containing Pseudomonas hybrid, Chakrabarty obtained "his" novel strain. Though the process was, in many senses, creative and his own, the novel organism was not *his creature.*

Even in true compositions of matter, that is, when chemicals are placed together to produce a new mixture or compound, nature is commanded only as she is obeyed. The potentialities of given matter may be exploited, but they cannot be artfully created. The laws of nature permit prediction and control of phenomena, but they too are not of our making and cannot be transgressed. One might say, what nature's God keeps asunder, no man can put together. Man's ability to change nature is, in principle and in practice, always consistent with and limited by nature's unchanging ground.

Ironically, in its pride, human innovativeness also respects itself too little because it lacks self-understanding. It fails to appreciate its source in the permanent power of mind, given to human beings but not of their own making. Our inventiveness is not our invention; neither are the truths it discovers.

The Court acknowledges that "Einstein could not patent his celebrated law that $E = mc^2$; nor could Newton have patented the law of gravity." The reason given is curious: "Such discoveries are manifestations of . . . nature, free to all men and reserved exclusively to none." The Court fails to appreciate the deeper reason why a truth cannot be patented. Once it is published, it is shareable. To know it is to make it your own. But truth is your own in a very special way, unlike your other possessions. The greatest thinkers have understood that truths are neither private nor property, that they *come* unbidden to mind, mysteriously, and that insight is neither at one's disposal nor of one's own making. Homer, the greatest of the makers, assigns credit to the

Muse. Finally, the claim of "intellectual property" is unfounded, even for inventions.

In the ever-changing being that is given to living organisms, the two poles of natural permanence—mobile matter and sensitive awareness, culminating in mind—are bound together. In human beings, living nature at last becomes conscious of itself. If we are sober in our practice and mindful in our thought, it is given to us human beings to learn our place in the natural whole and to discover something of its distinctive beauty and mysterious ground. Without such self-knowledge, the project for mastery and possession of nature is a Faustian bargain. Reacquiring a respect for our relatives, the ever-changing living forms, could regain for us a much needed recognition and appreciation of the natural and unchanging source of all change.

PART II

Holding the Center
The Morality
of Medicine

CHAPTER SIX

The End of Medicine
and the Pursuit of Health

*And I said that the cure itself is a certain leaf, but in addition to the
drug, there is a certain charm, which if someone chants when he makes
use of it [the drug], the medicine altogether restores him to health;
but without the charm there is no profit from the leaf.*

—"Socrates," in Plato's *Charmides* (155e)

American medicine is not well. Though it remains the most widely
respected of professions, though it has never been more technically
competent, it is in trouble, both from without and from within.

The alleged causes are many; I will mention a few. Medical care is
very costly and not equitably available. The average doctor sees many
more patients than he should, yet many fewer than would like to be
seen. On the one hand, the physician's powers and prerogatives have
grown as a result of new technologies yielding new modes of diagnosis
and treatment and new ways to alter the workings of the body. His
responsibilities have grown as well, partly due to rising patient and
public demands for medical assistance in addressing behavioral and
social problems. All kinds of problems now roll to the doctor's door,
from sagging anatomies to suicides, from unwanted childlessness to
unwanted pregnancy, from marital difficulties to learning difficulties,
from genetic counseling to drug addiction, from laziness to crime. On
the other hand, the physician's new powers have brought new dilemmas, concern over which has led to new attempts to regulate and control his practices, including statutes, codes, professional review bodies,
ombudsmen, national commissions, and lawsuits brought by public-
interest law and consumer groups. More and more physicians are being
dragged before the bar, and medical malpractice insurance has become
both alarmingly scarce and exorbitantly expensive.

Health care has become an important political issue. A right to
health has been frequently claimed and embraced by politicians. The
Federal government is now directly in the life-saving business, obliged

to pay for kidney machines for anyone in need. And the National Health Insurance on the horizon will surely bring the medical profession even more under governmental control, at the very least, by defining what will count as health care through determining what will be paid for.

Last but not least, people both in and out of medicine have begun to wonder out loud whether and to what extent medicine is doing good. No longer simply charmed by the profession's diagnostic and therapeutic wizardry, some people are seriously asking whether the so-called health care delivery system really does—or can—deliver or foster improved health for the American people.

This last question points to a more fundamental cause of medicine's illness: medicine, as well as the community that supports it, appears to be perplexed regarding its purpose. It is ironic, but not accidental, that medicine's great technical power should arrive in tandem with great confusion about the standards and goals for guiding its use. When its powers were fewer, its purpose was clearer. Indeed, since antiquity, medicine has been regarded as the very model of an art, of a rational activity whose powers were all bent toward a clear and identifiable end. Today, though fully armed and eager to serve, the doctor finds that his target is no longer clear to him or to us. Sometimes it appears to be anything at which he can take aim; at other times it appears nowhere to be found. In fact, the very existence of a target is implicitly questioned by those who have begun to change the name of the doctor from "physician" to "member of the helping professions."

At what should the medical art aim? What is the proper end, or the proper ends, of medicine? Continued confusion about this matter could bring about, more directly than any other cause, the demise of the profession, even if there were to remain people with M.D. degrees whom their clients called "Doctor." For without a clear view of its end, medicine is at risk of becoming merely a set of powerful means, and the doctor at risk of becoming merely a technician and engineer of the body, a scalpel for hire, selling his services upon demand. There is a connection between the two meanings of "end" suggested by the title of this chapter. It is in part for this reason that I have chosen to inquire regarding the end, or purpose, of medicine, with the hope that we might more seriously regard—that is, look back at, pay attention to, and, finally, esteem—the end or purpose of the medical art. Moreover, only by again attaining clarity about the goal of medicine can we intelligently evalute efforts to attain it and wisely plan for their improvement. Otherwise, for all our good intentions, our health policies will be mere tinkerings in the dark, at great risk of doing more harm than good.

The End of Medicine

I trust it will shock no one if I say that I am rather inclined to the old-fashioned view that health—or if you prefer, the healthy human being—is the end of the physician's art. That health is a goal of medicine few would deny. The trouble is, so I am told, that health is not the only possible and reasonable goal of medicine, since there are other prizes for which medical technique can be put in harness. Yet I regard these other goals, even where I accept their goodness as goals, as false goals for medicine, and their pursuit as perversions of the art.

Let us examine some of the false goals that tempt today's physicians. First, there is what is usually called "happiness" in its sadly shrunken meaning, but which might best be called "pleasure"—that is, gratifying or satisfying patient desires, producing contentment. This temptation arises largely because of the open-ended character of some contemporary notions of mental health, in which frustration or anxiety or any unsatisfied desires, no matter how questionable, are taken as marks of ill health, requiring a remedy.

Some examples of gratification may be helpful. A woman gets a surgeon to remove a normal breast because it interfered with her golf swing. An obstetrician is asked to perform amniocentesis, and then abortion, if the former procedure shows the fetus to be of the undesired gender. "Dr. Feelgood" devotes his entire practice to administering amphetamine injections to people seeking elevations of mood. To these real but admittedly extreme examples one could add, among others, the now generally accepted practices of performing artificial insemination or arranging adoptions, performing vasectomies and abortions* for nonmedical (i.e., family planning) reasons, dispensing antibiotics or other medicines simply because the patient wants to take something, and some activities of psychiatrists and many of cosmetic surgeons (e.g., where the surgery does not aim to correct inborn or acquired abnormality or deformity). I would also add the practice, now being advocated more and more, of directly and painlessly killing a patient who wants to die.

All of these practices, the worthy and the unworthy alike, aim *not* at the patient's health but rather at satisfying his, albeit in some cases reasonable, wishes. They are acts not of medicine but of indulgence or gratification in that they aim at pleasure or convenience or at the satisfaction of some other desire, and not at health. Now, some indulgence may be necessary in the *service* of healing, as a useful means to

*Abortion—nearly all of it nontherapeutic in this sense—is now the most common surgical procedure in the United States. Over 1.5 million abortions were performed in 1980 (a year in which there were 3.6 million live births).

the proper end: I see nothing wrong in sweetening bad tasting medicine. But to serve the desires of patients as consumers should be the task of agents other than doctors, if and when it should be the task of anyone.

Even in its fuller sense, happiness is a false goal for medicine. By gerrymandering the definition of health to comprise "a state of complete physical, mental, and social well-being," the World Health Organization has in effect maintained that happiness is the doctor's business (even if he needs outside partners in this enterprise). Complete mental well-bing—not to speak of the more elusive and ambiguous "social well-being" that will certainly mean different things to the pope, President Reagan, and Comrade Chernenko—goes well beyond the medical province of sanity, depending as it does on the successful and satisfying exercise of intelligence, awareness, imagination, taste, prudence, good sense, and fellow feeling, for whose cultivation medicine can do little. (That *happiness*, even in its full sense, is different from *health* can be seen in considering whether it would ever make sense to say, "Call no man healthy until he is dead.")

A second false goal for medicine is social adjustment or obedience, or more ambitiously, civic or moral virtue. The prevention of crime or war, the taming of juvenile delinquents, the relief of poverty and racial discrimination, the reduction of laziness and philandering, the rearing of decent and moral men and women—all worthy goals in my opinion—are none of the doctor's business, except as the doctor is also a human being and a citizen. These are jobs for parents, policemen, legislators, clergymen, teachers, judges, and the community as a whole—not to speak of the individual citizens themselves.* It is doubtful that the physician has the authority and competence, as physician, to serve these goals with his skills and techniques.

The difficulty is, of course, that only doctors are able and legally entitled to manipulate the body; hence the temptation to lend his licensed skill to any social cause. This temptation is bound to increase as we learn more about the biological contributions to behavior. In an increasing number of circumstances, the biological contribution will be seen as most accessible to intervention and most amenable to change. Hence, biological manipulation will often hold out the promise of dramatic and immediate results. Brain surgery and behavior-modifying drugs already have their advocates in the battles against criminal and other so-called antisocial behavior, and, for better or for worse,

---

*Improvements in public order and private virtue may, of course, lead secondarily to better health, as with the reduction of crimes or drunkenness. (This theme will be discussed more fully below.) Conversely, medicine and its attendant institutions, including programs of health insurance, may have secondary consequences for society and morals, say, for the the redistribution of income or the sense of personal responsibility for one's state of health.

there is good reason for believing that these techniques may be effective at least in some cases some of the time. But even assuming that we should accept, for example, psychosurgery for some men committing frequent crimes of violence, or the dispensing of drugs in schools for some restless children, or (at some future time) genetic screening to detect genotypes that may predispose to violent behavior, I doubt that it is the proper business of medicine to conduct these practices, even though, on balance, there may be overriding prudential reasons for not establishing a separate profession of biobehavioral conditioners.

I reject, next, in passing, the claim that the alteration of human nature, or of some human natures, is a proper end for medicine, whether it be a proposal by a psychologist for pills to reduce human aggressiveness—especially in our political leaders, the suggestions of some geneticists for eugenic uses of artificial insemination, or the more futuristic and radical visions of man-machine hybrids, laboratory-grown optimum babies, and pharmacologically induced peace of mind. Also to be resisted is that temptation first dangled by Descartes (and repeated in various forms by others many times since) who wrote, in praise of the prospects for a new medicine based on his new physics: "For the mind depends so much on the temperament and disposition of the bodily organs that, if it is possible to find a means of rendering men wiser and cleverer than they have hitherto been, I believe that it is in medicine that it must be sought." My difficulty with such proposals rests partly on my skepticism concerning whether some of the proposed improvements would indeed be improvements, and partly on whether these goals are indeed realizable by the use of the proposed biomedical techniques. But in addition—and, for the present purpose, this is decisive—I would argue that these goals are not proper goals for the healing profession.

I skip over the much discussed question of whether the physician should be also a seeker after scientific truth, and whether and to what extent he may or should conduct research on patients not for their immediate benefit. Insofar as the knowledge sought is pertinent to the art of healing, its pursuit is a necessary means to the end of medicine and cannot be ruled out of bounds on that score, though serious and difficult moral questions remain whenever human beings are used as means, regardless of the end served.* There may be good practical

---

*These questions are, I believe, resolvable, at least in principle, along the following lines: By knowingly and freely consenting to serve as an experimental subject, the patient is not serving *merely* as a means; he becomes, as it were, a co-inquirer, and the obligation to secure his consent explicitly acknowledges that he is not to be regarded merely as a means. Nevertheless, a whole nest of theoretical and practical questions remain, ranging from the meaning and limits of "consent," "informed," and "free" to the design of procedures that would adequately protect the subject against risk and abuse without undermining the freedom to inquire.

reasons to keep clearly delineated the activities of the physician as healer and the physician as student of health and disease, all the more so where research done by doctors is not clearly and directly in the service of the health of their patients. But as the art depends upon knowledge, so the search for knowledge cannot be excluded from the art.

Let me, with some misgivings, suggest one more false goal of medicine: the prolongation of life, or the prevention of death. It is not so clear that this is a false goal, especially as it is so intimately connected with the medical art and so often acclaimed as the first goal of medicine, or, at least, its most beneficial product. Yet *to be alive* and *to be healthy* are not the same, though the first is both a condition of the second and, up to a point, a consequence. One might well ask whether we desire to live in order to live healthily and well, or whether we desire to be healthy and virtuous merely in order to stay alive. But no matter how desirable life may be—and clearly to be alive *is* a good, and a condition of all the other human goods—for the moment let us notice that the prolongation of life is ultimately an impossible, or rather an unattainable, goal for medicine. For we are all born with those twin inherited and inescapable "diseases," aging and mortality. To be sure, we can still achieve further reductions in *premature* deaths, but it often seems doubtful from our words and deeds that we ever regard any death as anything other than premature, as a failure of today's medicine, but, if we are diligent, avoidable by tomorrow's.

If medicine takes aim at death prevention, rather than at health, then the medical ideal, ever more closely to be approximated, must be bodily immortality. Strange as it may sound, this goal really *is* implied in the way we as a community evaluate medical progress and medical needs. We go after the diseases that are the leading causes of death, rather than the leading causes of ill health. We evaluate medical progress, and compare medicine in different nations, in terms of mortality statistics. We ignore the fact that we are largely merely changing one set of fatal illnesses or conditions for another, and not necessarily for milder or more tolerable ones. We rarely stop to wonder of what and how we will die, and in what condition of body and mind we shall spend our last years, once we are able to cure cancer, heart disease, and stroke.

I am not suggesting that we cease investigating the causes of these diseases. On the contrary, medicine *should* be interested in preventing these diseases or, failing that, in restoring their victims to as healthy a condition as is possible. But it is precisely because they are causes of *unhealth*, and only secondarily because they are killers, that we should be interested in preventing or combating them. That their prevention and treatment may enable the prospective or actual victims to live longer may be deemed, in many cases, an added good, though we should not expect too much on this score. The complete eradication of heart disease, cancer, and stroke—currently the major mortal diseases—would, according to some calculations, extend the average life expectancy at birth only by approximately six or seven years, and at age sixty-five by no more than one and a half to two years. In the United States medicine's contribution to longer life has nearly reached its natural limit.*

By challenging prolongation of life as a true goal of medicine, I may be challenging less what is done by practicing clinicians and more how we think and speak about it. Consider a concrete case: An elderly woman, still active in community affairs and family life, has a serious heart attack and suffers congestive heart failure. The doctor orders, among other things, oxygen, morphine, and diuretics and connects her to a cardiac monitor, with pacemaker and defibrillator handy. What is the doctor's goal in treatment? To be sure, his actions, if successful, will help to keep her alive. But his immediate intention is to restore her circulatory functions as near to their healthy condition as possible; more distantly, his goal is to see her able to return to her premorbid activities. Should the natural compensating and healing processes succeed, with his assistance, should the cardiac wound heal and the circulation recover, the patient will keep herself alive.

We all are familiar with those sad cases in which a patient's life has been prolonged well beyond the time at which there is reasonable hope of returning him to a reasonably healthy state. And in such cases—say a long comatose patient or a patient with end-stage respiratory failure—a sensible physician will acknowledge that there is no longer any realizable therapeutic or medical goal, and will not take the mere preservation of life as his objective. Sometimes he may justify further life-prolonging activities in terms of a hope for a new remedy or some dramatic turn of events. But when reasonable hope of recovery is gone, he acts rather to comfort the patient and to keep him company, as a friend and not especially or uniquely as a physician.

I do not want to be misunderstood. Mine is not an argument to permit or to condone medical callousness, or euthanasia practiced by physicians. Rather, it is a suggestion that doctors keep their eye on their main business, restoring and correcting what can be corrected and restored, always acknowledging that death will and must come, that health is a mortal good, and that as embodied beings we are fragile beings that must snap sooner or later, medicine or no medicine. To

*From 1900 to 1970 the average overall life expectancy of white males in the United States increased about twenty-two years; for those who reached age sixty-five the increase was only one and a half years (see Chapter 12, "Mortality and Morality: The Virtues of Finitude").

keep the strings in tune, not to stretch them out of shape attempting to make them last forever, is the doctor's primary and proper goal.

To sum up: Health is different from pleasure, happiness, civil peace and order, virtue, wisdom, and truth. Health is possible only for mortal beings, and we must seek it knowing and accepting, as much as we are able to know and accept, the transience of health and of the beings who are healthy. To serve health and only health is a worthy profession, no less worthy because it does not serve all other goods as well.

What Is "Health"?

There was a time when the argument might have ended here, and we could have proceeded immediately to ask how the goal of health may be attained, and what the character of public policy toward health should be. But since there is nowadays much confusion about the nature and meaning of health, we may have made but little progress by identifying health as the proper purpose of medicine.

If the previous section might be viewed as an argument against creeping medical imperialism expanding under a view of health that is much too broad, there remains a need to confront the implications of a medical isolationism and agnosticism that reduces its province under a view of health that is much too narrow. Indeed, the tendency to expand the notion of health to include happiness and good citizenship is, ironically, a consequence of, or reaction to, the opposite and more fundamental tendency—namely, to treat health as merely the absence of known disease entities and, more radically, to insist that health as such is nothing more than a word.*

We are thus obliged, before turning to the question of what we can do to become healthier, to examine the question, What is health? For what was once self-evident, now requires an argument. I begin with some of the important difficulties that confound the search for the meaning of health.

1. What is the domain of health? Is it body, or body and soul?

---

*Claude Bernard opens his book *An Introduction to the Study of Experimental Medicine,* held by some to be a founding document of our scientific medicine, with the following sentence: *"To conserve health and to cure disease:* medicine is still pursuing a scientific solution of this problem, which has confronted it from the first." Yet he says in Chapter 1 of Part II, "[N]either physiologists nor physicians need imagine it their task to seek the cause of life or the essence of disease. That would be entirely wasting one's time in pursuing a phantom. The words life, death, health, disease, have no objective reality."[1]

Can only individuals be healthy, or can we speak univocally, and not analogically, about a healthy marriage, a healthy family, a healthy city, or a healthy society, meaning by these references something more than collections of healthy individuals? I think not. In its strict sense, health refers to individual organisms—plants and animals, no less than humans—and only analogically or metaphorically to larger groupings. I will set aside the question of whether only bodies or also souls are or can be healthy, since it appears difficult enough to discover what health is even for body. While there is disagreement about the existence of a standard of health for the soul—or, if you prefer, about whether there is "psychic health"—no one I think denies that if health exists at all, it exists as a condition at least of bodies. For the sake of simplicity, then, we shall confine our investigation in the present context to somatic or bodily health.*

2. Health appears to be a matter of more and less, a matter of degree, and standards of health seem to be relative to persons, and also relative to time of life in each person. Almost everyone's state of health could be better, and most of us—even those of us free of overt disease—can remember being healthier than we are now. Yet, as Aristotle long ago pointed out, "health admits of degrees without being indeterminate." In this respect, health is like pleasure, strength, or justice, and unlike "being pregnant" or "being dead."

3. Is health a positive quality or condition, or merely the absence of some negative quality or condition? Is one necessarily healthy if one is not ill or diseased? One might infer from modern medical practice that health is simply the absence of all known diseases. The classic textbook, *Harrison's Principles of Internal Medicine,* is a compendium of disease and, apart from the remedies for specific diseases, contains no discussion of regimens for gaining and keeping health. Indeed, health is never explicitly discussed, and the term "health" does not even occur in the index.

Clinical medicine's emphasis on disease and its cure is readily understood. It is the sick, and not the well, who seek out medical advice. The doctor has long been concerned with restoration and remedy, not with promotion and maintenance, originally the responsibilities of gymnastic (physical fitness programs) and dietetics. This orientation has been encouraged by the analytic and reductive approach of modern medical science and by the proliferation of known diseases and treatments—both leading to a highly specialized but highly frag-

---

*In doing so, we are supported by a sensible tradition which held that health, like beauty or strength, was an excellence of the body, whereas moderation, wisdom, and courage were excellences of soul. While excluding these latter goods from the goal of medicine, I do not mean to deny to a more minimal state of psychic health, namely, sanity or "emotional equilibrium," a possible place among the true ends of medicine.

mented medicine. Doctors are too busy fighting disease to be bothered much about health, and, up to a point, this makes sense.

Yet among pediatricians, with their well-baby clinics and their concern for normal growth and development, we can in fact see medicine clearly pointing to an overall good rather than away from particular evils. The same goal also informs the practices of gymnastic and dietetics. Together, these examples provide a provisional ground for the claim that health is a good in its own right, not merely a privation of one or all evils. Though we may be led to *think* about health and to discover its existence only through discovering and reflecting on *departures* from health, health would seem to be the primary notion. Moreover, as I hope will become clear, disease, as the generic name for the cluster of symptoms and identifiable pathological conditions of the body, is not a notion symmetrical with, or opposite to, health. Health and *un*health (i.e., health and falling short of health) are symmetrical, not so health and disease.

4. Who is the best judge of health, the doctor or the patient? On the surface this looks like, and has increasingly been treated as, a question about power and the locus of authority, connected with the rise of consumerism and suspicion about all kinds of expertise, and fostered by loose talk about health as a commodity, as something money can guarantee, as something determined by expressed need of patients and delivered or served on demand by doctors. But the question has deeper roots and more important implications.

If medicine is an art that aims at health, and if an art implies knowledge of ends and means, then the physician is a knower. As unnatural as it may seem that someone else should know better than I whether or not I am healthy—after all, it is my body and my pain and not the doctor's—still the doctor as a knower, *should* know what health and healthy functioning are, and how to restore and preserve them. In principle, at least, and to a great extent in practice, doctors *are* experts; they are men who know not only how we feel about, and what we wish for, our bodies, but how our bodies work and how they should work. This alone justifies their prescribing bad tasting medicine, or their mutilating a healthy abdominal wall to remove an inflamed appendix or even a nonsymptomatic ovarian cyst; this alone justifies, but surely it *does* justify, doctors giving orders and patients obeying them.*

*Several recent commentators have made much of this "giving orders" in their attempts to prove the tyrannical character of medicine and physicians. But they mistakenly treat doctor's orders as categorical imperatives, when in fact they are only hypothetical ones: "*If* you want again to be healthy, take this." In most cases it is fair to construe the patient's desire to become healthy from his self-presentation to the physician. One must, however, concede that doctor's orders are sometimes given with inadequate exploration of the patient's needs and aspirations, and with inadequate explanation and instruction regarding the propriety of the ordered therapy or regimen.

Yet the case for health as an objective condition, in principle recognizable by an expert, and independent of patient wishes and opinions, needs to be qualified. Health and unhealth, as well as all diseases, occur only in particular living beings, each experiencing *inward* manifestations of health or its absence. The patient's feelings of illness or well-being must be reckoned with, not only because the patient insists, but because they are pertinent signs in the assessment of health. To be sure, there are people who feel fine but harbor unbeknownst to themselves a fatal illness (e.g., the vigorous athlete whose routine blood count shows early leukemia). Still, when a patient complains of headaches or backaches, funny noises in his ears, fatigue, weakness, palpitations on exertion, pains or cramps in the abdomen, or dizziness, *he is not healthy,* even if he looks and acts healthy and even if the doctor fails "to find anything wrong" (i.e., fails to discover a cause for the symptoms). A *negative* report by the patient always, or almost always, counts.

There need be no discordance between the "objective" and "subjective" manifestations of health and unhealth. For the most part, they do correspond. The individual's state of health shows itself both to him and to the outsider, including the expert. If the doctor were more the expert on health than on particular diseases, he would be less quick to dismiss as insignificant the whole range of complaints that he cannot ascribe to known diseases and causes. They may not be significant of disease, but they are signs of unhealth.

5. Health is said to be relative not only to the age of the person but also to external circumstances, both natural and cultural. A person with hay fever can be well in the absence of ragweed pollen or cats, and incapacitated in their presence. The hereditary deficiency of the enzyme glucose-6-phosphate-dehydrogenase results in serious illness for the individual who eats fava beans or takes certain drugs, but is otherwise without known consequence. Eyeglasses, it is said, make myopia no longer a disability. Paraplegia may not interfere much with the life of a theoretical physicist or a president of the United States, whereas an ingrown toenail could cripple the career of a ballerina. If various functions and activities are the measure of health, and if functions are affected by and relative to circumstances, then health, too, so the argument goes, is relative.

Yet all these points, however valid, do not prove the relativity of health and unhealth. They show, rather, the relativity of the *importance* of health and unhealth. The person without hay fever, enzyme deficiency, myopia, paraplegia, and ingrown toenails is, other things being equal, *healthier* than those *with* these conditions. To be sure, various absences of health can be ignored, and others overcome by change of circumstance, while still others, even if severe, can be rendered less incapacitating. But none of this affects the fact that they *are* absences of health, or undermines the possibility that health is something in its own right.

The most radical version of the relativist argument challenges the claim that health is a *natural* norm. According to this view, what is healthy is dependent not only on time and circumstance, but even more on custom and convention, on human valuation. To apply the concept or construct "healthy" is to throw our "judgment of value" onto a factual, value-neutral condition of the body; without human judgment, there is no health and no illness. A recent commentator, Peter Sedgwick, argues that "all sickness is essentially deviancy" and that illness and disease, health and treatment are "social constructions":

> All departments of nature below the level of mankind are exempt both from disease and from treatment. The blight that strikes at corn or at potatoes is a *human invention,* for if man wished to cultivate parasites rather than potatoes (or corn) there would be no "blight" but simply the necessary foddering of the parasite-crop. Animals do not have diseases either, prior to the presence of man in a meaningful relation with them. . . Outside the significances that man voluntarily attaches to certain conditions, *there are no illnesses or diseases in nature. . . .* Out of his anthropocentric self-interest, man has chosen to consider as "illnesses" or "diseases" those natural circumstances which precipitate the death (or the failure to function according to certain values) of a limited number of biological species: man himself, his pets and other cherished livestock, and the plant-varieties he cultivates for gain or pleasure. . . . Children and cattle may fall ill, have diseases, and seem as sick; but who has ever imagined that spiders or lizards can be sick or diseased? . . . The medical enterprise is from its inception value-loaded; it is not simply an applied biology, but a biology applied in accordance with the dictates of social interest.[2]

Insofar as one considers only disease, there is something to be said for this position—but not much. Disease-entities may in some cases be constructs, but the departures from health and the symptoms they group together are not. Moreover, health, although certainly a good, is not therefore a good whose goodness exists by convention or by human decree. Health, illness, and unhealth all may exist even if not discovered or attributed. That human beings don't *worry* about the health of lizards and spiders implies nothing about whether or not lizards and spiders *are* healthy, and any experienced student of spiders and lizards can discover—and not merely invent—abnormal structures and functionings of these animals. Human indifference is merely that. Deer can be healthy or full of cancer, a partially eaten butterfly escaping from a blue-jay is not healthy but defective, and even the corn used to nourish parasites becomes abnormal corn, to the parasite-grower's delight.

Sedgwick must be partly forgiven for his confusion, for he has no doubt been influenced by a medicine that focuses on disease-entities and not on health, by a biology that does not consider wholes except as mere aggregates, and by that conventional wisdom of today's social science which holds that *all* goods are good because they are valued, and *all* values are in turn mere conventions, wholly tied to the culture or the individual that invents them. To be sure, different cultures have different taxonomies of diseases, and differing notions of their cause. But the fact that *some* form of medicine is *everywhere* practiced—whether by medicine men and faith healers or by trained neurosurgeons—is far more significant than the differences in nosology and explanation: It strongly suggests that healers do not fabricate the difference between being healthy and being unhealthy; they only try to learn about it, each in his own way. It also suggests that health is everywhere valued *because* it is good.

I turn next away from these difficulties to the constructive part of the search for health. To begin with, I should say that I am not seeking a precise definition of health. I am rather inclined to believe that it is not possible to say definitively what health is, any more than it is possible to say wholly and precisely what "livingness" or "light" or "knowledge" or "human excellence" are. What I hope to show more clearly is what *sort* of a "thing" health is, so that we can be more secure in recognizing and promoting it, even if we are unable to capture it in speech. The boundaries or limits of health—and hence, its de*fin*ition—may always remain hazy; but our search may help us to recognize, and thus hold, the center.

First, I note that in ordinary speech we generally use the terms "health" and "healthy" as if we know what we are talking about. The term may become problematic when we decide to investigate it, but we—and I would guess people in most cultures—speak of health as if the term intends a real condition, found in and displayed by real people, including themselves. When military questionnaires or civil service applications ask about our state of health, we are not at a loss as to what is being inquired about, even if we may not have a simple or ready answer; the twin tendencies to exaggerate or to deny illness in answering such questionnaires prove all the more that we regard the question as meaningful and the answer as important. Even those cases in which someone feels and acts "fit as a fiddle" but harbors a fatal disease give us no difficulty: We say that the appearance of health was deceptive. The possibility of making such an error, far from undermining the existence of a true condition of health, in fact presupposes it; appearances can only be deceptive if there are realities with a view to which we discover deception.

Various idioms and expressions also support our contention that health is recognizable. Have we not heard it said of someone that he is the picture of health? In these and other expressions, we point to certain exemplary individuals as standards, suggesting that healthiness shines forth and makes itself known.

Etymological investigations may provide some clues for what we recognize when we recognize health. The English word "health" literally means "wholeness," and "to heal" means "to make whole." (Both words go back to the Old English *hal* and the Old High German *heil*, as does the English word "whole.") To be whole is to be healthy, and to be healthy is to be whole. Ancient Greek has two etymologically distinct words translatable as "health," *hygieia* and *euexia. Hygieia*, the source of our word "hygiene," apparently stands for the Indo-European *sugwiges*, which means "living well," or more precisely, a "well way of living." *Euexia* means, literally, "well-habited-ness," and in this context, "good habit of body."

Two observations are worth noting: (1) Both the Greek and the English words for health are totally unrelated to all the words for disease, illness, sickness. (This is also true in German, Latin, and Hebrew.) The Greek words for health, unlike the English, are completely unrelated also to all the verbs of healing: Health is a state or condition unrelated to, and prior to, both illness and physicians. (2) The English emphasis on "wholeness" or "completeness" is comparatively static and structural, and the notion of a whole distinct from all else and complete in itself carries connotations of self-containedness, self-sufficiency, and independence. In contrast, both Greek terms stress the *functioning* and *activity* of the whole, and not only its working, but its working well.*

Aided by these etymological reflections, we turn now from words to things in search of instances of *wholeness* and of *working-well* in nature. We shall look, of course, only at part of what is today called nature. We are not tempted to seek health in mountains or rocks or hurricanes, for these are surely not organic wholes. We look only at *animate* nature, at plants, animals, and man—true wholes, if any there be.

---

*The Greek terms suggest that health is connected with the way we live and imply perhaps that health has largely an inner cause. Indeed, it seems reasonable to think of health understood as "living well" or "well-habited" as the cause of itself. Just as courage is the cause of courageous action and hence also of courage, so "living well" *is* health, is the *cause* of health, and is *caused by* health. The activities that in English usage we might be inclined to see as *signs* or *effects* of health, might in the Greek usage appear as the *essence* of health.

Related to this, the Greek seems to imply that to stay healthy requires effort and care, that however much nature makes health possible, human attention and habit are required to maintain and preserve it. Health is neither given nor usually taken away from the outside, nor is it the gratuitously expected state of affairs.

But are plants and animals authentic wholes, or are they mere aggregates masquerading as wholes? In Chapter 10, "Teleology, Darwinism, and the Place of Man: Beyond Chance and Necessity?," I will try to show at greater length why living things cannot even be looked at, much less understood, except as wholes—and in this sense at least, as teleological beings—regardless of whether or not the species originally came to be by nonteleological processes. I will here present only some of the evidence.

First, consider the generation of living things. Each organism comes to be not at random, but in an orderly manner, starting from some relatively undifferentiated but nevertheless specific seed or zygote produced by parents of the same species, and developing, unfolding itself from within, in successive stages that tend toward and reach a limit, itself, the fully formed organism. The adult that emerges from the process of self-development and growth is no mere outcome, but a completion, an end, a whole.

Second, a fully formed mature organism is an organic whole, an articulated whole, composed of parts. It is a structure and not a heap. The parts of an organism have specific functions that define their nature as parts: the bone marrow for making red blood cells; the lungs for exchange of oxygen and carbon dioxide; the heart for pumping the blood. Even at a biochemical level, every molecule can be characterized in terms of its function. The parts, both macroscopic and microscopic, contribute to the maintenance and functioning of the other parts, and make possible the maintenance and functioning of the whole.

But perhaps the best evidence that organisms are wholes, and that their wholeness and their healthiness correspond, is the remarkable power of self-healing. In hydra, planaria, and many plants, the power to restore wholeness shows itself in amazing degree, in the form of adult regeneration. A plant cutting will regrow the missing roots, a hydra regrows amputated tentacles, and each half of a divided planarian will regenerate the missing half. In human beings, various organs and tissues (e.g., skin, the epithelia of the digestive tract, liver, bone marrow, and lymph nodes) have comparable regenerative powers. More generally, nearly all living things heal wounds or breaks and tend to restore wholeness. Foreign bodies are engulfed and extruded by amoebas and by man. This tendency to maintain wholeness by rejecting *additions* to the whole becomes marvelously elaborate in the immune system of higher animals, which sensitively recognizes and combats the entry of alien elements, whether infectious agents, tumors, or grafted tissue.

The highly complex phenomenon of pain is also a sign that organisms are wholes. Pain provides an advance warning, or is an accompanying sign, of a threat to bodily integrity. Yet its presence is as much a sign of wholeness as its opposite, for pain, in normal circumstances, attests to a healthy nervous system detecting, and at the same time representing as an insistent sign, the presence of the threat of unhealth. (Here we see again a connection between experienced bodily feeling and actual conditions of the body.)

So far my examination of wholeness has been largely, or at least explicitly, structural and static, in keeping with a view of health as capturable in a picture of health. Yet can one capture healthiness in a photograph? Don't we need at least a movie camera?

One way to examine this claim is to ask, Is being healthy compatible with being asleep? In a way, and up to a point, the answer must be yes. If we are healthy, we do not cease being healthy when we sleep. Sleep is necessary to stay healthy, and insomnia is sometimes a symptom of illness. Digestion, respiration, circulation, and metabolism continue quite normally while we sleep, but only if and because we do not sleep for long. Even this vegetative activity requires periodic wakefulness, at least enough to bite, chew and swallow. Moreover, continued sleep would rapidly produce feebleness and atrophy of bones and muscles, as well as more gradual losses of other functions. And even if none of these disasters were to befall us, ours would be a sleepy kind of wholeness; the sleeping Rip Van Winkle might not have been sick, but he was hardly healthy. The wholeness of a man is not the wholeness of a statue of a man, but a wholeness-in-action, or working-well of the work done by the body of a man.

What constitutes well-working? The answers will vary from species to species—among other things, web-spinning for a spider, flight for some birds, swimming for others. For a given species, there will be some variations among individuals, increasingly so as functions are dissected into smaller and smaller subfunctions. For certain functions, the norm will be a mean between excess and deficiency—for example, blood pressure can be too high or too low, as can blood sugar or blood calcium; blood can clot too quickly or too slowly; body temperature can be too high or too low. And while there is some arbitrariness in our deciding on the lower and upper limits of the so-called normal range in all these cases, this indistinctness of the margins does not indicate nature's arbitrariness or indifference about the norm. For we note that the body has elaborate mechanisms to keep these properties balanced, often very precisely, between excess and deficiency, to preserve homeostasis.

There is also well-working and ill-working of the activities and powers that watch over bodily wholeness—including wound-healing, immunity, and pain. A certain excess in wound-healing common among individuals of African origin leads to piled up scars known as keloids, whereas people with diabetes or Cushing's syndrome heal cuts very badly and slowly. In some individuals, the immune system sometimes makes a mistake, like an overzealous guardian, and attacks various native tissues as if they were alien, thereby producing the so-called autoimmune diseases; yet in others, antibody formation is, by inheritance, defective. Finally, certain neurological conditions can affect the pain-receiving pathways to make nonthreatening impingements seem painful, or in other cases, to allow threatening ones to do their damage painlessly.

Yet it is at the whole animal that one should finally look for the measure of well-working, for the well-working of the whole. That there are mechanisms for restoring well-working at this level can be seen by considering the case of a dog missing one hind leg. Such a dog still runs—though certainly not as well as when he had four legs—by positioning his remaining hind leg as close as he can to the midline of his body, to become a more-balanced tripod, and he does this without being taught or without previous experience in three-legged running. There appear to be "rules of rightness," as Polanyi calls them, unique to each level of bodily organization, whose rightness is not explicable in terms of the lower levels, even though failure at the lower levels can cause failure at the higher—for example, a broken wing can prevent flight, but two intact wings, good chest muscles, and hollow bones don't add up to flight. Think about trying to give a mechanical account of the rules of rightness for the well-functioning that is riding a bicycle or swimming or speaking.

Thus, it is ultimately to the workings of the whole animal that we must turn to discover its healthiness. What, for example, is a healthy squirrel? Not a picture of a squirrel, not really or fully the sleeping squirrel, not even the aggregate of his normal blood pressure, serum calcium, total body zinc, normal digestion, fertility, and the like. Rather, the healthy squirrel is a bushy-tailed fellow who looks and acts like a squirrel; who leaps through the trees with great daring; who gathers, buries, covers but later uncovers and recovers his acorns; who perches out on a limb cracking his nuts, sniffing the air for smells of danger, alert, cautious, with his tail beating rhythmically; who chatters and plays and courts and mates, and rears his young in large improbable-looking homes at the tops of trees; who fights with vigor and forages with cunning; who shows spiritedness, even anger, and more prudence than many human beings.

To sum up: Health is a natural standard or norm—not a moral norm, not a "value" as opposed to a "fact," not an obligation—a state of being that reveals itself in activity as a standard of bodily excellence or fitness, relative to each species and to some extent to individuals, recognizable if not definable, and to some extent attainable.

If you prefer a more simple formulation, I would say that health is "the well-working of the organism as a whole," or again, "an activity of the living body in accordance with its specific excellences."*

The Pursuit of Health

The foregoing inquiry into the nature of health, though obviously incomplete and in need of refinement, has, I hope, accomplished two things: first, to make at least plausible the claim that somatic health is a finite and intelligible norm, which is the true goal of medicine; and second, by displaying something of the character of healthiness, to provide a basis for considering how it might be better attained. *Curiously, it will soon become apparent that even if we have found the end of medicine, we may have to go beyond medicine in order to find the best means for attaining it.*

Though health is a natural norm, and though nature provides us with powerful inborn means of preserving and maintaining a well-working wholeness, it is wrong to assume that health is the simply given and spontaneous condition of human beings, and unhealth the result largely of accident or of external invasion. In the case of non-human animals, such a view could perhaps be defended. Other animals instinctively eat the right foods (when available) and act in such a way as to maintain their natually given state of health and vigor. Other animals do not overeat, undersleep, knowingly ingest toxic substances, or permit their bodies to fall into disuse through sloth, watching television and riding in automobiles, transacting business, or writing articles about health. For us human beings, however, even a healthy nature must be nurtured, and maintained by effort and discipline if it is not to become soft and weak and prone to illness, and certain excesses and stresses must be avoided if this softness is not to spawn overt unhealth and disease. One should not, of course, underestimate the role of germs and other hostile agents working from without; but I strongly suspect that the germ theory of disease has been oversold, and that the state of "host resistance," and in partic-

---

*Whatever progress we may have made in our search for health, large questions still remain, which I defer to another occasion. These questions include: What activities of the living body should be considered, and are all of them of equal rank? What are the specific excellences or fitnesses of various organisms, and can one hope to articulate these standards for a being as complex as man, whose activities are so highly diversified and differentiated? What is a living body, and what a specifically *human* living body? Finally, what is the relation of health of body to psychic health? I touch on some of these questions in later chapters (see, e.g., pp. 201–203, and Chapter 11, "Thinking About the Body").

ular the immunity systems, will become increasingly prominent in our understanding of both health and disease.

Once the distinction is made between health nurture and maintenance on the one hand, and disease prevention and treatment on the other, it becomes immediately clear that bodily health does not depend only on the body and its parts. It depends decisively on the psyche with which the body associates and cooperates. A few examples will make this clear, if it is not already obvious. Some disorders of body are caused, at least in part, by primary problems of soul (psyche); the range goes from the transitory bodily effects of simple nervousness and tension headaches, through the often severe somatic symptoms of depression (e.g., weight loss, insomnia, constipation, impotence), to ulcers and rheumatoid arthritis. Other diseases are due specifically to some aspect of the patient's way of life: cirrhosis in alcoholics, hepatitis in drug addicts, special lung diseases in coal miners, venereal disease in prostitutes.

But the dependence goes much farther than these clear psycho- and sociosomatic interactions. In a most far-reaching way, our health is influenced by our temperament, our character, our habits, our whole way of life. This fact was once better appreciated than it is today.

In a very early discussion of this question, in the Platonic dialogue *Charmides,* Socrates criticizes Greek physicians for foolishly neglecting the whole when attempting to heal a part. He argues that "just as one must not attempt to cure the eyes without the head or the head without the body, so neither the body without the soul."[3] In fact, one must care "first and most" for the soul if one intends for the body to be healthy. If the soul is moderate and sensible, it will not be difficult to effect health in the body; if not, health will be difficult to procure. Greek medicine fails, it is charged, because men try to be physicians of health and moderation separately.

Socrates does not say that excellence of soul and excellence of body are one and the same; indeed, health is clearly distinguished from moderation. Rather, the claim is that health is at least in large part affected by or dependent upon virtue, that being well in body has much to do with living well, with good habits not only of body but of life.

Now Socrates certainly knew, perhaps better than we, that accident and fortune can bring harm and ill health even to well-ordered bodies and souls. He knew about inborn diseases and seasonal maladies and wounds sustained in battle. He knew that health, though demanding care and discipline, and requiring a certain control of our bodily desires, was no sure sign of virtue—and that moderation is not all of virtue. He knew, too, as we know, human beings whose healthiness was the best thing about them, and he knew also that to be preoccupied with health is either a sign or a cause of a shrunken human life.

Yet he also knew what we are today altogether too willing to forget: that *we are in an important way responsible for our state of health,* that carelessness, gluttony, drunkenness, and sloth take some of their wages in illness. At a deeper level, he knew that there was a connection between the fact that the human soul aspires beyond mere self-preservation and the fact that men, unlike animals, can make themselves sick and feverish. He knew, therefore, that health in human beings depends not only on natural gifts, but also on taming and moderating the admirable yet dangerous human desire to live better than sows and squirrels.

Today we are beginning again to consider that Socrates was possibly right, that our way of life is a major key to our sickness and our health. I would myself guess that more than half the visits to American doctors are occasioned by deviations from health for which the patient, or his way of life, is in some important way responsible. Most chronic lung diseases, much cardiovascular disease, most cirrhosis of the liver, many gastrointestinal disorders (from indigestion to ulcers), numerous muscular and skeletal complaints (from low-back pain to flat feet), venereal disease, nutritional deficiencies, obesity and its consequences, and certain kinds of renal and skin infections are in important measure self-induced or self-caused—and contributed to by smoking, overeating, overdrinking, eating the wrong foods, inadequate rest and exercise, and poor hygiene. To these conditions must be added the results of trauma—including automobile accidents—in which drunkenness plays a leading part, and suicide attempts, as well as accidental poisonings, drug abuse, and many burns. I leave out of the reckoning the as yet poorly studied contributions to unhealth of all varieties made by the special stresses of modern urban life.

There are even indications that cancer is in some measure a disease of how we live, even beyond the clear correlations of lung cancer with smoking and cancer of the cervix with sexual promiscuity and poor sexual hygiene. If the incidence of each kind of cancer could be reduced to the level at which it occurs in the population in which its incidence is lowest, there would be 90 percent less cancer. Studies have shown that cancers of all sorts—not only cancers clearly correlated with smoking and drinking—occur less frequently among the clean-living Mormons and Seventh Day Adventists.

The foregoing, it will be noted, speaks largely about disease and unhealth, and about the role of our excesses and deficiencies in bringing them about. Unfortunately, we know less about what contributes to healthiness, as nearly all epidemiological studies have been studies of disease. But in the past dozen years, there have appeared published reports of a most fascinating and important series of epidemiological studies on health, conducted by Dean Lester Breslow and his colleagues at the UCLA School of Public Health. Having first developed a method for quantifying, albeit crudely, one's state of health and well-functioning, they investigated the effect of various health practices on physical health status. They have discovered, empirically, seven independent "rules" for good health, which correlate very well with healthiness, and also with longevity. People who follow all seven rules are healthier and live longer than those who follow six, six more than five, and so on, in perfect order. Let me report two of their more dramatic findings: The physical health status of people over seventy-five who followed all the rules was about the same as those aged thirty-five to forty-four who followed fewer than three; and a person who follows at least six of the seven rules has a life expectancy eleven years longer at age forty-five than someone who has followed less than four. Moreover, these differences in health connected with health practices persisted at all economic levels and, except at the very lowest incomes, appeared largely independent of income.[4]

The seven rules are: (1) Don't smoke cigarettes. (2) Get seven hours of sleep. (3) Eat breakfast. (4) Keep your weight down. (5) Drink moderately. (6) Exercise daily. (7) Don't eat between meals. ("Visit your doctor" is not on the list, though I must confess that I cannot find out if this variable was investigated.) It seems that Socrates, and also Grandmother, may have been on the right track.

One feels, I must admit, a bit foolish, in the latter half of the twentieth century—which boasts the cracking of the genetic code, kidney machines, and heart transplants—to be suggesting the quaint formula "Eat right, exercise, and be moderate, for tomorrow you will be healthy." But quaint formulae need not have been proven false to be ignored, and we will look far more foolish if Breslow and his colleagues are onto something that, in our sophistication, we choose to overlook.

Implications for Policy

What might all this point to for medicine and for public policy regarding health? Let me try to sketch for you the implications of the preceding sections, which, as a point of departure, I would summarize in this way: Health and only health is the doctor's proper business; but health, understood as well-working wholeness, is not the business only of doctors. Health is, in different ways, everyone's business, and it is best pursued if everyone regards and minds his *own* business— each of us our own health, the doctor the health of his patient, public health officials and legislators the health of the citizens.

With respect to the medical profession itself, there is a clear need to articulate and delimit the physician's domain and responsibilities, to protect against expansion and contraction. The more obvious and perhaps greater danger seems to be expansion, given the growing technological powers that can serve nontherapeutic ends and the rising demands to put these powers to nonmedical uses. The medical profession must take the initiative in establishing and policing the necessary boundaries. The American Medical Association, the state and county medical societies, and the various specialty organizations would do well to examine current practices and to anticipate new technologies with a view to offering guidance to their members amidst these dangers. In some cases, they might well try to discourage or proscribe certain quasimedical or extramedical uses of medical technique. The American College of Obstetrics and Gynecology, for example, should consider regulations barring its members from helping prospective parents determine or select the sex of their child-to-be, or the American Association of Neurological Surgeons could establish strict guidelines for the permissible uses, if any, of destructive brain surgery for the sake of modifying behavior.

It is true that such guidelines can always be violated in the privacy of an examining room or operating theater—but what rule cannot?—and it is also true that the decentralized character of American medicine makes professional self-regulation more difficult than in, say, Britain. Still, the profession has heretofore not concerned itself with the problem, and it would be foolish to declare inefficacious a remedy not even contemplated because the disease itself had yet to be recognized.

Medical licensure provides an alternative device for drawing boundaries. It would be worthwhile to reconsider the criteria for medical licensure and the privileges and prerogatives that it is meant to confer. The current system of licensing was designed largely to protect the public, and the reputation of the profession, against incompetents and charlatans. Yet this license to practice healing is now *de facto* a license to conduct research on human subjects, as well as a license to employ biomedical technique in the service of any willing client, private or public, for almost any purpose not forbidden by law. Because these various techniques involve direct physical or chemical intervention into the human body, and because the practice of such interventions has been restricted to those who know about and can protect the human body, a medical license has been regarded as a necessary condition for all these extramedical activities, but it has also come to be regarded as a *sufficient* condition.

Some have argued that changes in licensing be made to clearly distinguish the healing profession, and to require special (and additional) licensing to engage in clinical research, practice various forms of biomedical indulgence, or serve purposes of social reform and social control. In some cases, people have called for completely separate professions—of, say, abortionists, artificial inseminators, mercy killers, surgical beautifiers, mood elevators, and eugenic counselors. This approach is recommended not only because it keeps the boundaries neat, but because it prevents the poor use of medical expertise and training, since at least some of these procedures and practices—including first-trimester abortion—could be mastered by moderately intelligent and dextrous high school graduates given six months of technical training.

On the other hand, since the demand for these extramedical services is unlikely to disappear, it might be dangerous to separate them from the practice of medicine. Keeping the various functions and "professions" mixed together under the medical umbrella might cover them all with the long-standing ethical standards of the traditional medical profession, a protection that might not readily be provided, or even sought by, the "younger professions" if they were to be separated or expelled from the healing profession. Those who hold this view are willing to tolerate some confusion of purpose in exchange for what they believe will help produce necessary restraints. But whether the restraint would in fact be forthcoming is an open question.

The greatest difficulty is how to protect the boundaries of the medical domain against unreasonable *external* demands for expansion. The public's misperception of medicine is ultimately more dangerous than the doctor's misperception of himself. The movement toward consumer control of medicine, the call for doctors to provide therapy for social deviants and criminal offenders, and the increasing governmental regulation of medical practice all run the risk of transforming the physician into a mere public servant, into a technician or helper for hire. Granted, the doctor must not be allowed to be a tyrant, but neither must he become a servant. Rather, he must remain a leader and a teacher. The community must respect the fact that medicine is an art and that the doctor rightly is a man of expert knowledge, deserving of more than an equal voice in deciding what his business is. Though one may rightly suspect *some* of the motives behind the medical profession's fear of governmental intrusion, one must acknowledge the justice of at least this concern: Once the definition of health care and the standards of medical practice are made by outsiders—and governmental health insurance schemes all tend in this direction—the physician becomes a mere technician.

Yet if the medical profession wants to retain the right to set its own limits, it must not only improve its immunity against foreign additions to its domain, it must also work to restore its own wholeness. The profession must again concern itself with health, with wholeness, with well-working, and not solely with the cure of disease. The doctor must attend to health maintenance, and not only treatment or even prevention of specific diseases. He should no longer look befuddled when a patient asks him, "Doctor, what regimen do you suggest in order that I may remain healthy?" This implies, of course, changes in medical orientation, which in turn imply changes in medical education, both difficult to design in detail and not easy to institute in practice. But again, we have not seriously thought about how to do this because we have not seen that it was something that might need doing. To recognize and identify this defect is to take the first, and thus the biggest, step toward its amelioration.

I am not saying that doctors should cease to be concerned about disease, or that they should keep us in hospitals and clinics until we become fully healthy. I do suggest, however, that physicians should be more interested than they are in finding ways to keep us from their doors. Though medicine must remain in large part restorative and re-medial, greater attention to healthy functioning and to regimens for becoming and remaining healthy could be very salutary, even toward the limited goal of reducing the incidence of disease. Little intelligence and imagination have thus far been expended by members of the profession, or by health insurance companies, to devise incentive schemes that would reward such a shift in emphasis (e.g., that would reward financially both patient and physician if the patient stays free of the need for his services). I invite people cleverer than I to make such efforts, especially in conjunction with the likely changes in the financing of medical care.

Moving beyond the doctor-patient setting to the area of medical research, the major point is the importance of epidemiological research on *healthiness*. We need to devise better indices of healthiness than mortality and morbidity statistics, which, I have argued, are in fact not indices of health at all. The studies like those of Breslow and his collaborators are a step in the right direction and should be encouraged. Only with better measures of healthiness can we really evaluate the results of our various health practices and policies.

We also need large-scale epidemiological research into health maintenance in order to learn more about what promotes, and what undermines, health. More sophisticated studies in nutrition, bodily exercise, rest and sleep, relaxation, and responses to stress could be very useful, as could expanded research into personal habits of health and hygiene and their effects on general healthiness, overall resistance to disease, and specific resistance to specific diseases. We need to identify and learn about healthy subgroups in the community, such as the Mormons, and discover what accounts for their success.

All of these things are probably obvious, and most of them have been championed for years by people in the fields of public health and preventive medicine—though they, too, have placed greater emphasis on disease prevention than on health maintenance. Their long-ignored advice is finally beginning to be heeded, with promising results. For example, studies report a surprising downturn (after a twenty-five year climb) in the death rate from heart attacks among middle-age men, attributed in part to changes in smoking and eating habits, exercise, and to new treatments for high blood pressure.* Yet this approach will always seem banal and pedestrian in comparison with the glamorous and dramatic style of high-technology therapeutics, with the doctor locked in combat with overt disease, displaying his marvelous and magical powers. My high regard for these powers cannot stifle the question as to whether the men who first suggested adding chlorine to drinking water or invented indoor plumbing didn't contribute more to healthiness than the Nobel Prize winners in medicine and physiology who discovered the chemical wonders of enzyme structure or of vision. It might be worthwhile to consider by what kinds of incentives and rewards the National Institutes of Health or the AMA might encourage more and better research into health maintenance and disease prevention.

However, as has been repeatedly emphasized, doctors and public health officials have only limited powers to improve our health. Health is not a commodity that can be delivered. Medicine can only help those who help themselves. Discovering what will promote and maintain health is only half the battle; we must also find ways to promote and inculcate good health habits and to increase personal responsibility for health. This problem is, no doubt, the most fundamental and also the most difficult task. It is but one more instance of that age-old challenge: How to get people to do what is good for them without tyrannizing them. The principles of freedom and of wisdom do not always—shall I say, do not often?—lead in the same direction.

Because this is not a new problem we have some experience in how to think about it. Consider the problem of getting people to obey the law. Policemen and judges are clearly needed to handle the major crimes and criminals, but it would be foolish to propose, and dangerous to provide, even that degree of police surveillance and interference required to prevent only the most serious lawbreaking. Though justice is the business of the policeman and the judge, it is not their business alone. Education—at home, in schools, in civic and religious institutions—can teach lawabidingness far better than policemen, and where the former is successful, there is less need of the latter.

Yet even without considering the limitations of this analogy, the limits of the power of teachers—and of policemen as well—to produce

---

*There are, regrettably, other trends: cancer of the lungs in women is on the rise, clearly connected with increased smoking among women. Cancer of the cervix is also increasing.

lawabidingness are all too apparent. And when one considers that fear of immediate, identifiable punishment probably deters lawbreaking more than fear of unhealth deters sloth and gluttony, we see that we face no simple task. The wages of poor health habits during youth are only paid much later, so much later that it is difficult to establish the relation of cause and effect, let alone make it vivid enough to influence people's actions. If it isn't likely to rain for twenty years, few of us are likely to repair our leaky roofs.

This is not a counsel of despair. On the contrary, I am much impressed with the growing interest in health and health education in recent years, including the greater concern for proper nutrition, adequate exercise, dental hygiene, and the hazards of smoking and the evidence that, at least among some groups, this attention is bearing fruit. Nevertheless, when we consider the numerous impediments to setting in order our lives and our communities, I think we should retain a healthy doubt about just how healthy we Americans are likely, as a community, to become.

This skepticism is rather lacking in most political pronouncements and policies regarding health. Making unwarranted inferences from medicine's past successes against infectious disease, being excessively impressed with the technological brilliance of big hospital medicine, mobilizing crusades and crash programs against cancer and heart disease, the health politicians speak as if more money, more targeted research, better distribution of services, more doctors and hospitals, and bigger and better cobalt machines, lasers, and artificial organs will bring the medical millennium to every American citizen. Going along with all this is a lack of attention to health maintenance and patient responsibility. While it would surely be difficult for the Federal government to teach responsibility, we should not be pleased when its actions in fact discourage responsibility.

One step in this direction is the growing endorsement of the so-called right to health, beyond the already ambiguous and dubious right to health care. Here is a typical argument:

> The right to *health* is a fundamental right. It expresses the profound truth that a person's autonomy and freedom rest upon his ability to function physically and psychologically. It asserts that no other person can, with moral justification, deprive him of that ability. The right to *health care* or the right to *medical care,* on the other hand, are qualified rights. They flow from the fundamental right, but are implemented in institutions and practices only when such are possible and reasonable and only when other rights are not thereby impeded.[5]

If the right to health means only the right not to have one's health destroyed by another, then it is a reasonable but rather impotent claim in the health care arena; the right to health care or medical care could hardly flow from a right to health, unless the right to health meant also and mainly the right to become and to be kept healthy. But if health is what we say it is, it is an unlikely subject of a right in either sense. Health is a state of being, not something that can be given, and only in indirect ways something that can be taken away or undermined by other human beings. It no more makes sense to claim a right to health than a right to wisdom or courage. These excellences of body and of soul require natural gift, attention, effort, and discipline on the part of each person who desires them. To make my health someone else's duty is not only unfair, it imposes a duty impossible to fulfill. Though I am not particularly attracted by the language of rights and duties in regard to health, I would lean much more in the direction, once traditional, of saying that health is a *duty,* that one has an obligation to preserve one's own health. The theory of a right to health flies in the face of good sense, serves to undermine personal responsibility, and in addition places obligation where it cannot help but be unfulfillable.

Similarly, the amendment to the Medicare legislation that provides payment for kidney-machine treatment for all in need, at a cost of from $10,000 to $40,000 per patient, is, for all its good intentions, a questionable step. First of all, it establishes the principle that the Federal government is the savior of last resort—or as is more likely at this price tag, the savior of first resort—for specific persons with specific diseases. In effect, the government has said that it is in the national interest for the government to pay, disease by disease, life by life, for life-saving measures for all its citizens. The justice of providing benefits of this magnitude solely to people with kidney disease has been loudly questioned, and hemophilia organizations, among others, are pressing for government financing of equally expensive treatment. Others have called attention to the impossible financial burden that the just extension of this coverage would entail. Finally, this measure gives governmental endorsement, in a most dramatic and visible way, to the high-cost, technological, therapy-oriented approach to health. This approach has been challenged, on the basis of a searching analysis of this kidney-machine legislation, in a report by a panel of the Institute of Medicine of the National Academy of Sciences, which, with admirable self-restraint, comments:

> One wonders how many billions of dollars the nation would now be spending on iron lungs if research for the cure of polio had not been done.[6]

This is not to say that, in the special case of the kidney machines under the special circumstances in which the legislation was passed, a persuasive case was not made on the other side. Clearly, it was hoped that perfection of kidney transplantation or future prevention of kidney disease would make this high-cost insurance obsolete before too long. Moreover, no one wishes to appear to be, or indeed to be, callous about the loss of life, especially preventable premature loss of life. Still the dangers of the kidney-machine legislation must be acknowledged.

One might even go so far as to suggest that prudent and wise legislators and policymakers must in the future resist (in a way that no private doctor should be premitted to resist) the temptation to let compassion for individual calamities and general sentimentality rule in these matters. Pursuing the best health policy for the American people—that is, a policy to encourage and support the best possible health for the American people—may indeed mean *not* taking certain measures that would prevent known deaths. Only by focusing on health and how one gets it, and by taking a more long-range view, can our health policy measure up in deed to its good intentions.

The proposals for a National Health Insurance seem also to raise difficulties of this sort, and more. Medical care is certainly very expensive, and therefore, for this reason alone, not equally available to all. The economic problems are profound and genuine, and there are few dispassionate observers who are not convinced that something needs to be done. Many technical questions have been debated and discussed, including the range of coverage, and the sources of financing, and organized medicine has voiced its usual concern regarding governmental interference, a concern that I have already indicated I share in regard to the delimination of the doctor's role and the scope of health care. But some of the most serious issues have received all too little attention.

The proposals for National Health Insurance take for granted the wisdom of our current approaches to the pursuit of health and thereby insure that in the future we will get more of the same. These proposals will simply make available to the noninsured what the privately insured now get: hospital-centered, highly technological, disease-oriented, therapy-centered medical care. The proposals have entirely ignored the question of whether what we now do in health is what we *should* be doing. They not only endorse the status quo, but fail to take advantage of the rare opportunity that financial crises provide to reexamine basic questions and directions. The irony is that economizing in health care is probably possible only by radically reorienting the pursuit of health.

One cannot help getting the impression that it is economic *equality,* not health, and not even economizing, that is the primary aim of these proposals. At a seminar in which I participated, a career official of HHS informally expressed irritation at those who are questioning whether the health-care delivery system is really making us healthier and suggested that their main goal was to undermine liberal health programs enacted in recent years. Yet this official went on to say that even if the evidence conclusively showed that all the government's health programs in no way actually improved health, the programs ought to be continued for their extramedical (i.e., social and economic) benefits. For myself, I confess that I would prefer as my public health official the cold-hearted, even mean-spirited, fellow who is interested in health and knows how to promote it.

All the proposals for National Health Insurance embrace, without qualification, the no-fault principle. They therefore choose to ignore, or to treat as irrelevant, the importance of personal responsibility for the state of one's health. As a result, they pass up an opportunity to build both positive and negative inducements into the insurance payment plan, by measures such as refusing or reducing benefits for chronic respiratory disease care to persons who continue to smoke.

There are, of course, complicated questions of justice raised, and even to suggest that the sick be in any way penalized, implying responsibility and fault, flies in the face of current custom and ways of thinking. Yet one need not be a Calvinist or a Spartan to see merit in the words of a wise physician, Robert S. Morison, writing on much the same subject:

> In the perspectives of today, cardiovascular illness in middle age not only runs the risk of depriving families of their support, or society of certain kinds of services; it increasingly places on society the obligation to spend thousands of dollars on medical care to rescue an individual from the results of a faulty living pattern. Under these conditions, one wonders how much longer we can go on talking about a right to health without some balancing talk about the individual's responsibility to keep healthy.
>
> I am told that Thorstein Veblen used to deplore the fact that in California they taxed the poor to send the rich to college. One wonders how he would react to a system which taxes the virtuous to send the improvident to hospital.[7]

But even leaving aside questions of justice and looking only at the pursuit of health, one has reason to fear that the new insurance plan, whichever one it turns out to be, may actually contribute to a worsening rather than an improvement in our nation's health, especially if there is no balancing program to encourage individual responsibility for health maintenance.

One final word. Despite all that I have said, I would also emphasize that health, while a good, cannot be the greatest good, either for an individual or for a community. Politically, an excessive preoccupation with health can conflict with the pursuit of other important social and economic goals (e.g., when cancer-phobia leads to government regulations that can unreasonably restrict industrial activity or personal freedom). But more fundamentally, it is not mere life, or even a healthy life, but rather a good and worthy life for which we must aim. And while poor health may weaken our efforts, good health alone is an insufficient condition or sign of a worthy human life. Indeed, though there is no such thing as being too healthy, there is such a thing as being too concerned about health. To be preoccupied with the body is to neglect the soul, for which we should indeed care "first and most," and more than we now do. We must strike a proper balance, a balance that can only be furthered if the approach to health also concentrates on our habits of life.

CHAPTER SEVEN

Practicing Prudently
Ethical Dilemmas in Caring for the Ill

Nor is prudence a knowledge only of general principles, but it must also know the particulars; for it is practical and action is always about particulars.

—Aristotle, *Nicomachean Ethics* (1141b14–17)

To fully deliver the diseased from their sufferings and to blunt the violence of their diseases, and not to begin to treat those who are overmastered by their diseases, knowing that in such cases medicine is powerless.

—Hippocrates, *The Art*

What should the doctor do? How should the doctor act? These are questions increasingly raised by both physician and layman alike, often in relation to the growing incidence and variety of difficult dilemmas in the care for the seriously ill. Formal discussion of ethical matters now occurs in most medical schools, in medical journals, and at professional meetings, while the popular press and television give full coverage to tantalizing cases and vexed questions of policy. To some extent, perhaps, this is the pendulum's return to ethics after a generation or two of silent confidence in the belief that one could get on quite well—and not only in medicine—without moral reflection. However, there are good reasons to think that that confidence no longer can or will prevail. First, there are the notorious dilemmas about abortion, drug abuse, informed consent, clinical trials, genetic counseling, malpractice, and the like, as well as enduring problems in the allocation of scarce resources. But there are also more central matters. Medicine's great success against acute infectious disease, as well as some nutritional and endocrine deficiencies, now brings it increasingly face-to-face with chronic and incurable illness, against which it has no magic bullets, and also increasingly face-to-face with irreversible loss of mentation and with terminal illness, against which *there can be no*

magic bullets. Frustrated in our desires to fully master our bodily fate,
we are forced to rediscover the limits of medicine. For the universal
goal of health—which we have argued in the previous chapter is the
prime goal of medicine—is attainable and maintainable only provisionally and temporarily, manifested necessarily always in particular
frail and perishable beings. Regarding in this chapter more closely the
practice of medicine, we see that the physician will with nearly every
patient sooner or later find himself in an ethical dilemma, poised between his service to the natural norm of health, toward which the healing powers both of his art and the patient's body are pointed, and the
particular needs and sufferings of frail human beings, incapable of full
or lasting wholeness and well-working, and burdened also by fear,
shame, and general anxiety. What should the doctor do? is not a new
or transient question.

Experienced and seasoned physicians, I suspect, know all of this
already. They have for years confronted concrete, troublesome, and
often poignant dilemmas in their day-to-day practice. They now look
with not a little suspicion and skepticism at the efforts of eager-bea-
vers who are busily devising guidelines, rules, laws, and procedures
that they hope will solve one or another of these medical-ethical dilemmas. Experienced physicians have reason to be skeptical: They know,
even if only tacitly, that their deepest dilemmas—like most other large
human questions—are not to be *solved*, but only *faced*.

The doctor's ability to face these dilemmas is unlikely to be improved by exposure to the remote, abstract, and scholarly argumentation practiced by academic "ethicists." Exhortations to be good and
sermons to avoid evil, while perhaps useful on some occasions, are usually too simplistic to touch either the troublesome questions or the
questionable practices. Discussions of shocking and bizarre cases usually do little to illuminate the myriad ordinary practical difficulties.
And judicial opinions or legislative and executive guidelines respond-
ing to these cases usually lack the understanding, subtlety, and flexibility that good practice necessarily requires. How does one usefully
address the ethical aspects of medical practice?

In more than fifteen years of discussing questions of medical ethics with physicians, I have been impressed by their reluctance to generalize the principles of their conduct. They counter philosophical
argument of principles with anecdotal accounts of cases. "Every case
is altogether unique," they frequently insist. For several years, I must
confess, I was impatient with this approach. It seemed to me then that
my physician interlocutors were either too lazy or thoughtless to articulate the tacit premises of their conduct (premises that seemed, to
me at least, readily accessible through analysis of their cases) or else
too frightened to subject those premises to careful scrutiny and crit-

icism. Moreover, it just is not true that the ethical aspects of each case are in *every* respect unique, any more than are the medical aspects. Why not seek the same clarity and precision in thinking about medical ethics that we seek about disease, which, after all, also manifests itself only in particular cases, each in some sense unique?

This chapter might be said to be a reflection on this question and a defense of the peculiar antipathy of physicians toward formal and abstract ethical reasoning. I have come in large measure to appreciate the practitioners' point of view. Indeed, I increasingly believe that the attempt to replace the often inarticulate yet prudent judgments of discerning physicians with explicit rules or procedures will not lead to better decisions. It is an old story—easily verified by living in a university—that there is no necessary connection between sophistication in ethical argumentation and good conduct.

Yet the superiority of trusting in the prudence or discernment of the practitioner in *specific* cases presupposes that the practitioner understands *in general* what his practice is and what it is for. His ability to exercise prudence in particular cases requires his greater self-consciousness and thoughtfulness about questions such as: Who and what is the physician? Whom and what does he serve? What is his relation to his patient and society? To contribute to such general reflections on the nature of medicine, I shall here consider certain general and unavoidable perplexities regarding the doctor's proper business, rather than the more usually discussed specific ethical dilemmas (e.g., abortion, confidentiality, or the allocation of scarce resources). I begin with some general observations about ethics and medicine.

Some General Remarks About Ethics and Medicine

What is an ethical dilemma? What makes an ethical dilemma ethical? Identifying an ethical dilemma is sometimes as difficult as resolving it, as the following story makes clear.

A shabbily dressed elderly man has come to consult the professor of philosophy with a question about business ethics. "Me and my partner, we have a confectionary store in the Bronx. Last week in comes a young man, very distracted, probably in love, asks for a package of cigarettes. Staring dreamily at the ceiling, he puts down a ten-dollar bill, takes his cigarettes, and starts out of the store, leaving his change on the counter. Now, professor, comes a question, business ethics. Should I or should I not tell my partner?"

Identifying ethical dilemmas depends not only on the perceived presence or absence of relationships that may give rise to obligations of honesty or justice. It depends even more on our perceptions and opinions about human nature and the nature of morality. For some of us, morality is primarily conceived of in terms of right and wrong, good and evil, purity and sin, and issues forth in rules designed to guide but primarily to restrain wayward humanity, seen as destined toward the bad in the absence of such restraints. For others, morality is primarily conceived of in terms of benefits and harms, and issues forth in calculations aimed at optimizing pleasure, safety, and comfort, and minimizing pain, danger, and disease, mankind being seen as already correctly disposed by nature toward the former and away from the latter. Still others talk the moral language of rights and correlative duties, a morality that issues forth in procedural and institutional arrangements that seek to maintain and extend a precarious individual autonomy, threatened by the fact that everyone is assumed to be at best selfishly indifferent to the good of unrelated others. Finally, others talk not of sins, costs and benefits, or rights, but of virtue and vice, of the dispositions or stands of human beings regarding fear or bodily pleasure, or wealth or honor, an approach to ethics that emphasizes character rather than rules of conduct and that seeks to nourish by habituation in good deeds the latent, yet otherwise feeble, human tendency toward goodness, thought to be present in most human beings.

I bother you about these different approaches to ethics—the approaches that emphasize (1) moral rules of thou shalts and shalt nots, (2) calculations of benefits and harms, (3) procedures to protect freedom and autonomy, and (4) the cultivation of fine character—for two reasons. First, because it is important to recognize that the very formulation and definition of ethical problems is itself subject to broad variation and debate, indeed, about extremely fundamental matters. Second, and this is my underlying thesis, to which I will return briefly later, I suspect that some of the difficulties medicine now faces stem from the fact that the profession's traditional perception of the ethical universe differs sharply from the prevailing American culture on this very matter. Medicine, long a profession of cultivated gentlemen and (more recently) ladies, and always a profession of action, has relied largely on the character of physicians and their esteemed virtues of tact, gentleness, gravity, patience, modesty, justice, humanity, and, above all, prudence or practical wisdom to cope with the myriad human situations they would encounter in practice. Though there have been codes of ethics for physicians—something like rules of conduct— these have not until recently been thought desirable or necessary for regulating the behavior of the well-intentioned and well-brought-up practitioner, who must in any case be allowed to exercise the prudent judgment of the man on the spot that practice requires and for which rules cannot be given.

In contrast, the ethos of our community has different sources: the great religious traditions, with their commandments and rules, some embodied also in our criminal law; the Anglo-American libertarian tradition, with its freedoms and rights, and, to protect these liberties, institutional arrangements meant to supply the defect of the want of better motives; and the utilitarian tradition of comfort and safety, fueled by technological progress, which strives to make life less poor, nasty, brutish, and short. If this thesis is correct, then it is small wonder that medicine should find itself somewhat at odds with the broader community even in the formulation of what constitutes an ethical dilemma, let alone what one should do about it. Doctors who have long taken for granted their benevolent intentions toward their patients and prided themselves on their ability to judge the "just right thing to do in the circumstances" will not see the need for, and understandably will bridle at, the proposals of moralists and legislators to establish fixed rules of conduct, of economists and health policy planners who want to convert everything into cost-effectiveness data, or of lawyers and patient advocates who insist on establishing a patient bill of rights or laying down an explicit contract between the so-called consumers and providers of health services. The profession jealously, and, I think, by and large justifiably, guards its own view of its professional responsibility and, more important, its own ability to discern, describe, and face its ethical dilemmas.

Yet, however reasonable it is to see medicine as an autonomous profession with its own ethical views and practices, it cannot be denied that individual physicians are members of the broader community, and the institution of medicine is subject to law and enmeshed in the broader social matrix. It is the dilemmas caused by these external relations that I want to consider next, as I turn to the first of the three larger perplexities I wish to consider.

Whom Does the Doctor Serve?

Whom does the doctor serve? He serves his patients: the ill, the diseased, the dying, and also the worried-well who might be ill, diseased, or dying. This is a truism and has been so since the beginning of medicine. The Hippocratic Oath states, "I will apply dietetic measures for the benefit of the sick according to my ability and judgment; I will keep them from harm and injustice," and again, "Into whatever houses I may enter, I will come for the benefit of the sick."

Yet this is not the last word. There are, and have always been, other potential beneficiaries of the physician's services. And I do not mean only the physician himself, whose self-interest is notoriously not always congruent with those of his patients, whether in matters of time or money. The patient's family, other unrelated patients, community institutions such as schools or corporations, and the broader sociopolitical community sometimes bid for, or lay claim to, the physician's services and attention. Sometimes the service demanded conflicts with the physician's service to individual patients.

This is not a new problem. Since antiquity, physicians attached to armies on the battlefield have served to maximize the army's fighting strength. Today, triage medicine is justifiably practiced in battle, where the overarching purpose of the common defense and community survival justifies overriding the usual presumption to help those most in need in favor of restoring those most able to fight. In peacetime, for reasons of public health and safety, physicians are legally obliged to report cases of venereal disease or gunshot wounds, and, if the *Tarasoff* decision in California* establishes a precedent, may be *required* to violate patient confidentiality to inform other people who may be at risk for serious harm from potentially homicidal maniacs. In practices that have not received the critical ethical scrutiny they deserve, psychiatrists now serve as officers of the court in civil commitment proceedings, and many physicians are in the employ of businesses or athletic teams, often, at best, with divided loyalties, and sometimes, in fact, undertreating or mistreating employees or athletes for the sake of the organization's purpose. The rise of family medicine and family psychiatry has acknowledged that the well-being of families is intimately connected with that of the ill, thus overcoming a sometimes unfortunate exaggeration of the focus on ailing persons, as when, for example, psychiatrists encourage patients to undertake or continue extramarital affairs, indifferent to the effect on the patient's spouse and children.

Though the problem of divided loyalty is old, it has been greatly exacerbated in recent years, and threatens to become much worse, for several related reasons. First, there are changes in the character of medical care. Some fruits of technological progress, such as renal dialysis or transplantation or the intensive care units, raise explicit questions of patient selection for services in short supply, often placing a physician in a bind between loyalty to one patient and another or between loyalty to patient and institutional policy. Second, the

---

*A graduate student confided to his psychotherapist that he intended to kill a young woman who had rejected him, and he later did so. The girl's parents (Tarasoff) sued, among others, the therapist and his employer, the University of California, for failure to warn them, the girl, or others who might have alerted the girl and for failure to confine the student. The Supreme Court of California held that when a therapist predicts that his patient is a danger to someone else, he has a duty to warn the prospective victim. A majority of the court also held that if the therapist, on the basis of professional expertise, *should have* been able to predict danger, "he bears a duty to exercise reasonable care to protect the foreseeable victim of that danger."[1]

enormous costs of medical care, in large measure the result of its technological sophistication, have meant new roles for third parties. What insurance companies will pay for often determines what patients will receive, and this frequently means unnecessary hospitalizations and other measures not clearly in the patient's best interest. The actuarial and statistical approach that necessarily governs the practices and policies of third parties is frequently at cross purposes with the best treatment or procedure for a given patient with special needs or problems. Third, and probably most important, is the massive role of the Federal government in its hospital building programs, Medicare and Medicaid, kidney-machine legislation, health maintenance organizations (HMOs), Professional Standards Review Organizations (PSROs), human experimentation committees, food and drug regulations, and equal opportunity employment regulations, and other species of intervention too numerous to mention. Finally, there is the rise of the public health movement, with its focus on whole populations rather than the individually ill, a focus supported also by the statistically inclined cost-effectiveness orientation of our leading health economists and policymakers.

One conclusion is obvious. The physician's attempt to benefit sick patients is constrained and shaped decisively on all sides by many nonmedical considerations. The individual doctor-patient relationship occurs in institutional settings and under ruling practices that have little whatsoever to do with the needs of the individually ill, and the powerful interests of these larger organizations, from hospitals to the U.S. Department of Health and Human Services, in many cases take precedence. Looking to the future, we can expect this tendency to increase—not just in medicine, but in education and many other activities—and some have called on us to abandon our preference for individual attention and opt for what one thoughtful physician has called "statistical compassion."

What are physicians to think about all this? Can the medical profession retain its integrity if it abandons its patient-centered orientation? The American medical profession has traditionally said no. Yet there are signs of change—to me, at least, disturbing signs of change. The text of the new Principles of Medical Ethics of the American Medical Association is much more ambiguous on this subject than one might wish. Only one of the seven proposed articles deals with the physician's primary business, namely, patient care, and moreover, strange to say, it makes no mention of the patient: "A physician shall be dedicated to providing competent medical service with compassion and respect for human dignity."[2] Competent service to whom and for whose benefit? And why not "respect for the dignity of his patients" or even "of human persons," rather than the more abstract "respect for human dignity"? The other provisions occasionally speak of the patient, but their themes are the physician's relationship to the profession, the law, rights, learning and consulting and informing the public, the physician's freedom "to choose whom to serve," and his responsibilities as a member of society. Surprisingly, there is nothing in the proposed code equal to the shiningly clear statement of the World Medical Association's 1948 Declaration of Geneva: "The health of my patient will be my first consideration."[3]

Now there is no question but that medicine cannot be a simply autonomous profession. In truth, it never has been. As Aristotle long ago observed, the political is most authoritative and architectonic; politics (i.e., the ruling laws, institutions, and customs of a regime) "ordains which of the sciences are to be in the cities . . . and legislates as to what people shall do and what things they shall refrain from doing."[4] Moreover, as health has become a major public business— now 8 percent of our gross national product is spent on health care— it is to be expected and even welcomed that the political community as a whole, at its various levels, deliberate about how best to promote the nation's health, given that resources will always be limited. The allocation of scarce resources has always been ultimately a political matter, here and elsewhere, to be resolved by the political process— even when that political process decided that government not interfere in a free-market allocation. Medicine may seek to influence the outcome of political deliberation—and no doubt we would have better health policies if our health planners were more intimately familiar with the actual practice of medicine in hospital, home, and examining room—but the ruling character of the political must be conceded, even if with fear and trembling.

The crucial question lies elsewhere: Can these broad social and political considerations be allowed to enter within the bounds of the doctor-patient relationship? Here, perhaps, lies the greatest threat to the traditional ethic of the profession: the intrusion of the political or social perspective into the day-to-day care of single patients.

Consider this case in point. In an article in *The Hastings Center Report*, a physician and a philosophy professor discussed a case that turned on whether an elderly woman, with a recurrent episode of severe respiratory failure, previously successfully treated, should again be treated with a respirator, a decision to treat requiring that she occupy the last vacant bed in the intensive care unit (ICU). The young physician wrote not only to oppose treatment but also to oppose leaving the decision either to the patient and her family or to the physician:

> Although Mrs. A's family has the right to refuse the ICU transfer and the respirator, they do not have the right to demand it. Self-determination

with regard to medical care is one thing, but a just allocation of scarce resources is another. The first right belongs to the patient, but the second is a claim which society makes upon the use of its health care resources.

Shall the physician decide? This would be an odd way to decide such questions (although in fact many are decided by physicians). Physicians, like families, have vested interests in some patients compared to others. And little in their educational background qualifies them as experts *in the just allocation of scarce life-saving medical resources.*"[5]

But the proper question before the doctor was not one of just allocation of medical resources (though it might have become one if he had two patients both of whom needed the last bed). The question was whether his patient needed help and whether she could benefit from the treatment—and never mind if it meant filling the empty last bed in the ICU, which, after all, is there to be used. The questions of distributive justice have their place, say, in the decision to build or expand the ICU. But once the facility is available, these considerations are irrelevant to the physician's judgment that therapy will be beneficial, and should not prevent his offering it to the patient if the patient is willing. Within the broader limits set by social and institutional policy, the physician must practice unswervingly the virtues of loyalty and fidelity to his patient.

This is no mere sentimentality or moralizing. The principles involved are easy to state. They go to the very heart of the medical activity. Loyalty to the patient must be paramount because the mysterious activity of healing depends on trust and confidence, which is lodged by the vulnerable and dependent patient with the physician in the very act of submitting to his care. (By the way, would it not be unethical to undermine or betray such trust, even were it not indispensable for the healing activity?) One who is ill submits to the care of the healer in the expectation that the healer will care for his well-being as much as he does himself and that the healer will pursue it to the best of his ability. Even more fundamentally, the legitimate claim for individual patient-centered medicine can *never* be obliterated, even come the revolution, for the obvious, but sometimes forgotten, reason that illness almost always happens to bodies, and nothing is more individuating and private than body. My pain, my weakness, my illness—unlike my *thoughts* on these matters—are absolutely unshareable. Even the socialization of medicine could not socialize the radically private property of body. An ultimately patient-centered and individualized orientation is necessarily constitutive for the healing profession under any political order.*

---

*This thesis will be qualified in Chapter 9, "Is There a Medical Ethic?: The Hippocratic Oath and the Sources of Ethical Medicine," pp. 240–246.

What Does the Doctor Serve?

Having established provisionally that the physician serves first and foremost the patient, we face the second perplexity: *What* about the patient is served by the physician? The dilemma may be put this way: Does the physician serve the patient's *needs* or the patient's *desires* and *wishes?* Or again, does the physician serve the patient's *good* or his *rights?* Can it be simply true, without qualification, that the good physician is the *servant* of the *patient?*

Let me illustrate with some examples. When cosmetic surgeons lift faces, inflate bosoms, and straighten noses, are they serving needs or desires? When the obstetrician agrees to determine the sex of the unborn child by amniocentesis and abort fetuses of the unwanted gender, does he serve need or desire? When the internist gives in to a mother's request to tranquilize her teenage daughter's anxiety before her first cello recital, does he serve need or wish? And what of the psychiatrist who has sexual relations with his patients in the office as part of his treatment of frigidity?

Honoring so-called patient's rights may also conflict with serving the patient's good. It is now claimed that patients have a right to know the whole truth about their diagnosis and treatment, including an account, in detail, of all possible complications and untoward consequences of proposed therapies. There is said to be a right of access to medical records, a right to refuse treatment as well as a right to obtain treatment (especially for persons involuntarily committed for mental illness), a right to health care, a right to determine the fate of one's body, even a right to die or to be mercifully killed. Some of these claims seem to me dubious, while others touch on important matters that could be accommodated without resorting to the uncompromising and contentious talk of "rights." But be that as it may, the main point here is that physicians increasingly face the uncomfortable choice of either risking harm to patients by catering to their rights or risking suit from patients by ignoring their rights in order to serve their good.

One suspects that these difficulties and tensions will increase, for they seem to be related to powerful tendencies of modern life. The increasingly sophisticated means for intervening in the human body and mind have also produced new ends for the use of biomedical technique. And it has been noted often that the triumph of technique, fueled initially by rather modest and unexceptionable goals, itself gives rise to an inflation and proliferation of desires, which in turn breed the further growth of the technology that is needed for their satisfaction. This is especially true under conditions of great freedom and prosperity, and in cultures such as our own that esteem comfort and safety, and in which a highly literate and demanding populace comes to expect a technical solution to all of life's difficulties.

The swelling of demands and desires for goods and services is paralleled by the rising stress on personal autonomy, intensified by a declining influence of the authority of tradition and traditional authorities and, often, by a frank attack on *all* authority, including that of the professions, medicine among them.

There is no doubt that physicians will have to make their peace with these tendencies requiring them to serve patient desires and respect patient rights. The AMA's new Principles of Medical Ethics have, perhaps unwittingly, made clear concessions in that direction. They say, for instance, that "A physician shall be dedicated to providing competent medical service"[6] without specifying anywhere the end to be served by medical competence. Indeed, the word "health" occurs in the Principles only twice, both times as an adjective in the phrase "other health professionals." The Principles also say that "A physician shall respect the rights of patients" without specifying what rights patients are deemed to have, and who will determine them.[7] (It would be most instructive, if we had time, to compare the understanding of the doctor-patient relationship implied by the notion "respect the *rights* of the patient" with the more traditional understanding implied by the Hippocratic Oath's "I will keep them from harm and injustice," the former emphasizing the patient's autonomy, prerogatives, and rights, the latter the patient's neediness and the overarching norms of good and right; the former presupposing a physician who needs to be exhorted not to violate his patients, the latter presupposing a benevolent healer who must keep the oft intemperate patient from violating himself. See Chapter 9, "Is There a Medical Ethic?: The Hippocratic Oath and the Sources of Ethical Medicine.")

But the medical profession should be wary of conceding too much in making its peace. For, as with the first perplexity, there are some fundamental matters at stake, matters that again strike at the heart of the healing relation. If we could recover a deeper understanding of this relation and of the art of medicine generally, we might be more alert to the dangers that threaten it. One danger is contained in the pressures to treat medicine not as an art or a profession that is practiced but as a technical service that is delivered, like auto repair or plumbing, in this case, a service "provided" by physicians or by the "health care delivery system" and "consumed" by patients (see Chapter 8, "Professing Medically: The Place of Ethics in Defining Medicine"). This new understanding shifts the focus to the buying and selling, to the relation of exchange, and away from the mysterious activity of healing and the crucial and incomparable interpersonal bonds of the healing relation—far beyond what the psychiatrists call transference and countertransference.

Many in the so-called consumer movement, not particularly sensitive to fine distinctions, would replace an understanding of the healing relationship founded on ideas of covenant, vocation, philanthropy, fidelity and trust, or devotion to the art with a new view: a relation of contract, with explicitly defined items contracted for exchange. Such a notion both presupposes a lack of trust and further exacerbates it, thus interfering with the healing relation. Indeed, it simply misunderstands and hence denies the essence of the healing activity, which depends on hope and confidence, care and trust, no less than on technique. It is an activity whose nature and effectiveness are shrouded in mystery, and whose outcome is always so uncertain and subject to the vicissitudes of conditions and circumstances as to make contractual promises unreasonable. It is an activity much like child-rearing or teaching, activities also grossly misunderstood by those who would reduce them as well to a species of contract.

A second mortal danger is contained in the now popular notion that a person has a right over his body, a right that allows him to do whatever he wants to it or with it. Civil libertarians may applaud such a notion, as an arguably logical expansion of the right of privacy, of the right to be free from unwanted or offensive touchings of one's body. But for a physician, the idea must be unacceptable. No physician worthy of the name would honor a patient's request to pluck out his eye if it offends him or lop off a breast to improve a lady's golf swing. Medicine violates the body only to heal it. Doctors respect the integrity of the body not only because and if the patient wants or allows them to. They respect and minister to bodily wholeness because they recognize, at least tacitly, what a wonderful and awe-inspiring—not to say sacred—thing the healthy living human body is. They know or should know—for they practice a profession whose very foundation presupposes—the precariousness of human life and the dependence of all good things on a well-working body, whose great powers and frailty both command respect and modesty. No doctor who understands the profession should be guilty of the contempt for the body or arrogance of the will that declares "my body" to be a mere thing to be disposed of or carved up at "my will." (I commend to the reader's attention on the subject of the body the writings of Richard Selzer, M.D., *Mortal Lessons* and *Confessions of a Knife*.[8] Also see Chapter 11, "Thinking About the Body.")

Now, to be fair, one must concede that medicine has not itself been sufficiently mindful of these matters. The assertion of patient's rights and the move toward increasing patient autonomy, it must be ac-

knowledged, are in part in response to *excessive* authoritarianism and
mystification—even arrogance—on the part of physicians. If patients
stridently insist on being treated as persons, demand that their wishes
be honored, and occasionally show contempt for the body, it is perhaps
because physicians have all too often shown a forgetfulness of the soul,
of the human aspects of patients and patient care. Sad to say, one can
even learn contempt for the body from some physicians, who in their
eagerness to treat this or that abnormality forget about the well-working of the body as a whole, or prescribe dangerous drugs to remedy
trifling complaints. Thus, while it is true that the physician has been
rightly committed more to patient good than to patient rights, to patient need than to patient wish, it is also true that physicians frequently now hold too narrow a view of need and of good, too shrunken
a view of the integrity of the human organism, and almost no view at
all of the riches and mysteries of the human soul. Modern science and
modern medicine have not taught our culture well on most of these
matters.

Can we find a way out of this perplexity regarding the doctor-patient relationship and the object of medical service, and the dilemmas
it creates? Even in thought, there is no simple answer. It is not always
easy to distinguish a need from a desire, a reasonable desire from an
unreasonable one, or to decide which clearly reasonable desires and
wishes deserve the services of a physician. Patients do have a need
for respectful treatment and for by-and-large truthful and patient
counsel, and it is good for them to take as large a role in their own
health-maintenance as is possible—even if these needs and goods are
not owed them because they claim them as rights. There is truth in
the slogan that patients are first of all persons, and only secondarily
patients to the physician's ministrations.

Practically speaking, there are likewise no simple rules for balancing these considerations in deciding what to do in patient care. Certain virtues seem to be required: moderation in the physician's view
of what he can and cannot accomplish; gravity before the awesome
mysteries of human being; understanding of the human aspects of the
lives, hopes, and fears of the ill; courage to resist unwarranted demands for pills or procedures; and prudent judgment to discern the
warranted from the unwarranted. These are not the virtues of the servile. And though it is true that the doctor must not seek to be a master, so also must he not stoop to be a slave.

Indeed, to close this section, it is worth reconsidering whether we
have not been mistaken in the very posing of our questions: Whom
and what does the physician serve? Is the doctor really a *servant?* In
a way he is, providing we remember that one can serve not only people
but also ideals, not only a worldly master but also a noble calling. One

way of stating the conclusion of this part of my argument is to assert
that the physician's loyalty to his patient must be decisively qualified
by the physician's loyalty to his art and to its norms and goals, or
again, that the physician serves not the patient simply but, rather,
the *good* of the patient. At a deeper level, one could say that in healing
the human body the physician is also assisting and serving that innate
power of nature that is manifested in each patient but greater than
all of them, a power that all ancient peoples acknowledged in regarding medicine as a sacred or holy art. (Consider in this connection the
beginning of the Hippocratic Oath [see Chapter 9], the discussion of
healing in Leviticus 13 and 14, or the Greek deification of the founders
of medicine.)

Yet the language of service is also not quite right when applied to
the relation between doctor and patient, for service implies mastery
or lordship, and the doctor is, in truth, neither a master nor a servant
of the patient. With respect to the patient's body, he is a helper, a co-
worker with nature and with the patient himself, in providing the ill
body its proper aid. With respect to the patient as person, one of his
main functions, oft neglected, is to be a leader and a teacher, one who
leads the activities of healing and one who teaches patients and the
community about regaining and maintaining healthy functioning. The
word "doctor" literally means "teacher," from the Latin verb *docere,*
"to teach," in this case one who teaches the wisdom and wonders of
the body to patient and pupil alike.

What Is the Patient's Good?

In reflecting on the nature of the medical vocation, we have come via
a somewhat labyrinthine path to a not very startling assertion: The
doctor, by teaching and technique, through patient understanding and
astute judgment, promotes the patient's good. Our third perplexity
follows directly: What is the patient's good? More precisely, of the
many things that are good for the patient, which is it the physician's
business to promote? We are no longer asking about medicine's foreign or domestic relations, but about its very constitution: What is
the purpose of medicine?

This is, in a way, a strange question to be asking, since most of
the time physicians can go about unreflectively *doing* their proper
business, tacitly if silently clear about their goals. But because new
technological powers permit physicians to serve multiple ends, and
because much of medical practice is so fragmented and specialized,
there is today some confusion and uncertainty about the nature and
limits of the purposes of medicine. In the previous chapter I argued
that this confusion was a serious cause of the malaise of American

medicine. Against the growing tendency to enlarge the medical mission, I argued that *health* was *the* proper end of medicine, whereas other albeit worthy goals—such as pleasure, contentment, happiness, civil peace and order, virtue, wisdom, and truth—were false goals for the healing art. Against the narrow perspective of the high-technology, highly specialized therapy-centered predilections of recent decades, I also tried to outline a functional notion of health and argued that health, understood as the "well-working" of the organism as a whole, is not just the absence of disease, but a positive good and the proper norm for medical practice, one that implies that there is more to healing illness than curing disease. I will not repeat that part of the argument here and will trust that few physicians would deny that health is the main purpose of medicine or that some tacit notion of the norm is latent in every attempt at healing—"healing" meaning literally, "a making whole."

Instead, I want to look at difficulties that arise when healing is not fully possible, say, in the case of chronic and irreversible illness, or, especially, with those who are dying. But first, it will be useful to extend and refine the argument about the meaning of healthiness, by dealing briefly with an obvious and previously unaddressed problem with the conclusion itself—or with *any* proposed functional definition of health—namely, which well-working activities or functions of the living body should be considered? Is it possible to articulate, let alone define, the specific bodily excellences and fitnesses for a being as complex as man, whose activities are so highly diversified and differentiated? In short, when seeking a functional definition of health, which functions count? The answer of the medical profession and its followers will not always be the same as that of the so-called unenlightened layman, as the following story reveals:

> Two friends meet on the street. Says one: "Did you hear? Smith died."
> "Smith? Which Smith?"
> "You know Smith. He had terrible ulcers for almost forty years."
> "Smith? With ulcers?"
> "You know Smith, he had chronic emphysema."
> "Smith? Ulcers, emphysema? No, I don't know him."
> "But you *do* know him; Smith, with psoriasis and a pacemaker put in five years ago."
> "Smith? Ulcers, emphysema . . . psoriasis . . . pacemaker?" He shrugs his shoulders.
> "Dammit, you know Smith, he walked with a limp and had a tumor on his left shoulder the size of a grapefruit."
> "Tumor? Limp? . . . pacemaker? Psoriasis? Emphysema? Ulcers? Nope, not one of my Smiths."
> "Smith. Smith the lawyer! Downtown office. A terrific pianist. Wife's a painter. Lived on Lake Shore Drive near Belmont!"
> "*That* Smith? He died? My God, he was as healthy as a horse!"

It is likely that the second man was simply a poor observer. But maybe not. In what sense might Smith *in fact* have been "healthy," if not "as a horse," then at least healthy *enough* to have his death come as a complete shock? What signs of healthy functioning are we looking at?

Do we look mainly at his high gastric acidity, diminished vital capacity, elevated blood pressure, scaly skin, reduced mobility of knees and hips, etc.? Or do we look mainly at his continuing practice of law; his joyful relationships with spouse, children, and friends; his ongoing interest in music, art, sports, and politics; his regular (if slowed-down) walks along the lake front; his eager mind and alert aesthetic sense; and his desire to continue to go on living?

Which are the functions of the well-working human organism *as a whole,* in relation to its *proper human environment?* Do we not see that the ordinary medical viewpoint, while emphatically useful in diagnosis and treatment of specific disease, is really rather partial—both biased and incomplete—when it comes to healthiness? Is it not true that the examining room and hospital provide a rather restricted environment in which to discern the level of *human* functioning—that the patient's powers and desires for work, friendship, love, learning, awareness, mobility, thought and memory, self-command, and the sheer enjoyment of life fail to receive their proper evaluation there?

To be sure, these powers and appetites are hard to evaluate and defy quantitative measure. Moreover, specifically medical techniques may have little to contribute to their improvement or maintenance. Still, if it is ultimately *these* functions that medicine seeks to serve and preserve—if its concern with liver function, pulmonary function, and other vital functions is ultimately in the service of supporting the ability and desire of the patient to function *in the conduct of his life*—then doctors might do well to take the broader view. Such a view seems especially relevant to caring for the chronically and irreversibly ill. It permits a more hopeful and enterprising outlook on what can be done for people who no doubt have incurably compromised functionings of their parts, but who may nevertheless display nearly intact, or at least readily supportable, functionings as wholes.

Concretely, this means that *supporting* the healthy life-functioning of the chronically diseased becomes less a matter of fighting disease and more a matter of physical therapy and rehabilitation; of hearing aids, dentures, and glasses; and especially of addressing habits of life and patterns of social arrangements. These are now *ancillary* services of disease-battling medicine, but they need to become central to the care of the chronically ill, who though manifesting incurable and even progressive disease, may continue to be able to function more or less healthily as wholes for some time to come.

I might add that greater attention by physicians to these life-functions and life-situations might even improve their care for the acute illnesses of their patients. Take the case of a man of seventy with an acute episode of bacterial pneumonia. Of course, the doctor isolates and treats the sensitive organism with appropriate antibiotics. But should he not also be interested in the fact that the man is recently a widower, that he lives alone in a fourth-floor walk-up, that he has difficulty climbing stairs because of painful osteoarthritis of one knee, that he is therefore rarely out of doors or in contact with other people, that he therefore is eating poorly and is growing isolated and more depressed—in short, that his pneumonia is perhaps only a sign of, maybe in part even a result of, a decline in his functioning as a human being? Would not a *good* doctor also look to the knee, and consider analgesics, or physiotherapy, or mechanical supports? Might he not explore with the patient the possibility of moving his residence to a ground-floor apartment, perhaps closer to shopping, children, and friends? Might he not deal also with his isolation and depression?

Now I know that these matters are sometimes as intractable to treatment as chronic disease of liver or lung. But we will never know unless we try. And we will be unlikely even to think of trying if we adhere to a notion of healthy functioning that looks only to parts and parts of parts, rather than at wholes, and which thinks of health only as the absence of known disease.

When Health Is No Longer Possible

There remains to be considered a quandary that arises even if we all agree not only that health is the physician's primary business, but also on the meaning of healthiness: What to think and what to do in the face of incurable disease and unhealable illness? What good does the physician seek to promote when healthy functioning is out of the question?

To be sure, some level of healthy functioning is never out of the question, this side of death. The degree of health and departures from health form a continuum, and even people in the so-called persistent vegetative state must have healthy vegetative functions—respiration, metabolism, and excretion. But few of us would accept the preservation of such a reduced level of function as a proper *goal* for medicine, even though we sadly accept it as an unfortunate and unforeseen *result* of treatment that had higher aspirations, and even if we refuse actively to cause such vegetative life to cease. So the question remains, not only regarding the comatose, but also the terminally ill, the irretrievably dying, the irreversibly deteriorating, especially in mind and awareness, and the otherwise anguished and miserable: What should medicine's goal be for them?

To be sure, easing of pain and relief of suffering, along with supporting and comforting speech, are always in order, all the more so in the presence of incurable and progressive illness. Relief of suffering stands, next to health, as a crucial part of the medical goal, and medicine has always sought to comfort where it cannot heal. The real quandaries concern activities for prolongation or preservation of life. When, if ever, is it appropriate to withhold or interrupt treatments that might be life-preserving or life-prolonging?

This, too, is an old quandary, faced by physicians even before the modern era of antibiotics, respirators, and defibrillators, say, in decisions about whether or not to amputate a gangrenous limb in an elderly patient. But there is no doubt that such dilemmas now arise with vastly greater frequency and have generated a burgeoning professional and public concern. The main causes include: (1) the effectiveness of life-sustaining technologies and the zeal of young physicians in their use; (2) the existence of many more "incurables" who, particularly in hospital settings, are especially impotent in expressing their wishes to be left alone, many of whom have no enduring relationship with a physician who could knowingly act on their behalf; and (3) the physician's fear of criminal and civil liability, partly in the face of ambiguity regarding the law (e.g., whether turning off a respirator is a proximate cause of death). Yet recognizing these sources of our dilemmas in no way diminishes them. Nearly everyone agrees that the practice of intervening and prolonging has gotten out of hand in many, many cases.

How to think about this problem? I confess I find it enormously difficult. Indeed, perhaps the most important lesson to be driven home by the medical profession and to its members is just how complicated a matter this is. It simply will not yield to simple formulae such as "death with dignity" or "life is sacred" or "dispense with extraordinary means." Such terms as "incurable," "dying," "terminal," and "hopeless" are notoriously vague, not to speak of "dignity." "Ordinary" and "extraordinary" sometimes means only "customary" and "unusual," sometimes are relativized to the particular circumstances of each patient, so that what would be ordinary for one patient becomes extraordinary for another. Even the word "treatment" is ambiguous, sometimes implying effective therapy, sometimes merely naming the applied procedure or intervention. Measures that can be said to be life-preserving span a continuum from respirators and dialysis machines, through antibiotics and insulin, to intravenous glucose and water, even to food and drink. The distinction between an action and an omission, which I find useful in some cases, is neither always obvious nor obvious to all, especially if judged only from the result. There are notorious problems in discerning the patient's own judgment regarding his state of being and suffering or his desire to be treated or not. For at least these reasons, the attempt to solve these dilemmas—by generalizations embodied in rules, guidelines, or statutes, or even by court decisions—seems fraught with dangers, some of which I will indicate soon. In no other area is there a greater need for sober and prudent judgment of the man-on-the-spot and yet, in no other area is there more reluctance and resistance to leave matters simply to prudence. When it comes to the utter finality of the end of a life, we all long to proceed with certitude and clear conscience.

The easy way out is to adhere always to the principle "Sustain life regardless," and there is much to be said for the sentiment involved (if not for the practice), reverence for life being a constitutive yet fragile principle of decent human community. Nevertheless, it seems to me that a true reverence for life might include permitting it to end, free from further assaults on life's sanctity committed in its name. This, I submit, has been the traditional view of the medical profession, and, I might add, of Christian religious tradition.[9]

Illuminating this view is in part the recognition of human finitude. Even the most healthy human being must someday die, despite all efforts of the most competent physician. Such it is to be a mortal being, and such it is to have but extremely limited powers. Physicians need to accept these lessons no less than anyone else. If medicine takes aim at death prevention, rather than at health and relief of suffering, if it regards every death as premature, as a failure of today's medicine, but avoidable by tomorrow's, then it is tacitly asserting that its true goal is bodily immortality. Once it is put that way, it should be clear that physicians must teach themselves and their patients to make their peace with finitude. As I have argued earlier (pages 163–164), physicians should try to:

> [K]eep their eye on their main business, restoring and correcting what can be corrected and restored, always acknowledging that death will and must come, that health is a mortal good, and that as embodied beings we are fragile beings that must snap sooner or later, medicine or no medicine. To keep the strings in tune, not to stretch them out of shape attempting to make them last forever, is the doctor's primary and proper goal.

Medicine has traditionally refused to make prolongation of life its goal, not only because the goal was finally unreachable, but also because it recognized that efforts in that direction often produced more harm than good—in pain and discomfort as well as anguish and anxiety. (Many have observed that anxiety about death from disease has never been higher, and it would be a hollow victory indeed if the extra time medicine provides us is squandered in worry. Of the deeper ills of a society dedicated to staying alive rather than to living well, we are but dimly aware. See Chapter 12, "Mortality and Morality: The Virtues of Finitude.")

These thoughts suggest a useful beginning for thinking about the concrete cases in which healing is impossible and interventions are being contemplated. The first question should not be, Will this intervention prolong life? but, Will this intervention increase or decrease this patient's discomfort, pain, and suffering? The concern of the physician should be the condition of the life to be sustained and not especially its duration. This means, for example, being willing, if the circumstances are correct, to give high doses of narcotics if needed, even at the risk of respiratory depression, or to forgo resuscitation or antibiotics in the face of underlying terminal illness or severe debilitation of mind. To judge if the circumstances are correct, here and now, is the work of prudence or practical wisdom or discernment—a virtue not teachable in medical school and not replaceable by computer programs. It is the cardinal virtue of experienced and seasoned practical men.

Now, against this view there are important arguments. Some will suggest that I am urging physicians to begin to make judgments of so-called quality of life. It may be argued that such a move also invites considerations of "social worthiness" or other alien matters to contaminate medical decisions, with not only individual lives but our very reverence for life in jeopardy. The consideration in medicine of quality of life, it is correctly said, was the fundamental error of the Nazi physicians. These are serious concerns, not to be taken lightly. I, too, am made uncomfortable by breezy talk about quality of life, and even by the terms in which the sanitized notion of "quality" is substituted for "goodness," with the attendant implication that quality is a readily measurable matter. Nevertheless, I do believe consideration of the condition of the individual patient's health, activity, and state of mind must enter these decisions, if the decision is indeed to be for the patient's good. I think one can walk between the extremes of vitalism and "quality control," and uphold in so doing the respect that life itself commands for itself. For life is to be revered not only as manifested in physiological powers, but also as these powers are organized in the form of a life, with its beginning, middle, and end. Thus, life can be revered not only in its preservation, but also in the manner in which we allow a given life to reach its terminus. For physicians to adhere to efforts at indefinite prolongation not only reduces them to slavish technicians without any intelligible goal, but also degrades and assaults the gravity and solemnity of a life in its close.

Fortunately, protection against the dangers of callous indifference, patient neglect, and the intrusion of alien considerations is available, if the physician assesses the patient's condition from the patient's point of view. This usually means open and frank discussion with the patient, where that is possible. Delicate questions of truth-telling emerge—how much to say, and when and how—never to be settled by an inflexible adherence to principle. It is even more difficult to discern truly from the patient's words what he actually believes and feels. In a recent article in the *New England Journal of Medicine,* Drs. David Jackson and Stuart Younger show how patient ambivalence, reactive depression, displacement, fear, and ignorance, or family misperceptions of patient attitudes, can lead astray a physician who relies too much on a verbalized request for cessation of treatment.[10] Discerning the patient's true sentiments and outlook is, again, often a matter of tact and care and subtlety. Physicians will soon need to be sensitive in detecting pressures applied to patients because of the emerging climate of opinion that trumpets cessation of treatment, that romanticizes "death with dignity,"[11] or that trivializes death's meaning by turning dying into a management problem for death experts called "thanatologists," just as they should have been sensitive in the past to how their own bias for "prolongation regardless" also subtly manipulated patient attitudes. The physician must know each patient if he is to teach him appropriately.

To be frank, the record of the profession as a whole in its treatment of the untreatable has left many dissatisfied, and there are growing movements, from several quarters, to remove the matter from the discretion of individual doctors and to bring practice under some form of external and uniform guidelines. We shall probably see more of this in the future, and the prospect is not encouraging. Consider, for example, the California Natural Death Act, passed in 1976.[12] This well-intentioned statute permits a person prospectively to execute a written directive to his physician to withhold or withdraw life-sustaining procedures in the event of a "terminal condition," but the operative clause indicates that such instruction is valid *only* in cases in which "death is imminent, *whether or not* such procedures are utilized" (emphasis added). In other words, the law permits the patient only to instruct the physician to desist from *useless* procedures, a halt that should be regarded by physicians not as a legally granted privilege but rather as a professionally based duty. By implication, the law seems to cast some doubt on whether withholding treatment under any other condition is acceptable, or whether a patient who has not written such a directive must be assumed to desire full-scale and recurrent prolongation efforts. Could the physician decline to resuscitate a senile woman who has crippling arthritis, but no terminal illness, either with or without such a directive? Does the law silently forbid— or discourage—what it does not explicitly permit?

Or consider the *Saikewicz* case in Massachusetts, which has been followed by a spate of other cases, brought to court by physicians who mistakenly believed that they now need a judicial directive to cease treatment, if the patient is incompetent.[13] To be sure, they need a judicial opinion in advance if they wish to obtain immunity against possible legal action for their decision, but what are we to think of the self-understanding of the profession if it will practice only under the promise of immunity for its errors? Moreover, the risk of penalty is probably greatly exaggerated. Before 1981, no American physician had been sued or prosecuted for terminating treatment of a dying patient, competent or incompetent, for whom there was no hope of cure. The results of a recent landmark case strongly support the physician's prerogative to discontinue useless medical interventions. Two California physicians, Drs. Robert Nejdl and Neil Barber, had been accused (by a hospital nurse) of murder for removing a respirator and then a feeding-tube from an irreversibly comatose patient, in fact, at the family's request. In a remarkably sensible and thoughtful opinion, the California Court of Appeals—following the recommendations of the President's Commission's report, *Deciding to Forego Life Sustaining Treatment*[14]—acquitted the physicians, and affirmed that there are limits on the physician's duty to prolong life through medical interventions: "There is no duty to continue [life-sustaining machinery] once it has become futile in the opinion of qualified medical personnel."[15]

Guidelines set by the profession may also needlessly tie physician's hands. A widely discussed article in the *New England Journal of Medicine*, "Orders Not to Resuscitate," proposes that orders not to resuscitate are appropriate if the disease is irreversible, the physiological status of the patient is irreparable, and death is imminent, "in the sense that in the ordinary course of events, death probably will occur within a period not exceeding two weeks."[16] Are there no conditions of the patient other than imminent death within two weeks that could justify a refusal to defibrillate? And can a doctor now no longer write orders not to resuscitate such patients?

While there are pitfalls at every turn, sure footing will not be had as a result of statutes, court decisions, or professional guidelines and regulations. Indeed, here is an area where dilemmas cannot be neatly solved but only soberly faced, where the desire for certainty and the cleanest conscience must give way to the satisfaction that a grave decision was conscientiously made, with utmost seriousness and adequate consultation, and in the interest of doing the most good and causing the least suffering to the individual patient, as that patient

himself would wish. Written directives, rules, and guidelines are no
substitute for sobriety, common sense, and discernment in the search
for the patient's good.

Yet I would make one qualification, and propose one rule, well
stated in the Hippocratic Oath: "I will neither give a deadly drug to
anybody if asked for it, nor will I make a suggestion to this effect."
There *is* a difference between permitting to die and directly killing, if
not in the result for the patient, then certainly in understanding the
activity of the agent. This age-old rule against mercy killing by physicians, though supported by Judaism and Christianity, and by Anglo-
American law, has its true roots in the very idea of the physician as
healer. Hans Jonas has put the matter well:

> As to outright hastening death by a lethal drug, the doctor cannot fairly
> be asked to make any of his ministrations with this *purpose*, nor the hos-
> pital staff to connive by looking the other way if someone else provides
> the patient the means. The law forbids it, but more so (the law being
> changeable) it is prohibited by the innermost meaning of the medical vo-
> cation, which should never cast the physician in the role of a dispenser
> of death, even at the subject's request.[17]

Coda

I close with a few general observations, again about ethics and medicine, as these emerge from my reflections. Though it is perhaps nothing to brag about, I know that I have said little that could help a
practicing physician "solve" any ethical dilemma now waiting for him
in the hospital. I have chosen to concentrate on some basic elements
of the profession: Who and what is the physician? What is his relation
to his patient and society? What are the goals of his art? I believe that
what is most needful at present is self-conscious reflection by physicians regarding themselves and the nature of their professional com-
mitments. To be sure, through some procedures and rules, we can try
to "solve" or at least cope with this and that ethical dilemma, but it
is through clarity about his ends and prudence and moderation in his
choice of means that a self-understanding physician is able to stand
well before all dilemmas.

There is notion abroad that there is, or that there can and should
be, a science of medical right and wrong or, at least, of proper decision
making to which doctors can turn for expert help to solve medicine's
ethical dilemmas. This is worse than an illusion. It represents a declaration of moral bankruptcy on the part of the profession, which once
understood the ethical as integral to the medical, and which never sup-
posed that the "dilemmas of caring for the ill" could be neatly solved.

The call for rules, guidelines, and procedures; the convening of ethics committees; and the encouragement of statutory regulations are a search for yet one more technical solution—this time a technical *ethical* solution—for problems produced by our already foolish tendency to seek technical medical solutions for the weighty difficulties of human life. If a doctor would be a physician and not merely a body technician, he must also be a knower of souls, those of his patients and, not least, his own.

CHAPTER EIGHT

Professing Medically
The Place of Ethics in Defining Medicine

Then you also understand that sick people in the cities, slaves and free, are treated differently. The slaves are for the most part treated by slaves, who either go on rounds or remain at their dispensaries. None of these latter doctors gives or receives any account of each malady afflicting each domestic slave. Instead, he gives him orders on the basis of the opinions he has derived from experience. Claiming to know with precision, he gives his commands just like a headstrong tyrant and hurries off to some other sick domestic slave. In this way he relieves his master of the trouble of caring for the sick.

The free doctor mostly cares for and looks after the maladies of free men. He investigates these from their beginning and according to nature, communing with the patient himself and his friends, and he both learns something from the invalids and, as much as he can, teaches the one who is sick. He doesn't give orders until he has in some sense persuaded; when he has on each occasion tamed the sick person with persuasion, he attempts to succeed in leading him back to health.

—"Athenian Stranger," in Plato's *Laws* (702b–e)

Not least among the myriad confusions and uncertainties of our time are those attending efforts to discern and articulate the goals, limits, and essential characteristics of the medical profession. To certain time-honored perplexities (e.g., Is medicine a science or an art? How is health of body related to health of soul?) have been added new questions, arising from changes in medical practice (e.g., specialization, the shift from home-care to hospitals, and amazing new techniques for diagnosis and treatment) and from changes in the broader community (e.g., secularization, egalitarianism, and a growing distrust of authority). Within the medical profession, there are uncertainties not only about difficult ethical dilemmas, but also about the nature and basis of the doctor-patient bond and even about the goals and limits of medical ministrations (see Chapters 6, "The End of Medicine and the Pursuit of Health," and 7, "Practicing Prudently: Ethical Dilemmas in Caring for the Ill"). Outside the profession, there are various efforts to cut medicine down to size: not only widespread malpractice litigation and massive governmental regulation, but also various attempts, by consumer groups and others, to redefine medicine as a *trade* rather than a *profession* and the physician as a morally neutered technician for hire under contract.

Recently, even the Federal Trade Commission and the courts have joined the attack on medicine's professional standing. In 1979, the FTC found that the AMA and other medical associations were engaging in illegal conspiracies to restrict competition by ethical bans on advertising and "contract practice" arrangements for physicians' services. The FTC order was modified and sustained by the U.S. Court of Appeals (in 1980) and affirmed by a split vote of the U.S. Supreme Court in 1982.[1] Medicine's adversaries are delighted by this new challenge to the profession's status:

> [T]he current antitrust attack on professional power seems to provide most of the tools necessary to break down excessive public and private deference to the medical profession. . . . By exposing the hollowness of the profession's claims to special virtue, antitrust activity will contribute to the spreading perception of medicine as just another interest group scrambling for a place at the public trough.[2]

Those of us who find these tendencies disturbing must acknowledge that organized medicine and its contemporary practitioners may have brought some of these troubles on themselves. Too many doctors have acted like tradesmen; too few have taken the trouble to clarify and affirm what they as physicians profess. Nevertheless, new follies will not remedy old ones, and care must be taken lest the baby be thrown out in efforts to purify its bath. There must be a limit to efforts at reformation and redefinition if we are to preserve a vigorous profession of healing. Though unfriendly and unwelcome, the present attacks on medicine as a profession provide the occasion for long-overdue reflection and reassertion of the meaning of professing medically. Despite the skill and energy of the critics, the defense can, I am confident, be solidly built. For if healing the sick is always the heart of the physician's business, and if, as I suspect, the *essential* features of the healing relationship between the physician and the ill have not been, indeed cannot be, altered by technological advance or social change, then medicine must remain, *at its core*, what it always has been: a very special profession, with its own goals, means, and intrinsic norms of conduct. The present inquiry is part of a search for that core, with special attention to the place of ethics in defining the medical profession.

Two prefatory remarks are in order, one about how I shall proceed, the other about where I shall come out. My approach is deliberately simple, though I hope not thereby simple-minded. To provide some anchorage against the drift of recent discussions, I would like to put down some elementary yet, I believe, weighty truths about medicine—things about which we need more to be reminded than instructed. These I hope to find partly by general reflections about professions, trades, and arts; partly by contrasting medicine with other human activities; and partly by looking directly at the ethical implications of certain features of medicine disclosed by the previous comparisons. I conclude that medicine necessarily remains a unique and intrinsically moral profession, with its own special norms and duties flowing from (1) the dignity and precariousness of the goal—health; (2) the human meaning of illness; and (3) the mutually shared self-consciousness of the doctor-patient relationship, an asymmetric human relationship explicitly defined relative to one person's illness and health and the other's professed devotion to healing and comforting, yet between persons who share equally in the gift of life and in its unavoidable finitude.[3]

Profession or Trade?

We begin with the notions of profession and trade. Why should physicians (or indeed, all sensible people) resist efforts to declare medicine a trade, and, for that matter, to have many trades considered as professions—as their practitioners now demand? What do we mean by this distinction? Actually, in certain broad usages—lexically derivative, but still venerable—"profession" and "trade" can and do mean the same thing: (1) *profession:* "In a wider sense: Any calling or occupation by which a person habitually earns his living (1576)"; (2) *trade:* "Anything practiced for a livelihood (1650)."[4] These shared meanings, which, as we shall see, bear in neither case any relationship to their etymological roots, collapse any distinction between profession and trade because both equally yield to their practitioners a livelihood. It is, I suspect, the considerable money-making powers of present-day physicians that leads the avant-garde to reclassify medicine as just another one among the many profitable occupations. But though it may surprise some laymen and even some physicians, the activity of healing is, both in principle and in fact, distinct from the activity of gaining remuneration for healing. A proof: Physicians sometimes treat for free. Every practitioner of any art or craft, profession or trade, may and often does practice the universal "art" of money-making, in addition to his particular activity. As the activity is often for the other fellow's good—doctoring, *as such*, benefits only the sick—justice requires wages or honor or some other form of benefit to the practitioner by way of reciprocation. And although it goes without saying that money-making, or, more properly, the love of gain, can distort the practice or corrupt the practitioner of any activity practiced for a livelihood, the activity itself has its own independent meaning and being. On these matters we expect more illumination from the original and more restricted meanings of "profession" and "trade."

The word "trade," derived from Germanic and Anglo-Saxon noun roots meaning "footstep" or "track," and verbs meaning "to tread," originally meant "the track or trail of a man or beast" or "a course or pathway," and derivatively, "the course, manner, or way of life; a regular or habitual course of action." From here one can see how "trade" could come to mean a habitual occupation, whose regularity is related to certain skills and crafts: "now usually applied to a mercantile occupation and to a skilled handicraft as distinguished from a profession, and *specifically* restricted to a skilled handicraft, as distinguished from a profession or mercantile occupation on the one hand, and from unskilled labour on the other (1546)"; It is a skilled handicraft—a manual art, informed by articulable know-how, acquired by habituation and practiced habitually—that enables the tradesman to tread regularly and reliably the same path.

If "trade" is etymologically grounded in the work of one's feet, "profession" traces home to an act of self-conscious and public—even confessional—speech. "To profess" preserves the meaning of its Latin source, *pro* + *fateor*, "to declare publicly; to own freely; to announce, affirm, avow." A profession, to speak precisely, would be an activity or occupation to which its practitioner publicly professes, that is, confesses, his devotion. But public announcement seems insufficient; it is unlikely that publicly declaring devotion to plumbing or auto repair would turn these trades into professions.

For some interpreters, learning and knowledge are the diagnostic signs of a profession. For reasons probably linked to the medieval university, with its tripartite division, the term "profession" has been applied especially to the so-called three *learned* professions—medicine, law, and theology—the practices of which are founded upon inquiry and knowledge rather than mere know-how—knowledge, respectively, of human nature, politics and ethics, and God. (It is worth noting that when medicine came to be considered a *learned* profession, it dealt with more than health and disease; books of medicine carried titles like *De Homine*). Knowledge or science remains crucial to the professions, all the more so in times of great scientific advance; few careers require as much schooling as does medicine. Moreover, the study of medicine properly belongs in a university, whereas carpentry and cosmetology do not. Yet the pursuit or acquisition of the knowledge does not alone make the knower a professional, one who is devoted to a way of life, albeit a way founded in part on that knowledge, artfully applied in practice. The knowledge makes the profession one of the *learned* variety; but its *professional* character is rooted in something else.

Some mistakenly seek to locate that something else in the prestige and honor accorded professionals by society, evidenced in their special titles and modes of dress and the special deference and privileges they receive. True, professionals are distinguished and honored members of their communities. True, the title "Doctor" and the white coat are prestigious marks of status to which others give deference. But these and other externalities do not constitute medicine a profession. The physician is not a professional because he is honored; rather, he is honored because of his profession. His title, his uniform, and the respect he is shown superficially signify and acknowledge something deeper: that he is a person of a professional sort, that is, one who knowingly and freely professes his devotion to a way of life worthy of such devotion. Just as the teacher devotes himself to assisting the learning of the young, looking up to truth and wisdom; just as the lawyer devotes himself to rectifying injustice for his client, looking up to what is lawful and right; just as the clergyman devotes himself to tending the spirit of his parishioners, looking up to the sacred and the divine; so the physician devotes himself to healing the sick, looking up to health and wholeness.

Being a professional is thus rooted in our moral nature and in that which warrants and impels making public professions or avowals of devotion to a way of life. It is a matter not only of the mind and hand but also of the heart, not only of intellect and skill but also of character. For it is only as a member of a community and as a being willing and able to devote himself to others and to serve some higher good that a man makes a public confession of his way of life. To profess is an ethical act, and it makes the professional *qua professional* a moral being who prospectively affirms also the moral nature of his activity.

Professing oneself a professional is an ethical act for many reasons. (1) It is an articulate public act—not merely a private and silent choice—a confession before others who are one's witnesses. (2) It freely promises continuing devotion, not merely announces present preferences, (3) to a way of life, not just a way to a livelihood; (4) in activity, not only in thought; (5) in service to some high good, which (6) is capable of calling forth devotion, because it is both good and high, but which (7) requires such devotion, because its service is most demanding and difficult, and (8) thereby engages one's character, not merely one's mind and hands. How the peculiar character of medicine warrants such professing and makes its own special demands on character is what we hope to articulate. First, though, we must deal with a possible objection.

Is the Art Morally Neutral?

Against our characterization of medicine as profession, inherently moral, it might be argued that medicine, like any other technique or skill or art, is morally neutral, does not itself bind character, and can be used for ill or good depending on the character of the one who uses the art or on the morality of his particular society. Especially as medicine is today explicitly studied and taught, it seems difficult to argue that medicine is inherently a moral activity. Our sciences of the workings of the body do not yield moral knowledge, or even the truth that health is good and disease bad; according to our scientific view of nature, disease is just as natural as health, both healthy and diseased processes equally "obeying" the same universal laws of nature. Moreover, the myriad changes in medicine and contemporary society have served to obscure the profession's ethical self-definition, at the very time that we see only too clearly the ethical dilemmas that medical advance has brought. Yet the question about medicine and morality goes back at least to classical antiquity and antedates the distinction of profession and trade.

In Ancient Greece, medicine was regarded as a *technē*, a term hard to translate, but perhaps best rendered as "art" if one remembers that it meant primarily the useful arts and handicrafts and only latterly, if at all, the fine arts. Shoemaking, carpentry, weaving, metalworking, and shipbuilding were *technai*, no less than navigation, strategics, legislation, or even divination. Although the arts generally had a manual element, this was in each case governed by a certain know-how, capable of being transmitted through teaching and learning. Unlike the fine arts of music and dance, the activity of the *technai* usually produced some work or product separate from the activity (e.g., a shoe, a ship, victory, health). Although the product was superior in dignity to the activity that produced it, and served as the *end* of the activity, the end was in turn useful to some further good. Moreover, this end was *given to* and presupposed *by* the *technē* and by the man of *technē* in his making: On this view, it is not the shoemaking *technē* that knows why shoes are good for feet or the medical *technē* that knows why health is good. It would follow that skill in the art could be divorced from character, that a bad man might be a good architect or physician, or, in Socrates's famous example in Plato's *Republic*, if justice were an art, then the clever guardian would equally be the clever thief. Virtuosity and virtue were not—and are not—the same thing.

There is certainly merit in these claims. For one thing, they highlight the difference between the mere presence of know-how and the disposition to use it and to use it beneficently. For another, they show how these purposive human activities serve goals other than themselves, and thus derive their meaning and worth finally from the contributions they make to the ultimate goods of human life. Thus, even if the physician as physician could discern that and why health is good, he may not be able, as physician, to assess how health stands relative to justice, learning, pleasure, or freedom, especially in concrete cases where these goods are in conflict. Still, while any *techne*, understood as dexterity governed by know-how, may be equally adapted to good or ill, and the medical *techne* to poisoning and curing, the *man* of *techne*, the *practitioner* of *techne*, is more than a technician. For one thing, he needs some semblance of virtue, even in order to acquire virtuosity. Conscientiousness, order, and industry, and some pride of workmanship, not to speak of elementary sobriety or temperance, are required: the lazy, slovenly, or dissolute rarely acquire or practice any art well. Moreover, the practitioner of any *techne* must generally be regarded as at least minimally dependable, trustworthy, honest, and fair if people are to continue to seek his services. Indeed, the *technites* or practitioner, in the *strict* sense, is understood to serve the good or interest of *others*. Strictly speaking, not even the practices of shoemaking and housebuilding are morally neutral, not to speak of medicine (which, by the way, is never one of the examples cited by Socrates to show how *techne*, as mere know-how, can be divorced from knowledge of and devotion to the goodness of its goal).

Yet even if we come to expect honesty, industry, reliability, and a disposition to serve our good from men of every *techne*, we have come no closer to identifying any special ethical features of medicine. These virtues are required in all the trades and professions—albeit perhaps in varying degrees and certainly with varying degrees of harm possibly resulting from their absence. Are there no standards of conduct—specific actions of commission or omission, specific duties or forbearances, specific virtues, specific ethical postures—especially or peculiarly fitting to the *healing* art? To approach this question, we need first to discern features of the medical art—both regarding ends and means. This we attempt by some simple comparisons.

Medicine: A Special Activity

Whatever might be said about the role of medicine in health-promotion and health maintenance (and I have supported these objectives in Chapter 6), medicine is first of all a corrective or restorative art. To heal means to make whole that which is impaired or less than whole. How does medicine differ from other "fix-it" arts, say, auto repair?

The auto mechanic, like the doctor, seeks to restore his "patient" to a well-working wholeness. His diagnoses depend on a good history and physical examination, backed by increasingly sophisticated "laboratory" tests of partial functions. Yet the "patient," though highly organized, is not an organism; although it yields automobility to its owner, it itself is not self-moving. The patient, and therefore also the goal of "therapy," is entirely man-made. Also, the patient is other than, is indeed a mere lifeless servant to, the client, whose in principle arbitrary will governs the goal of automotive manufacture as well as repair.

The veterinarian, unlike the auto-healer and like the physician, serves a *natural* end, the biologic well-working of his animal patient. Moreover, although he will often intervene to "violate" an animal's body, causing pain and further distress, the veterinarian even when he does so generally understands himself to be cooperating with the immanent powers of self-healing that characterize all living things. True, in some cases, the animal's "wholeness" is partially defined relative to the needs and desires of its human owner; different restorative treatments may be in order to permit one horse to run a race and another to pull the plow. However, an animal is also a self-moving and self-maintaining being, in and for itself. There are limits to the arbitrary human additions to what is basically a natural standard of animal wholeness and well-functioning, relative, of course, to species and, to some extent, to age, sex, and so on. Techniques to restore the health of a dog or a horse or a sheep may be invented by veterinary science, but animal health is in itself not a human invention. We can try to discover its meaning; we can defend it or ignore it, but it is a natural given, which, to a great degree, is the immanent though unconscious aspiration of the self-maintaining and self-healing powers and activity of every living thing. These two facts—the at least partial givenness of the norm of health and the existence of immanent powers that tend toward wholeness—are, or ought to be, minimally a restraint on arbitrary willfulness, but maximally a cause of awe and respect, for veterinarians no less than physicians. How these might function in professing medicine will be a topic for the end.

However veterinary and human medicine may be alike, they are decisively different in one major respect: The horse or dog does not bring himself to the healer. True, neither does a newborn baby nor a comatose adult; and the care of such patients sometimes presents knotty moral problems for this very reason. But we do best to take our bearings from the ordinary, not the unusual. In most human cases, it is not only illness but the *self-consciousness* of illness and the desire for health, as well as the *voluntary turning* for aid voluntarily professed, that establishes the relationship between patient and physician and gives it its special ethical dimensions. The physician, who is a knower of health and the numerous forms of its absence, who seeks to assist the healing powers in the human body, also must tend particular, necessitous human beings who, in addition to their symptoms, suffer from self-concern and often fear and shame—about weakness and vulnerability, neediness and dependence, loss of self-esteem, and the fragility of all that matters to them. We approach these features by looking briefly at other arts that minister to the client as willing patient.

Beauticians and barbers also work on the human body, albeit superficially. To both we allow the invasion of our private space; we have forgotten how as children we had to be coaxed to overcome a natural reluctance to allow strangers to touch and manipulate even our lifeless and unfeeling hair. Yet without depreciating the importance of grooming and appearance to self-esteem, and without denying that the self-exposure witnessed by some beauticians might call for discretion and protection of privacy, no one would claim for these arts the gravity of medicine or the intimacy appropriate to the doctor-patient bond. Closer to medicine, the same can be said for the dentist, the optometrist, or the podiatrist, and even within medicine, for some subspecialties, such as radiology. By and large, neither our survival nor our general possibilities for healthy functioning are threatened by dental caries, nearsightedness, or corns; moreover, for most sensible people, body image and sense of integrity are not especially bound up with teeth, retina, toes, or skeleton—all of which are ordinarily hidden from view, one's own and others', and all of which can be examined without the indignity of removing one's clothing.

When, in contrast, we come to the physician of the whole body—to the internist, the general practitioner, the primary care physician, or the family doctor—we come tacitly acknowledging the meaning of illness and its potential threats to present function, future aspiration, and ultimately all that we hold dear. We are willing to expose ourselves—not only our bodies but often certain intimate details of our lives—to this relative stranger, first, of course, because he has professed his ability and willingness to help (and we believe him), but also because we expect that he understands the meaning of our presenting ourselves for his examination and counsel. The physician, though he is healthy, shares the patient's awareness of vulnerability and mortality, not only because the *technē* of medicine rubs his nose in the multifarious fragilities of the flesh, but also because of his own humanity. The physician, too, is capable of self-consciousness, and recognizes himself in the other. He sees not only the *one-sided* relationship of benefactor and recipient appropriate to the patient's appeal for knowledgeable advice, but also the equal relationship grounded in their *mutual* and equal participation in the ever-precarious, necessarily finite, yet daringly aspiring and hope-filled venture called human life. Although he must keep his private feelings to himself, it is his own silent self-awareness that enables the physician to synthesize the abstract scientific knowledge of bodily workings with the concrete human experience of living in health and in sickness. As a result, he acquires the kind of understanding that permits him not only to fight disease but also by helpful speech to mediate between the patient's understanding and his own mysterious and silent body.

Other arts deal with parts of the body, but medicine deals—sometimes explicitly—with the fact of human embodiment, that is, with our strange and mysterious being comprising a grown-togetherness of soul (or perhaps only mind) and body. This mystery is reflected in our speech: We say both "I am my body" and "I have a body." We say both "my body" and "my mind." Which or what, then, is me? What constitutes my unity or wholeness?*

Ordinarily, when we are in health, this unity is never a question. One of the mysteries of the physical workings of our body is how little they intrude on our consciousness. For example, we must act to open our eyes, but we see effortlessly, totally unaware of the workings of our eye and brain. Even in touching, the body ordinarily silently suppresses the perception of the pressure on the finger; we "feel" instead only the hardness of the table. When all goes well, we are unaware of duality or contradiction. But pain, weakness, fever, or other symptoms disrupt our normal flow. We become aware of a certain doubleness: The pain is ours, but so is our desire and effort to combat it. We become self-conscious, and not only of our vulnerability. This self-attention and self-consciousness draws us away from activity, from immediacy, from free and easy relation to the world. And this disturbance of world-relation is not the least reason for seeking medical help: We want the duality silenced. Yet, paradoxically, to get the rift resolved, we must freely expose it to the physician.

All of these peculiarities of the doctor-patient relationship are manifest in the physical examination. In undressing ourselves for examination, we shed those conventional signs of our humanness and expose ourselves as we were when we came into the world: weak, insufficient, and unaccommodated. We give over our distinctively human upright posture as we lie down on the table to be examined. We subject ourselves to the objectifying gaze and touch of the physician, which necessarily abstract from our humanity and treat us as mere

*See Chapter 11, "Thinking About the Body."

body. Even if we are only getting a checkup, we cannot but be reminded of our vulnerability and mortality, and, in seeing ourselves
objectifiedly seen, we become objects to ourselves. Yet this act of self-
consciousness—all of these acts of self-recognition—are the signs not
of mere body, or of our lowliness, but rather of something very special,
something that—despite, or perhaps because of, its awareness of necessity—affirms life and wholeness and struggles gamely to preserve
and dignify them. In these efforts to face and meet necessity, sometimes with the physician's help, we are capable of rising above it. A
good physician makes the support of dignity an important part of his
professional task, offering hope without deception, behaving seriously
without solemnity, giving advice while respecting freedom. In so doing,
he tacitly recognizes and in deed demonstrates that we human beings
are beings mysteriously in-between yet pointed upward: necessitous
but freely able to combat necessity; vulnerable but resourceful, and
not only regarding vulnerability; isolated as body yet capable, through
soul and speech, of overcoming our isolation in self-conscious and passionate relations with others and with the world; perishable and subject to error and evil, yet full of longing and striving for the permanent
and the good.

Admittedly much of what I have said is merely implicit in the doctor-patient relationship and is rarely, if ever, present in this form to
the physician's consciousness. Still, many of the traditionally understood obligations and virtues requisite of a physician are derived or
derivable from this understanding of what is humanly at stake in the
medical practice. In the closing part of this chapter, I illustrate briefly
how certain ethical duties and stances follow from what I have suggested to be distinctive to the healing art: on the one side, a natural
good and immanent powers of self-healing; on the other, vulnerability,
shame, exposure, and neediness; surrounding both, a shared and self-
conscious awareness, however dim, of the meaning of the delicate and
dialectical tension between wholeness and necessary decay.

Special Moral Duties and Postures

First, let us look at certain self-imposed restraints on deeds and
speeches, long professed by physicians and stated already in the so-
called decorum paragraphs of the Hippocratic Oath: the forswearing
of sexual relations with patients and the forswearing of gossip and
breaches of confidentiality.*

---

*For a discussion of the entire Oath, including a more thorough treatment of these
passages, see Chapter 9, "Is There a Medical Ethic?: The Hippocratic Oath and the
Sources of Ethical Medicine," especially pp. 236–240.

Into whatever houses I may enter, I will come for the benefit of the sick, remaining clear of all voluntary injustice and of other mischief and of sexual deeds upon bodies of females and males, be they free or slave.

Things I may see or hear in the course of the treatment or even outside of treatment regarding the life of human beings, things which one should never divulge outside, I will keep to myself holding such things unutterable [or "shameful to be spoken"].

These professionally imposed restrictions on deed and speech grow out of the recognition that illness is inherently degrading and dehumanizing, and that it exposes and threatens the sick person's body, soul, and intimate relationships. The physician pledges himself not to aggravate degradation nor to exploit weakness. Understanding the meaning of exposure, he knows that it would be grossly unjust for one who must professionally uncover his patients' nakedness to take sexual advantage of their vulnerability and their need for his services. Understanding the unavoidable assaults on dignity that accompany illness—including the supine posture, the role of being patient rather than agent, and, especially, being regarded through the physician's reductive and abstract concepts of bodily function—the professing physician knows to set self-conscious and clear limits on his own reductive behavior. He also knows that it would be shameful to expose outside of the home or office what he has seen or heard as a privileged outsider invited into the private world of his patients. For he is obliged to protect the intimacy of the private against publicity and objectification, both of which threaten to produce not only explicit embarrassment and humiliation, but also—even in the absence of anything positively shameful—the corrosive kind of self-consciousness, which is the enemy of all spontaneity, immediacy, and intimacy. The perceived need to safeguard the patient's dignity, privacy, and self-esteem calls for the virtues of justice, self-restraint, and discretion.

Second, beyond these duties of restraint, there are positive duties, addressed especially to the patient's self-concern, to which I can here only point. Several have, in fact, already been anticipated or alluded to: cognitively, patient instruction about the nature and meaning of symptoms and disease; emotionally, support for the efforts to resist not only disease, but even more, fear, shame, and despair—compassion, by the way, is itself insufficient. From the healer, the patient needs more than sympathy for his frailty. If his higher powers to face up to and oppose disease and despair are to have a chance, he needs to be armed with knowledge, hope, and especially courage. Courage is what the doctor would order if he could; encouragement is what he can himself provide.

Instruction, support, and encouragement become all the more part of the doctor's professed business in the face of chronic illness and incurable disease. There are incurable conditions, but never untreatable patients. Concretely, this means that the physician is obliged to learn and advise about ways of *living* better with illness, through means not generally thought to be medical—involving advice about improved and more encouraging living situations, family support, alternative employment, transportation, etc. It also carries a strong presumption in favor of truthfulness, for the patient's dignity in facing his situation depends on knowing what it is and portends. And finally, it means being available—in principle, at all hours and in all places— albeit with limits set by the physician's need to preserve his own health and psychic well-being, a duty that has its own claims. When one professes medicine, one offers the healing, comforting, and encouraging hand, which, when it is grasped, may not be pulled away, at least not without providing for its replacement.

Finally, a concluding word about the physician's posture in the world. What should follow from his understanding of the medical art and the goals it serves? What should he make of the fact of self-healing, of the wondrous, even awe-inspiring powers of the living human body directed toward wholeness? What should he make of the realization that it is, in most cases, nature working within that is the true healer of disease, in whose service he is, as it were, but the physician's assistant? What should he think of our peculiarly human self-consciousness and the marvelous resilience of the human spirit in the face of adversity? What about the wondrous fact that the leaf of the fox glove contains the cure for congestive heart failure or the flower of the poppy the relief of pain—generally, that the world is full of all kinds of healing aids, actual and potential, lying around waiting, as it were, to be discovered and synthesized? What about his own powers as physician: that special power of mind, unique to human beings, capable of discovery, that can apprehend health and disease and find supports for the former and cures for the latter; and that special impulse of the soul, also unique to human beings, that beckons us to use our knowledge of wholeness to help make the wounded whole? A physician who could see through the trappings and equipment of his art to these native powers that make it possible would stand in the world neither as proud master nor as servile technician. He would stand self-consciously in-between, as one who professes, respectfully affirming and gratefully acknowledging the existence and support of powers not of his own or of any human making. As a professional, man bears witness to the being of something higher and more enduring, participation in which can only be called a blessing.

CHAPTER NINE

Is There a Medical Ethic?
The Hippocratic Oath
and the Sources
of Ethical Medicine

Between medicine and the love of wisdom there are no great differences; in fact medicine has all the things that lead toward wisdom: dislike of money, reverence, modesty, reserve, sound opinion, judgment, calm, steadfastness, purity, knowing speech, knowledge of things useful and necessary for life, the dispensing of that which cleanses, freedom from superstition, pre-eminence divine.

—Hippocrates, *Decorum*

The physician is only nature's assistant.

—Galen, *On the Humor*

Practiced diagnosticians have long been whispering in the corridors about the competing malignancies now eating away at American medicine and about the multiple strokes that have progressively paralyzed and anesthetized our public and private morality. But today they can pronounce from the rooftops the health and vigor of medical ethics. Medical ethics is flourishing, both as a field of intellectual activity and as an area of popular concern. In the past decade, we have seen a flood of new courses in colleges and medical schools; a spate of articles, journals, television programs, and books on medical ethics; the founding of several special institutes on ethics and the life sciences; new state and Federal statutes and regulations and numerous influential court cases on biomedical matters; a revision of the American Medical Association's Principles of Ethics; and no less than three national commissions to pronounce on ethical issues in biology and medicine. There is more here than fashion. Indeed, the rise of medical ethics—especially among its academic practitioners—can be understood as a response to the malaise both of American medicine and American morals, a response that holds out the promise of much needed therapy, not to say redemption.

Help from medical ethics is wanted for at least two **differentiable** purposes: to identify and curtail unethical or immoral **conduct or prac**tices by physicians or medical researchers, and to analyze **and resolve** difficult ethical dilemmas faced even by the most **upright physician,** say, in the proper uses of new biomedical technologies **or in the care** for the hopelessly ill. Though the medical profession has **increasingly** addressed itself both to medical misconduct and to the **genuine dilem**mas, the great rise of extraprofessional activity reflects, I **suspect,** a growing belief that medical ethics is too important to be left to doctors.

There are, to be sure, ample reasons today for public scrutiny of medical practices. In fact, at no time and in no place has medicine been a completely autonomous profession. Physicians have always been obliged to live under the law of their communities, even in those uncommon regimes like our own that have chosen liberally to leave the professions alone in most respects. This subordination of medicine is entirely proper: There is more to life than health. Sometimes the pursuit of health competes with the pursuit of other goods, and it has always been the task of the political community to order the different and competing ends. But medicine as a self-governing profession must today meet a new challenge, one that would deny autonomy to the profession because it denies the existence of a medical ethic as such. The challengers insist that what we call "medical ethics" or "the ethics of medicine" is but the application *to* medicine, as a particular but not unusual instance, of more universal norms of human conduct. For some, these more universal norms are derived from biblical religion, for others, from the dictates of reason (e.g., Kantian morality), while still others appeal only to positive law or to the self-determining, democratic procedures used to enact it. But all of the challengers agree that there are no special duties or principles of conduct unique to medicine, which arise directly from the nature or meaning of the medical art and the activity of healing the sick. So-called medical ethics is but that branch of universal ethics that casts its protective shade over the human relations between doctor and patient. Medical ethics does not grow out of medical roots.* Therefore, medicine cannot be self-governing.

Perhaps the deepest reason for looking beyond medicine to external sources of ethical principles is the widely shared belief that medicine—whether regarded as a science or as an art—is in its essence morally neutral, or as our jargon has it, value free. To begin with, the sciences of the workings of the body do not yield moral knowledge.

---

*Indeed, the rise of the *field* of medical ethics seems itself to spring from this uprooting of *medical* ethics.

Indeed, the body as studied by modern science does not even reveal that health is good and disease is bad. Both healthy and diseased processes "obey" the universal laws of nature; hence, according to our scientific view of nature, disease is just as natural as health. But even were we to reacquire a teleological understanding of nature, we would still have to face the ancient and enduring argument about the ethical neutrality of art.

If the art of medicine is defined primarily in terms of its powers, its techniques or skills, then the argument for its moral neutrality is hard to refute. Technique or know-how can be used for both good and bad ends. The man who best knows drugs is both the best pharmacotherapist and the best poisoner, and the man who best knows how to repair fractures knows best where and how to produce them. Only if an art is more than know-how, only if it is also know-what-and-what-for, and, further, only if that knowledge of ends informs the choice of means, could one begin to look to the art for its own inherent ethic. Most people today seem to deny to medicine, even in principle, such transtechnical and ruling knowledge of ends. Small wonder that they deny any medically based medical ethics.*

Are these prevailing opinions true? Is there no medical ethic? Must the ethical principles of the profession be imported from an outside source? Is the doctor morally bound not because he is a physician but only because he is a Christian or an American or a Kantian or utilitarian? Can none of the notions of good or right or duty or virtue pertinent to medicine be drawn from the meaning of medicine itself? Is medicine, and the knowledge on which it rests, utterly nonethical? Is our nature really indifferent as between health and disease? A full examination of these questions, and of the possibility of a medical ethic properly so-called, would require a thorough inquiry into both the nature of medicine and the nature of ethics, and ultimately into the relation of both to the nature of nature. I am not equal to the task, even were the reader willing. Instead, I intend to provoke ongoing reflection on these matters by an explication of the Hippocratic Oath.

---

*The argument about what we call the moral neutrality of the arts goes back to classical antiquity; for example, in Plato's *Republic*, Socrates argues that if justice is an art, the guardian skilled in that art will also be a clever thief. Though medicine is not treated in the Platonic dialogues as one of the *technai* that lacks knowledge of its end, the question is raised in one of the myths regarding Asclepius and Athena. Athena is said to have procured two powerful drugs in the form of blood taken from the Gorgon Medusa, the blood drawn from her left side providing protection against death, that from her right side a deadly poison. According to one version, Athena gave phials of both to the physician Ascelpius; according to the other version, she gave him only the beneficent drug, reserving the power of destruction to herself.[1]

Why the Hippocratic Oath

Of the numerous oaths and codes articulating principles of medical practice, the Hippocratic Oath is both the oldest and the most well known. Its authorship is unknown, and also its age—scholarly proposals for its source ranging anywhere from the sixth century B.C. to the first century A.D. For many centuries the Oath was taken as the very model of medical conduct; it informed the attitude of generations of physicians. Indeed, until the present generation, it was sworn to by students of many if not most of our nation's medical schools on the occasion of completing their studies and entering their chosen profession.

Yet its very antiquity is now increasingly a reason for its declining importance among us. Given the contemporary historicist prejudice, that is, given our belief that "truths" are simply relative to time-and-culture, and, therefore, that old documents bespeak and reflect only old cultures and beliefs, we increasingly regard the Oath merely historically, that is to say, not seriously. Skeptical of, and thus immune to, all claims to truth, we labor instead to *explain* the claim, in terms of its social context or the peculiarities of its claimants. The most influential article on the Oath in recent decades, written in 1943 by the eminent classicist and historian, Ludwig Edelstein,[2] seeks to establish, by means of a careful and illuminating analysis of the text itself, that the Oath was not even an expression of Greek medical thought in general but rather a product of a small sect of fourth-century (B.C.) Pythagorean physicians. Edelstein suggests that the Oath's influence in Western medicine is due to its later appropriation by Christian physicians who, unlike adherents of other pagan Greek sects, shared the Pythagoreans' concern for purity and holiness and their opposition to suicide and abortion. Having thus found what he takes to be its particularistic and historical origin, Edelstein feels entirely comfortable rejecting the ahistorical view that the Oath expresses "an ideal program designed without regard for any particular time or place."[3] He never raises for himself or for us the question of whether the Oath, despite its dated beginnings, might nevertheless speak truly and timelessly.

Not only our ideology, but also our changed circumstances contribute to our not taking the Oath seriously. Medicine has changed tremendously in our century. The institutions in which it is taught and practiced, the tools of diagnosis, the modes of treatment, even the principles for understanding the basic mechanisms of disease—all these would overwhelmingly surprise a nineteenth-century physician, let alone poor Hippocrates. To these changes in medical practice must be added changes in mores and beliefs in the community at large. For all these reasons, the Hippocratic Oath might seem to be an unlikely source of useful wisdom.

Yet even if our circumstances are inhospitable to the Oath's teachings, we should be wary of dismissing a document held by thoughtful men over many centuries to be *the* embodiment of medical wisdom. It is not one of nature's laws that wisdom is progressive or that opinions put forth long ago and by a small minority, even by Pythagoreans, cannot be superior to those of a majority or to those that replaced them. (On the historicist's own grounds, our historicism, which denies this possibility, might be nothing more than the ruling opinion of *our own* age.) Thus, even if Edelstein is right about the Pythagoreans, the Pythagoreans might have been right about medicine. And they might still be right if, as I believe, the essential activity of healing the sick is still the same, despite all the enormous changes in medical practice, that is, if to be healthy or whole still means largely what it did in ancient Greece, if the desire of the ill to be whole is no different, and if the healing relation between the physician and the one to be healed is *in essence* the same. The Oath could still teach truly, even if its teachings went unheard, drowned out by the noisy self-confidence of modernity. Not every lost cause deserves to lose.

With this as my defense for doing so, I turn to the Oath itself. I hope to show that it is internally coherent and comprehensive in its scope, that its various parts flow naturally from a profound understanding of what medicine is and must essentially be, and that it therefore stands as *an* expression of the medical ethic *in which the medical and ethical are as hard to separate as the concave and the convex.*

The Oath and Its Parts

I swear by Apollo Physician and Asclepius and Hygieia and Panaceia and all the gods and goddesses, making them my witnesses, that I will fulfil according to my ability and judgment this oath and this covenant:

To hold the one who has taught me this art as equal to my parents and to live my life in partnership with him, and if he is in need of money to give him a share of mine, and to regard his offspring as equal to my brothers in male lineage and to teach them this art—if they desire to learn it—without fee and covenant; to give a share of precepts and oral instruction and all the other learning to my sons and to the sons of him who has instructed me and to pupils who have signed the covenant and have taken an oath according to the medical law, but to no one else.

I will apply dietetic measures for the benefit of the sick according to my ability and judgment; I will keep them from harm and injustice.

I will neither give a deadly drug to anybody if asked for it, nor will I make a suggestion to this effect. Similarly I will not give to a woman an abortive remedy. In purity and holiness I will guard my life and my art.

I will not use the knife, not even on sufferers from stone, but will withdraw in favor of such men as are engaged in this work.

Into whatever houses I may enter, I will come for the benefit of the sick, remaining clear of all voluntary injustice and of other mischief and of sexual deeds upon bodies of females and males, be they free or slave.

Things I may see or hear in the course of the treatment or even outside of treatment regarding the life of human beings, things which one should never divulge outside, I will keep to myself holding such things unutterable [or "shameful to be spoken"].

If I fulfil this oath and do not violate it, may it be granted to me to enjoy life and art, being honored with fame among all men for all time to come; if I transgress it and swear falsely, may the opposite of all this be my lot.[4]

The Oath is divided into eight paragraphs, of which the first is the oath proper, an invocation and swearing, and the last something that resembles, but is not, a prayer. The covenant or promise made under oath is organized around two main subjects: conduct with respect to teachers and students (paragraph two) and conduct with respect to patients and their households (paragraphs three to seven). The latter is further subdivided into three paragraphs (three to five) that deal primarily with treatments, appropriate and inappropriate, and two paragraphs (six and seven) concerning decorum—though I hasten to add that this distinction between treatment and technique, on the one hand, and decorum and ethics, on the other, is foreign to the text, which treats each paragraph coordinately and equally to be sworn to. The Oath does not tell us where the merely technical ends and the ethical begins, properly so if all the provisions flow equally from an insight into the essence of medicine, understood not as ethically neutral technique but as activity constituted by a notion of good.

Conduct with respect to patients is further subdivided in an eminently sensible manner. The three so-called treatment paragraphs address ends and means. There is first a statement of the overall end, "the benefit of the sick," and an identification of the predominant means, "dietetic measures" (paragraph three). There follows a delimitation, by restriction, of the ends fittingly served by medically available means (paragraph four) and then a delimitation, by restriction, of the means properly used by physicians to attain medically desirable ends (paragraph five). The two so-called decorum paragraphs are exhaustively divided into deeds and speeches, both by way of self-restraint: what I will not do and what I will not say.

This purely formal analysis of the subject matter of the Oath is sufficient to raise an important question: Do these subjects adequately and precisely define the medical profession? Could there be a worthy definition of medicine or physicianship that did *not* address and delimit appropriate and inappropriate ends and means, deeds and speeches, as well in relation to teachers and students of the art as to patients? Are there important subjects omitted? Are there some that could be omitted without loss? I very much doubt it.

The new Principles of Ethics recently adopted by the American Medical Association (text on page 231) do in fact omit several of these subjects and add others.* They contain no statement of the end of medicine, saying only that a physician "shall be dedicated to providing competent medical service," nowhere defining or delimiting the goal being served. (Curiously, the word "health" occurs only as an adjective, in the phrase "other health professionals"; neither the sick nor the ill, sickness or illness, are mentioned, leaving us only the term "the patient." But "patient" is a term devoid of reference to the end. It is defined not relative to health, as are "the sick," but relative to the doctor as *active agent.* A patient is one who is patient or passive to the doctor's ministrations. By contrast, the word "patient" and its relatives do not occur in the Hippocratic Oath.) Moreover, the AMA Principles speak neither to limits on technique nor to duties to teachers or students (save an obligation to "deal honestly with patients and colleagues," and "to expose physicians deficient in character or competence, or who engage in fraud or deception"—the only deeds specifically ruled out of bounds).

Instead, the new AMA Principles deal with the physician's relations with society: respect for law, coupled with a responsibility to seek changes in laws bad for patients; respect for rights within constraints of law; and social responsibility of a physician "to participate in activities contributing to an improved community." I grant that these subjects are important, and despite my serious reservations about the particulars, I have no quarrel with appending principles on the subject of relations with society to a statement of professional

---

*The preamble to the seven principles claims that it is *"[a]s a member of this profession"* that a physician "must recognize responsibility not only to patients, but also to society," and that the standards of conduct it presents *"define the essentials of honorable behavior* for the physician" (emphasis added). My arguments in the sequel will challenge the suggestions that these principles do define the *essentials* of honorable physicianship and that the responsibilities enunciated flow from "member[ship] [in] this profession," rather than from duties as citizen or human being.

AMERICAN MEDICAL ASSOCIATION
PRINCIPLES OF MEDICAL ETHICS

Preamble:

The medical profession has long subscribed to a body of ethical statements developed primarily for the benefit of the patient. As a member of this profession, a physician must recognize responsibility not only to patients, but also to society, to other health professionals, and to self. The following Principles adopted by the American Medical Association are not laws, but standards of conduct which define the essentials of honorable behavior for the physician.

I. A physician shall be dedicated to providing competent medical service with compassion and respect for human dignity.

II. A physician shall deal honestly with patients and colleagues, and strive to expose those physicians deficient in character or competence, or who engage in fraud or deception.

III. A physician shall respect the law and also recognize a responsibility to seek changes in those requirements which are contrary to the best interests of the patient.

IV. A physician shall respect the rights of patients, of colleagues, and of other health professionals, and shall safeguard patient confidences within the constraints of the law.

V. A physician shall continue to study, apply and advance scientific knowledge, make relevant information available to patients, colleagues, and the public, obtain consultation, and use the talents of other health professionals when indicated.

VI. A physician shall, in the provision of appropriate patient care, except in emergencies, be free to choose whom to serve, with whom to associate, and the environment in which to provide medical services.

VII. A physician shall recognize a responsibility to participate in activities contributing to an improved community.[5]

principle. But they are at most peripheral. They are in themselves neither essentially medical nor essential to medicine. Moreover, those in the AMA Principles are peculiar to the relation of medicine to a modern, liberal, legalistic—not to say litigious—society. They would not accurately describe, much less constitute, medicine in traditional or premodern societies, in the People's Republic of China or the Soviet Union, or in most of the nations of the so-called Third World. However important, such politically determined principles must be foreign to medicine, if medicine *as medicine* is essentially the same, independent of regime. To make more plausible the suggestion that the Hippocratic Oath bespeaks the universal core of medicine, we must try to show not only that the *subjects* treated in the Oath exhaust the pertinent domain, but also that the *content* of each provision is not only defensible, but even somehow essential to, or constitutive of, the nature and meaning of the healing art.

The Content of the Oath: The Treatment Paragraphs

Let us begin, not illogically, with the passage that states the physician's main business: "I will apply dietetic measures for the benefit of the sick according to my ability and judgment; I will keep them from harm and injustice." This states unequivocally that it is sick individuals, not society or mankind or some abstract idea, who are the beneficiaries of the physician's activity. Moreover, the sick qualify for his services *because* they are sick, not because they have claims, desires, wishes, demands, or rights. The healer works with and for those who need to be healed, those who are not whole.*

"Dietetic measures," once the main staple of the physician's therapeutic offerings, now strikes us as anachronistic, what with our cornucopia of pharmaceuticals, machines, and surgical procedures. Still, there *are* numerous medical conditions—including diabetes, hypertension, ulcers, heart failure, and gout—in which a dietary regimen is a central part of therapy and many others in which we suspect that diet is important, both as a cause of illness and as a cure. Symptoms of malnutrition and starvation, as well as infections that feed on the malnourished, are prevalent over much of the globe; symptoms and diseases caused or exacerbated by obesity are common in the prosperous countries. And if we broaden "dietetic" to mean "ingestibles," so as to include alcohol and other toxic substances as well as contaminated or despoiled foods, we see that regulation of diet still plays—and will everywhere play—a decisive role in benefiting the sick. We are, in a sense, what we eat.

But the Oath's emphasis on dietetic measures bespeaks a deeper teaching about the nature of medicine. What, after all, is diet or nutrition? It is the steady provision of necessary materials steadily consumed for energy or transformed from other to same by the body in metabolism, by which the body maintains in organized equilibrium its own functioning integrity. The ancients, in their naiveté, spoke of a balance of the four elements or humors; Claude Bernard, in his sophistication, taught us of homeostasis of the *milieu intérieur,* the in-

---

*Arguments supporting this conclusion as a—or even *the*—constitutive principle of medicine are presented in Chapter 7, "Practicing Prudently: Ethical Dilemmas in Caring for the Ill," especially pp. 191-200. "An ultimately patient-centered and individualized orientation" was shown to be "necessarily constitutive for the healing profession under any political order" (p. 195). With the help of the Hippocratic Oath, we shall qualify the meaning of this "individualism" later in this chapter (pp. 240-246).

ternal environment. Each implied that the healthy state is a certain balance or harmony of parts or elements that—and this is the crucial point—*the healthy or harmonized body will produce on its own,* provided it acquires the right materials and is not obstructed, say, by superior invasive forces. The *body is its own healer,* and the physician a cooperative but subordinate partner who supplies the needed materials—whether it be protein or insulin, vitamin C or interferon, and, by extension, even antibiotics to help the body arrest invasive obstruction from without. I do not insist that all current treatments can be rationalized on this homely model of supplying the necessaries, nor do I mean to assert that health is homeostasis—though I do think the time is ripe for a return to these philosophical matters. Rather, I mean to emphasize the Hippocratic Oath's tacit assertion that medicine is a cooperative rather than a transforming art, and that the physician is but an assistant to nature working within, the body having its own powerful (even if not invincible) tendencies toward healing itself (e.g., wound healing and other regenerative activities, or the rejection of foreign bodies and the immune response). Though our current technical prowess tends to make us forget these matters, does not the Oath speak truly?

The inherent tendencies toward wholeness are much more precarious in human beings than in our animal friends and relations. Our dietary habits are not given by instinct but by convention, habit, and choice. Man is the animal most likely to make himself sick, in the very act whose purpose is to nourish and sustain. Sometimes we eat ourselves sick through ignorance of what is good for us, sometimes through incontinence and lack of self-restraint in the face of knowing better. In both these ways, we do ourselves unwitting harm and even injustice; that is, we treat ourselves unfittingly and worse than we deserve. The physician, according to the Oath, is not only one who brings a corrective diet to the already sick. He is also the one who seeks to prevent the ignorant and the self-indulgent from harming themselves. He has the knowledge needed to direct and inform the otherwise dangerously open and uninformed human appetites. It may at first seem strange to think that we human beings need such knowledgeable outsiders or that we do not know what is good for us, all the more so on the premise that our body is the primary healer and the doctor but the physician's assistant. But a body possessed of the power of reason, and hence also of choice, is a body whose possessor may lead it astray, owing to ignorance or wayward impulse. The physician, the ally of our body and of those inner powers working toward our own good, supplies needed knowledge, advice, and exhortation. He seeks to keep us from self-harm and injustice. The Oath's little paragraph on dietetics, properly unpacked, reveals the core of medicine.

The mention of the injustice that a person can do to himself becomes the link to the next provision, the one that forswears giving deadly drugs or abortifacients. This provision, I have already suggested, wisely implies that the true physician will not use his available means indiscriminately, to promote alien ends. The physician is no mere technician, selling his services on demand. But what is a proper and what an alien end? If the goal is to benefit the sick, what are the limits of "benefit"? The present passage shows us how to think about this, for the limits on the use of technique are derived from an understanding of the essence of the healing activity.

The forswearing of both giving a deadly drug if asked and making such a suggestion shows that the provision has suicide and active euthanasia in mind, rather than, say, a self-imposed restraint on murder or poisonings. The doctor refuses to participate directly in ending a life, whether one in the fullness of days or on the way to birth. To protect life, to maintain and support it, to restore it to wholeness, and certainly not to destroy it, is the common principle. To be sure, this principle gets outside support from religious teachings, pagan and Judeo-Christian. But, in our context, one sees that it can be derived from the inner meaning of medicine itself, especially if one remembers that the doctor is nature's cooperative ally and not its master. Is it not self-contradictory for the healing art to be the killing art—even in those cases in which we might all agree that killing is not murder (i.e., that it is legitimate)? Consider, for example, recent statutes in several states that, for reasons of humanity and economy, make injection of poison, rather than electrocution, the method of capital punishment. Put aside, for the moment, your reservations about capital punishment itself, and consider only whether physicians, if asked, ought to agree to administer the injections? Would that action be compatible with the medical vocation? I should say not. (Interestingly, at least for now, both the statutes and the medical profession agree with this conclusion: The statutes say that the injection shall not be made by a physician, and the AMA, I understand, has voted its official opposition to physician participation in such activities.)

Abortions are now performed by many doctors. I am aware that their uniform refusal to do so, were that to occur, would in practice nullify the "freedom to abort" that flowed from the Supreme Court's 1973 decision in *Roe* v. *Wade* (though, in fact, the overturning of legislative *prohibitions* on abortions did not *oblige* doctors to perform them). But the question before us here is not the morality or legality of abortion, but whether physicians who perform them are truly physicians. Set aside your opinions about abortion as such, and consider only whether the destruction of nascent life—life to which obstetricians give prenatal care, and whose needs are increasingly studied with a view to improving such care—is compatible with the inner meaning of medicine. If medicine is a technique neutral regarding the end, then the act of womb-emptying performed by obstetrical technique is equally properly used for birth and feticide. But if medicine is constituted by the task to assist living nature in human bodies to the work of maintenance and function and perpetuation, then one must wince at the monstrous because self-contradictory union that is the obstetrician-abortionist.* (I shall refrain from considering whether proper medical ministering to female humanity is finally compatible with assisting women in abortion, for reasons having to do not with fetal life but with the meaning of womanhood. The question, admittedly complex, is whether in opting for abortion a woman is doing harm or injustice to herself *as a woman* [i.e., by contradicting her generative nature] however much it may serve her wishes or aspirations as a [gender-free] human being.)

Perhaps I am mistaken in deriving these restraints strictly from the meaning of medicine. The end of this paragraph reads: "In purity and holiness I will guard my life and my art." What have purity and especially holiness to do with medicine? Moreover, the syntax might suggest that they are invoked as extrinsic principles, *outside* of "my life and my art," to guide how one lives one's life and how one uses one's art. It is especially this statement that led Ludwig Edelstein to conclude that the Oath is a Pythagorean and not simply a physician's oath, or, in other words, that the ethical teachings of the Oath are brought *to* medicine from some universalist (though sectarian) teaching rather than derived *from* the particular (though universal) nature of healing itself. I shall return to this question later.

The last of the three paragraphs on treatment forswears surgery, even in the face of the most excruciating pain. But it does so not, as in the previous case, because the relief of such suffering violates the meaning of medicine, but because the physician does not specialize in

---

*Though not central to the present argument, my own concern—and, as we shall see, the ultimate concern of the Oath—goes beyond maintaining the "purity" of the medical sect in a world otherwise impure. We must consider whether the insight into the goodness and vulnerability of life that permits a true physician to shun assisting in euthanasia, suicide, and abortion is an insight that ought to inform the entire community. Physicians as *physicians* might properly oppose death by injection or abortion on demand, and not merely their own participation in these practices. However this may be, they certainly ought to be concerned to oppose the apparent "medicalization" of human affairs (e.g., the treatment of crime or unwanted pregnancy on the model of disease) first, because something like "death by injection" will look superficially like medicine and will encourage the view that medicine is essentially but technical expertise; second, because the view of life's problems as "diseases" will increasingly make unreasonable demands on the medical profession; and ultimately, because treating human choices under the image of disease denies freedom and dignity to all human actions, finally also including that of healing the sick. (I am indebted to Ann Dudley Goldblatt and Martin Cook for these remarks.)

surgery. One might wonder whether the physicians of the Oath drew back from the knife because they regarded surgery as a mutilation or at least a violation of bodily wholeness, incompatible with their ministerial function. Surgery, however much it shares the goals of the medicine of diet and drug, is in its nature aggressive, invasive, and assaulting. Even today, we note differences between surgeons and internists in their stance and attitude toward the body. Nevertheless, it is unlikely that the Hippocratic physician forswears surgery as a violation of the medical spirit or, for that matter, of purity or holiness. For if this were his reason, he would not endorse the practice of surgery by those who do it. The physician is, rather, promising not to try himself to do what he cannot do, even in the face of a most severe suffering that might tempt his intervention. He willingly turns those in need over to someone competent in the necessary extramedical skill; the Oath thereby also teaches that we do not simply abandon those we cannot help ourselves. Cynical people may see here the beginning of specialization, trade restriction, fee-splitting, and the like, but the context favors a higher interpretation: Know your limits and let not your wishes to help exceed your competence to do so. Good intentions are dangerous if not coupled to competence. Who could disagree?

Today, it is true, surgery has been brought fully into the medical domain—though in Britain surgeons are still referred to as "mister" while other physicians are called "doctor," and a leading American medical school still distinguishes them in calling itself the College of Physicians and Surgeons. The knife for the stone is now a part of medicine. But today there are other areas and concerns pressing in on physicians, asking them to lend a hand. People oppressed in spirit or transgressing the law, people who beat their children or are beaten in love, people who fail in school or in life all ask the doctor to wield a new kind of knife he either does not have or cannot use with reasonable hope of success. With the proper updating, forswearing the knife still makes sense. It is only partly an excuse to say that there are no others who "are engaged in this work." Acting incompetently from good intentions is often worse than doing nothing, especially in the increasing number of instances in which the offer of pills and other forms of medical assistance for what are fundamentally problems of living fosters the false and enfeebling expectation that life itself has a technical solution.

The Decorum Paragraphs

"Into whatever houses I may enter, I will come for the benefit of the sick." The Oath continues with a reassertion of the end, "the benefit of the sick," this time not to define fitting therapeutic means but to delimit proper interpersonal conduct. Not the sick person's body but his humanity is now center stage, though, please note, the Oath does not need to remind us explicitly that the patient is a human being. The first words of this passage in the Greek text are "into houses" (*eis oikias*). The doctor visits the sick in their houses. The Hippocratic physician is less likely than his modern hospital-based descendant to forget that the sick have households, families, and lives of their own. The doctor enters the intimate life-world of the patient and must act accordingly. Medicine must and can be at home in the home. Today, the sick enter the less than intimate houses of medicine and they must act accordingly, leaving their own world behind.

We are moved to reflect on the meaning of the household, that nest and nursery of humanity—private, intimate, and vulnerable. Though its roots are the needs of bodily life—nurture, protection, reproduction, and then protection and nurture of the young—the household provides for more than the body. A richly woven fabric of nature and convention, it is established by law to nurture our nature. It is sustained by customs that humanize the human animal, engendering love and friendship, speech and education, choice and awareness, and shared beliefs and feelings.

This fragile network is protected by notions of the just and the shameful. Justice in the household is founded on clear demarcations of who belongs to whom, embodied first of all in the strict prohibitions against incest and adultery. Shame protects the intimacy and immediacy of family life—indeed, of all human relations. Intimacy and immediacy require protection against objectification caused by the presence of outsiders, in whose gaze we experience a breakdown of what Erwin Straus calls "protective shame,"[6] in the absence of which we feel a corrosive kind of self-consciousness.* At worst, the presence of strangers in the house threatens to efface altogether the wall between public and private should the strangers spread outside the house what is properly kept within.

Sickness is threatening not only to bodily life. The incapacity of a family member upsets the normal functioning of a household. More subtly, these disruptions force all to recognize the precariousness of the human. All our hopes and aspirations, our ways and achievements,

---

*Please note: It is the safeguarding of this protective shame that is the truest purpose of the so-called right of privacy, though, curiously, almost none of the defenders of privacy speak of this. Indeed, very few of them stop to consider *why* and *for what* privacy is good. We have here one of the many instances in which our dicta are severed from the only sources that can defend and nourish them and show them to be more than arbitrary prejudices. (For a thematic discussion of shame, see Chapter 13, "Looking Good: Nature and Nobility.")

and all our loves are at the mercy of this fragile body. This grim truth, as well as our particular degradations, are confirmed and ratified when the doctor enters a house. The viewing of this outsider makes it impossible for us to avert our eyes from the human meaning of our sickness. Only because the doctor brings hope of remedy do we risk exposing our bodies, our souls, and our instituted humanity, in all their vulnerability, to his objectifying gaze and manner.

The physician is both privileged and burdened by this exposure of vulnerability. He has the opportunity, rarely given to other human beings, to see without illusion the darker side of the human condition, and to see humanity, unprotected and stripped of pretense, struggling gamely to preserve itself. He is privileged to receive the trust implied by admission to the house, to the inner sanctum of the patient's life-world. Yet this trust is also a burden or at least a responsibility that it would be self-contradictory to violate. He is admitted to the private on condition that he keeps private what he sees and hears.

The Oath's specific restraints on deed and speech are now readily intelligible. They recognize the inherently degrading and dehumanizing aspects of illness and the vulnerability and exposedness of the sick person's body, soul, and intimate relationships. Accordingly, they address the danger that the physician might, in deed or speech, take advantage of the vulnerable and exposed, and that he might contribute further to the degradation and shame of those he seeks to lift up.

The absolute prohibition on sexual relations between the physician and his patients (or members of their households) follows directly from these considerations. Sex is proscribed not because of a general opposition to adultery or fornication nor because the physician is an ascetic who forswears bodily pleasures or who thinks sexuality as such is base or impure. The context, the Oath states clearly, is injustice, not wantonness or profligacy.* It is grossly unjust to take sexual advantage of the vulnerability of the sick, exploiting their grave need for one's service to obtain sexual favors.

But sexual relations between physician and patient are unjust in a much deeper sense. To study the illness of individuals, the physician must necessarily abstract, often greatly, not only from their individuality but even from their very humanity. Though the doctor knows better, he must often do things that imply, unjustly, that the human being is just body. Justice to our humanity, not to speak of the patient's self-respect, requires self-conscious and clear limits on this re-

---

*Though the Oath speaks of "*all*" voluntary [or "willing"; the Greek word, *ekousia*, does not imply deliberate choice] *injustice*," the emphasis is given to sexual matters. The word translated "mischief" has the connotation of seduction. It seems that it is in matters of sexuality, rather than money or property, that the physician *as physician* is most prone to injustice.

ductive behavior. Thus, even if patients are freely willing, it is profoundly unjust for a physician to treat sexually the bodies of those who, necessarily, must expose themselves to his gaze and touch. Both patient and physician tacitly understand and need to uphold the difference between medical and erotic gazing, and between palpation and caressing. Already reduced by illness, the patient whose nakedness is uncovered is medically reduced to body. The Oath speaks precisely to this matter of reduced condition: No sexual deeds "upon *bodies* of female and male persons, be they free or slave."*

Preserving the dignity of the patient requires not making his or her body an object of one's lust. Preserving the physician's view of the dignity of the patient requires also that he not succumb to the patient's sexual blandishments. Do these insights deserve the dustbin of history? Does not the Oath teach us today exactly what to think of the increasingly common practice of sexual relations between physician and patient, sometimes justified as part of the treatment for frigidity?

Just as it would be unjust to take advantage of the patient's exposed body, so it would be shameful for a physician to expose outside of the home—or consulting room—what he has seen or heard as a result of being invited into the private world of his patients. He will, no doubt, learn some embarrassing or downright disgraceful facts, which might, if publicized, be damaging to the reputation, honor, or self-esteem of the patient or family. But the promise of silence protects more than the skeletons in the closet. It insists on protecting the intimacy of the private against the intrusion of the stranger, albeit the invited stranger. The physician who is by necessity bidden to the inner circle

---

*A question might be raised whether the standards of justice that restrain the physician's deeds in this paragraph are strictly medical (i.e., accessible to the physician on the basis of his insights into the well-working wholeness of the human body) or whether they are political or social in origin. Indeed, I have already indicated that violations of a household cannot be simply violations of "natural" justice, given that the household is in part a conventional institution—even if in the service of our natural perfections as human beings. Still, I think a universal and transcultural standard of justice is implied here, witnessed first by the Oath's radical indifference to the *conventional* distinction between free and slave, and its insistence that the doctor, unlike the master, may not treat a slave as a mere body. Moreover, though different cultures define us differently, all acknowledge what the true physician knows immediately from his study of man: that man is more than his body, witness that he covers it up. Though different cultures dress us differently, all presuppose that man is by nature that animal which lives by custom. To overlook custom is to overlook the peculiarly human soul; to overlook the soul is to reduce man to body. The physician, even as he ministers to bodily wholeness, knows that the functioning wholeness of the human body goes well beyond the well-working of the physiological processes that make life possible. In short, it seems that the standards governing the physician's conduct in paragraphs six and seven flow from a physicianly insight into human nature, however much such insights are also accessible to nonphysicians or appropriated by the political community. One begins to see here how an ethic of medicine, rightly understood, could come to be the basis of ethics simply.

is thus brought under the constraint of protective shame that should govern the speech of family members in preserving all details of intimacy from public scrutiny. Publicity is the enemy of all intimacy, and the doctor, both as outsider and as objective observer, stands for publicity and objectification. He must move in and out with great discretion, and when outside must act as if he had never been inside. What is seen or heard inside is on no account to be divulged outside: "I will keep [them] to myself holding such things unutterable."

The decorum paragraphs not only address deeds and speeches. In speaking directly to the unjust and the disgraceful or shameful, the Oath implicitly articulates a notion of the just and the noble, of the right and the seemly, of the fair as equitable and the fair as fine. Those who understand ethics primarily in terms of character (the very word "ethics" comes from the Greek word "*ēthos,*" character), rather than in terms of rules of right and wrong, or of rights and duties, or of benefits and harms, have frequently divided the virtues or excellences of character into the noble and the just: the noble or the seemly or the fine, how a man displays on the surface in action, as beautiful or ugly, the inner state of his soul; the just, how a man does the fitting or owed deed to his fellow citizens. The Hippocratic Oath, while seemingly only a list of dos and don'ts, in fact reflects and embodies a deeper view about character and suggests, by implication, the virtues requisite for medicine: in addition to justice, also moderation and self-restraint, gravity, generosity, discretion, prudence, reverence. Can the same be said about the AMA's Principles of Ethics? Are its *feelings* of "compassion and respect for human dignity" (Principle I) an adequate substitute for those excellences of character? A further question: Where, according to the AMA's Principles, are doctors to learn these substitutes for virtue? How does the Oath answer this question?

Relations to Teachers and Students

The longest paragraph of the Oath, and the very first part of the covenant that is sworn, concerns teaching and learning, and, more importantly, relations to teachers and students. It goes far beyond the AMA's principle on learning: "A physician shall continue to study, apply and advance scientific knowledge, [and] make relevant information available to patients, colleagues, and the public. . . ." In the Oath, teaching and learning are placed in a human setting, with obligations and responsibilities to specific teachers and students, and not just to knowledge or to a generalized public.

Today, most readers of the Oath cannot understand why obligations to teachers of the art should be mentioned before obligations to patients. Changes in medical education are partly responsible; no one in this country learns medicine as an apprentice to a master physician.

Everyone now goes to school. Within these large institutions, which increasingly emphasize the sciences of medicine and aspire to the routinization of the art through more universalizable procedures of diagnosis and therapy, few of us can speak of individual teachers who taught us the art or whose understanding of the meaning of medicine we are following. For these reasons, this part of the Oath seems irrelevant.

The complaints against this paragraph sometimes go farther than the charge of anachronism. According to some, this provision is elitist and cultist, etiquette not ethics. It places duties and loyalties to fellow members before duties and loyalties to outsiders. It sets conditions on who is admitted to the guild. It in no way considers the wishes or opinions of outsiders. These critics assert that the Oath is trade-unionism at best, "a classical mystery-religion cult-initiation oath" at worst, with the "classical paternalistic, philanthropic obligations to benefit patients" not enunciated as "a promise made to patients."[7] Such critics have read neither carefully nor thoughtfully. They have given too little thought to how one learns proper conduct toward patients, and to the ultimate sources of character and morality. Let me illustrate.

The physician pledges to share the life of his teacher, to meet his needs if he is unable, to teach his teacher's sons without fee or contract (again, presumably, only if he were unable to do so himself). More generally, he covenants to hand down his knowledge to his own, his teacher's, and his, as it were, adopted sons*—in short, to all those who want to learn the art, and not just its technique, to those who will take the Oath and sign the covenant according to the medical *nomos* (law, custom). (We note, in passing, that the charge of nepotism and elitism is false: Anyone who will play by the rules is free to enter, and the physician pledges to teach him. A far cry from present medical school admissions!) All of these obligations flow from the beginning promise: "To hold the one who has taught me this art as equal to my parents." My parents gave me my life, my teacher gave me my art. Or again, my parents gave me life, my teacher a way of life, itself dedicated to upholding the life of all. It is no disrespect to one's parents to regard one's teacher in the art of medicine as their equal.

To see medicine in the context of fathers and sons, and one's art as analogous to one's life, permits many derivative insights. While esteeming both his life and his art, a physician who so reveres his profession is unlikely to take our embodiment as evidence that man

---

*Because I have undertaken to defend the wisdom of every part of the Oath, I should say here that I cannot defend the position, probably implied by the present passage, that medicine is a profession reserved for males, a suggestion somewhat (but only somewhat) undercut by the fact that two healer-*goddesses* are invoked at the beginning of the Oath. For myself, I have long thought, and even publicly stated, that medicine would be better practiced if more of its practitioners were women.

is finally intelligible as body. He will necessarily view both his patients and himself as part of some larger whole. Such a physician will understand that he is not a self-made or self-sufficient man, and that a belief in his own autonomy and independence is mistaken. He will appreciate that he owes both his life and his work to those who came before, that the art of medicine, like the rest of civilization, is a monument to the ancestors. By remembering his teacher and looking to his students, he will be kept aware of *his own* mortality, not only as a human being, but also as a practitioner of the art, while at the same time being called to remember the potential immortality of the art itself. Participating in this transcendent good while aware of the perishability of his own embodiment of it, the physician, like the parent or the teacher, discovers the importance of transmission and his duties to posterity. Through his concern for his own immediate descendants, he learns concern for posterity in general, and therewith a concern for unrelated others. Through his covenant to uphold his teacher and to teach the young, he sustains the art in community, for only through community can the art be transmitted.

The physician stands in the world not as one who claims his rights or demands his due; rather he stands gratefully, thankful for the existence of the art of medicine, for the devotion of his teacher, for the community of like-minded healers, and for the privilege of sharing in this noble work. From gratitude for the gift of the art can flow more readily the disposition to give to others, in the first instance to those near and dear, but soon also in an overflow to others. The disposition of philanthropy can come only to those who know the love of few. Just as the family, when it does its proper work, is the nursery of these ethical sentiments, opinions, and dispositions, so the family ties of the art inculcate what the technique alone cannot: that philanthropic and virtuous disposition which makes one eager to serve those in need for the sheer goodness of doing so. One cannot attach the man to the best precepts of the mind except by first winning his heart. Paragraph two of the Oath is thus indispensable to the meaning of medicine, and it belongs just where it is in the Oath, at the head of one's list of self-proclaimed duties, just as the relations it describes are the source of one's becoming dutiful.

This line of reasoning could also be applied to the order of the Ten Commandments. The so-called second table—with its proscription of murder, adultery, theft, bearing false witness, and coveting of what is thy neighbor's—is immediately preceded by the commandment to "Honor thy Father and Mother." This singular positive commandment of filial obligation stands ahead of one's duties to refrain from injuring one's fellow men. Does not this order of commandments— and, indeed, much of the rest of the biblical narrative—suggest that in filial piety are the seeds or source of human decency?

The seeds indeed, but finally not *the* source: The true source lies beyond. The commandment about parents is only number five. It is preceded by the commandment about the Sabbath, God's first sanctification, and by the three commandments about God Himself. In the list of the Commandments, filial piety and duty mediate between and connect duties to God and the created order and duties to man. In practice, filial piety is that conduit through which the source and content of what is good is transmitted down through the generations and distributed across to the rest of humanity. Concern with the *immediate* source of goodness leads back to the *ultimate* source of goodness, to the divine.

The Hippocratic Oath also begins with the divine. Not one's teacher but the gods have first mention. Faithfulness at least to the text obliges us to take seriously its beginnings.

Medicine and the Divine

The portions of the Oath so far discussed constitute a covenant, a binding promise, sanctioned by an oath sworn in the first paragraph. As in other such covenants under oath, appeal is made to a deity or deities, invoked not only to witness the making of the promise but also to observe subsequent conduct and to reward compliance and punish violation, according to the terms stipulated or implied in the swearing. In this case, at the end, the physician asks that a happy life and lasting honor and fame be granted him should he fulfill his oath, and that unhappiness, dishonor, and infamy be his lot should he violate and transgress it. The physician does not pray to God for favors or for charity or mercy. He asks instead for justice; he poses and exposes himself before the powers that be, hoping or wishing that he will receive from life and the world exactly what he deserves.

It is true, as some critics complain, that this covenant is unilateral; a person can bind no one but himself by oath. Moreover, it is not a promise made *to* patients. But it is not for these reasons either weak or self-serving. On the contrary, it obliges always and unconditionally the physician's full performance, regardless of the behavior of the patient or his ability to pay. Thus, the physician is bound much more stringently than he would be by an explicit contract with the patient, notwithstanding the impossibility of suit for nonperformance. Divine retribution, called for freely out of one's own mouth, makes a suit for breach of contract pale by comparison. Or so it would for those who believe in the gods by whom they swear.

To be sure, swearing an oath is no guarantee against perjury. In our courts we permit the cross-examination of witnesses under oath, and do not thereby impugn their piety. Nor do people always fulfill what they have in all sincerity sworn. The possibilities of swearing vainly and swearing falsely are as old as oath-taking. The further problems of invoking false gods or of falsely invoking the name of God add to the reasons why the taking of the Hippocratic Oath has become unpopular in our day: prospective surgeons or abortionists cannot honestly swear it; physicians who live under laws that will compel them sometimes to violate confidentiality may not be able to keep their promise; Jews and Christians will not swear by Apollo; and atheists will not swear at all, at least not solemnly.

Yet despite these unavoidable difficulties, we risk ignorance of medicine and ethics, and hence of our *own* self-understanding, if we adopt a condescending, outsider's attitude toward the oath of the Oath, and treat it merely as an expression of the Greek—or of some Greek—mind. We must try to see the Oath from the inside, yes, for the reason that such a view might perhaps be truer than our own. Unless we can understand it and see its power, we reject it in ignorance.

Four divinities are named by the Oath, but *all* the gods and goddesses are called to witness: The divine in its entirety is invoked. The physician swears by *all* that is sacred. His is the most powerful and solemn of swearings. The grave seriousness of the physician's business, and his considerable powers touching life and dignity—evident in our explication of the covenant—fully justify this gravity of swearing. From this, I trust, no one will dissent.

But the Oath suggests a more intimate connection between medicine and the divine. Apollo is invoked as "Apollo Physician," an unusual epithet for Apollo. Next comes Asclepius, the father of medicine, invoked as a god or at least a semidivinity. Asclepius was the offspring of Apollo—the god of light, truth, prophecy—and the nymph Coronis. Homer speaks of Asclepius as a skillful physician, but in later times he was worshipped as a god. He fittingly has two daughters, appropriately named Hygieia ("health," "living well") and Panaceia ("all heal"), the goddesses, respectively, of health maintenance and of curing illness, or, as we might say, of prevention and treatment. The divine father, his mortal son who became a god, and the two granddaughters who embody and carry on their father's work among human beings: We are shown a divine lineage and transmission, which not only serve as a model for the human perpetuation of the art, but teach us about the superhuman source of the human art of healing.*

*As the subsequent argument will suggest, medicine is, finally, not an autonomous profession because it depends on a larger whole: The powers of the living body and its standard of wholeness, the moderate hospitality of the natural order to human needs, and the remarkable givens (gifts) of intelligence, intelligibility, and insight necessary for the discoveries and practices of medicine all show the subordination of the art of healing to the natural whole. Its subordination also to the political whole, mentioned briefly in the introduction to this chapter, is an image of this larger relation.

How could anyone think that the source of medicine is other than human, let alone superhuman? I offer not a full argument but some questions and suggestions for your serious consideration. To begin with, who heals disease? Do doctors themselves make people whole, or do they mainly assist nature working within? To be sure, physicians increasingly know how to help, so much so that many are prone to forget or deny their subordinate and ministerial role. But any doctor who is honest with himself knows the limits of his powers, not only in cases where therapy fails but also in cases where we say *"it"* succeeded. The treatment may be necessary, but it is not therefore itself the sufficient or even major cause of healing.

In some cases the particular treatment is not even necessary; more precisely, not its pharmacological properties but the mere fact of its administration by a physician suffices to induce the cure, inert solutions "producing" the same benefits. This well-documented yet mysterious "placebo effect," which probably operates widely in medicine, reminds us of the complex and often uncanny interactions between *psychē* and *soma* (soul and body), so crucial for health and disease. Every doctor knows the therapeutic importance of "the will to live," and, conversely, the effects on health of grief and depression—for example, serious illnesses and death strike persons recently widowed or bereaved much more frequently than expected from their age and state of health. These "causes" are not under the control of the victims themselves, still less of their physicians. And even in its ordinary course, medicine does not know and cannot predict with assurance why some patients get better and some do not, some slowly and some quickly, some completely and some only partially. Perhaps these are matters that a more sophisticated science may in time explain. But they may *always* remain something of a mystery, and, in any case, not completely under human control.

However this may be, what are we to think of this fact of self-healing? Is it not simply wondrous, indeed awe-inspiring, that the living body contains an immanent power toward wholeness that acts on its own? Aren't we kidding ourselves when we think we can explain, and thereby explain away, that power in terms of the albeit indispensable chemical reactions it organizes? Further, is it necessary that the world be the kind of place in which living things are also self-healing things? If plants and animals, not to mention men, are truly the mere machines we claim they are, how come they are machines that can make and repair themselves?

Wonder and awe are also appropriate when we think about our environment, about nature without. How come the world is full of plants that serve our needs? It is only partly correct to say that if the world couldn't feed us, we would not have survived. The contingency of the beginnings and of the beginnings of perishable life still go unexplained: If persistence or survival is nature's sole concern, then the very existence of life is altogether implausible, inorganic matter being better suited to survive. But that aside, does the theory of natural selection satisfy our wonder at the fact that in the leaf of the foxglove is the cure for congestive heart failure or in the flower of the poppy the relief of pain? Is it *man's* doing that the world is full of all kinds of hidden healing aids, lying around to be discovered? The discoverer of a new drug, of course, deserves our praise. But what is owed for the fact that it was discover*able,* that is, both intelligible and in the world, though hidden, accessible to being discovered?

And finally, what about discovery itself? What are we to say of that special power of mind, unique to human beings, that is capable of discovery, that can apprehend health and disease and find supports for the former and cures for the latter? Do we know, can we know, what is responsible for insight, for that unique ability to be illuminated by the truths of the world—to be sure, only after we have worked hard to be able to see them? Hard work, scientific method, experimentation—of course. But insight? Intuition? Utterly mysterious. One can do much worse than to credit some higher-than-human power, not at our disposal or under our control, that strikes from afar, that, as it were, "flies in" with light and truth, that makes possible that correlation between receptive mind and intelligible object. One can do much worse than to credit Apollo who strikes from afar, himself a mysterious source of the dispelling of mystery, the source of clarity and form, harmony and order, light and truth. One can do much worse than to credit Apollo, source of purity and holiness no less than of light and truth, perhaps because with the light to see the truth, and the truth about the whole, Apollo—or whatever its name is—brings wholeness to mind and us to wholeness, and further, in the art of medicine, brings us beyond *awareness* of wholeness to a divinelike overflowing into action, permitting us to help make the wounded whole. Asclepius, or the true physician, directed by the Apollonian gift of insight into wholeness, brings human beings toward the whole. The doctor, divinely inspired healer of men, serves also the enduring whole by helping to heal the part. The insight that drew the holy, the healthy, and the whole from a common etymological root may point to the deepest wisdom, not only for medicine but perhaps also for how we are to live.

PART III

Deepening
the Ground
Nature Reconsidered

CHAPTER TEN

Teleology, Darwinism, and the Place of Man
Beyond Chance and Necessity?

[T]he race of birds was created out of innocent light-minded men who, although their minds were directed toward heaven, imagined in their simplicity, that the clearest demonstration of the things above was to be obtained by sight; these were remodeled and transformed into birds, and they grew feathers instead of hair. The race of wild pedestrian animals, again, came from those who had no philosophy in any of their thoughts, and never considered at all about the nature of the heavens, because they had ceased to use the courses of the head, but followed the guidance of those parts of the soul which are in the breast. In consequence of these habits of theirs they had their front legs and their heads resting upon the earth to which they were drawn by natural affinity, and the crowns of their heads were elongated and of all sorts of shapes, into which the courses of the soul were crushed by reason of disuse. And this was the reason why they were created quadrupeds and polypods. God gave the more senseless of them the more support that they might be more attracted to the earth. And the most foolish of them, who trail their bodies entirely upon the ground and have no longer any need of feet, he made without feet to crawl upon the earth. The fourth class were the inhabitants of the water; these were made out of the most entirely senseless and ignorant of all, whom the transformers did not think any longer worthy of pure respiration, because they possessed a soul which was made impure by all sorts of transgression, and instead of the subtle and pure medium of air, they gave them the deep and muddy sea to be their element of respiration. And hence arose the race of fishes and oysters, and other aquatic animals, which have received the most remote habitations as a punishment of their outlandish ignorance. These are the laws by which all animals pass into one another, now, as in the beginning, changing as they lose or gain wisdom and folly.

*"Timaeus," in Plato's Timaeus (91e–92c)

The wish to preserve oneself is the symptom of a condition of distress, of a limitation of the really fundamental instinct of life which aims at the expansion of power and, wishing for that, frequently risks and

249

*even sacrifices self-preservation. . . . [I]n nature it is not conditions of
distress that are* dominant *but overflow and squandering, even to the
point of absurdity. The struggle for existence is only an* exception, *a
temporary restriction of the will to life.*

—Nietzsche, *The Gay Science* (aph. 349)

Under the reigning orthodoxy that is modern natural science, to think
teleological thoughts about nature is to be guilty of heresy. We are
forbidden to give explanations in terms of final causes (ends, purposes,
or goals), to ask about the "why" or "what for" rather than the "how"
of things, and even, it seems, to wonder at or just to notice the seemingly purposive activities of all living beings. Things were not always
this way. In fact, modern science seized power precisely by overthrow-
ing an equally orthodox *teleological* natural science, whose origins
could somehow be traced back to Aristotle but whose character had
been altered and deformed almost beyond recognition by its connection with biblical religion and its solidification into Scholasticism. According to the founders of modern science, human beings may have
purposes or ends or goals, but nature and other natural beings do not.
Moreover, as Francis Bacon powerfully argued (in attacking the "Idols
of the Tribe"), the purposes of human beings regarding nature would
continue to be thwarted until we extirpate or neutralize the natural
human inclination for regarding nature teleologically:

> [F]or although the most general principles in nature ought to be held
> merely positive, as they are discovered, and cannot with truth be referred
> to a cause, nevertheless the human understanding being unable to rest
> still seeks something prior in the order of nature. And then it is that in
> struggling toward that which is further off it falls back upon that which
> is nearer at hand, namely, on *final causes, which have relation clearly to
> the nature of man rather than to the nature of the universe;* and from this
> source have strangely defiled philosophy.[1]

Rejecting, from the outset, all explanations in terms of ends or
purposes, and also notions of the causal status and dignity of *form*,
modern science began on its mechanistic and materialistic journey—
and, one must add, with astonishing success. Indeed, its very successes in explaining how nature works and in thereby providing power
to alter her workings have certainly justified—at least until now—science's antiteleological bent. True, the mechanistic account of nature,
including all living nature, left man rather isolated and estranged from
the rest, insofar as he was regarded the sole repository of truly purposive activity. True, some insisted that there could be no such spiritual island within the kingdom of corporeal nature, and sought—and

today still seek—to capture human nature wholly for mechanism. But proud of being able to wrest knowledge and power from nature, and for the time being secure in the dignity of their special place in the world ("the image of God"), men gathered all the more tightly to themselves the dignity of purpose and aspiration that they denied even to their closest animal neighbors.

These happy days may now be over. Two reasons—one practical, one philosophical—seem paramount. On the practical side, we have grown newly suspicious of the project for mastery of nature. So long as the blessings of technology were largely unmixed (or seen to be so), no one bothered much about the adequacy of the underlying theory. Its utility—and hence goodness for human life—was sufficient to attest to its truth. But now that technology has become a question, the science that supports it is increasingly subject to scrutiny and criticism, reaching in due course to its fundamental assumptions.

The more important reason is philosophical, and comes from an unlikely quarter: the theory of evolution. Man's evolutionary origin has made untenable the sharp divorce of man from the rest of nature. For man is, at least in origin, akin to nature. This discovery can, of course, lead one in opposite directions. Either man, too—his felt sense of inwardness, freedom, mind, consciousness, and purposiveness notwithstanding—is to be assimilated to the blind and dumb world of mechanism, or nature must again be seen more in the light of what common sense has always taken to be naturally human. The former path is being followed by, among others, the behaviorists and the neurophysiologists of mind (freely, purposefully, and with devotions to truth and the rightness of their theories that constitute a permanent embarrassment to—not to say a refutation of—these doctrines; on their own principles for explaining nature, their activity makes absolutely no sense at all). The latter path, which though less well traveled I find frankly more plausible, has been blazed by, among others, Hans Jonas:

> The *continuity* of descent now established between man and the animal world made it impossible any longer to regard his mind, and mental phenomena as such, as the abrupt ingression of an ontologically foreign principle at just this point of the total flow. With the last citadel of dualism there also fell the isolation of man, and his own evidence became available again for the interpretation of that to which he belongs. For if it was no longer possible to regard his mind as discontinuous with prehuman biological history, then by the same token no excuse was left for denying mind, in proportionate degrees, to the closer or remoter ancestral forms, and hence to any level of animality: common-sense evidence was reinstated through the sophistication of theory—against its own spirit, to be sure.... In the hue and cry over the indignity done to man's metaphysical status in the doctrine of his animal descent, it was overlooked that

by the same token some dignity had been restored to the realm of life as a whole. If man was the relative of animals, then animals were the relatives of man and in degrees bearers of that inwardness of which man, the most advanced of their kin, is conscious in himself.[2]

Where might this leave us regarding teleology? The matter is unclear. For even granted that the theory of evolution reconnects man and the animals, the Darwinian explanation of the primary *mechanism* of evolution, with its stress on chance and necessity, would seem to offer no support to teleologists and others seeking to find meaning in nature's ways or man's place in nature. And yet, like so many other aspects of the meaning of Darwinism, this, too, has been a controverted question.

On the one side, John Dewey praised Darwin for having done away with the need to think about nature in terms of final causes. Dewey claimed that, thanks to Darwin's discovery,

> philosophy forswears inquiry after absolute origins and absolute finalities in order to explore specific values and the specific conditions that generate them.[3]

Hans Jonas, reporting on the consequences of Darwinism for philosophy, appears to agree:

> It was the Darwinian theory of evolution, with its combination of chance variation and natural selection, which completed the extrusion of teleology from nature. Having become redundant even in the story of life, purpose retired wholly into subjectivity.[4]

Yet, on the other side, in 1874, Darwin's friend and colleague, Asa Gray, the great American botanist and evolutionist, wrote:

> Let us recognize Darwin's great service to Natural Science in bringing back to it teleology; so that, instead of Morphology *versus* Teleology, we shall have Morphology wedded to Teleology.[5]

Darwin replied to Gray:

> What you say about Teleology pleases me especially, and I do not think any one else has ever noticed the point.[6]

Who is right, Dewey and Jonas, or Gray and Darwin? The answer, of course, is that both are right. Should anyone protest that both *cannot* be right, we shall have to concede that he, too, is right. It seems we must ourselves take up the question afresh, beginning with what is first for us—on the one hand, living plants and animals; on the other, the confused term "teleology."

Organisms as Purposive Beings:
Elements of Teleological Analysis

Do plants and animals have purposes? Do they come to be and do they function "for the sake of something"? To say yes makes one a teleologist. But the genus "teleologist" has many species, and some are more fit than others.

I begin with several cruder kinds of teleology. While these will ultimately be of little interest for us, I mention them for two reasons: First, these are among the simple-minded teleological notions that have brought a bad name to any and all teleological thinking. Indeed, these are, for many people, the only teleological notions they know. Second, these notions were dominant in teleological thought prior to Darwin, and, as we shall see, played a major role in the genesis of his theory.

Many of these cruder views take their bearings from the fact that plants and animals live not only by and for themselves, but also and always in relation to other plants and animals. One animal's substance is another animal's food: Men are food for worms, worms are food for ducks, ducks are food for men. Mutual service, rather than predation, is sometimes the dominant relation: Bumblebees are the indispensable fertilizers of clover, and the nectar of clover becomes food for the bees. Plants fix and use carbon dioxide released by animal respiration and, in turn, give off oxygen consumed by animals. Also, organisms are structured as if designed for some of these external relations; the beak of a cardinal is a perfect nutcracker for sunflower seeds, the beak of a woodpecker a perfect drill to reach insect treasures stored under bark.

These very evident uses of one organism and its productions by another, and these very evident elements of structure that permit such exploitation, gave rise to the view that all of nature is one grand system, an organic whole, in which and for which every living thing plays its proper and necessary part. Each animal and plant has a worldly purpose, outside of itself, relative to and for others. In a common anthropocentric version of this external—or relative—teleology, all of nature is seen as being for the use and benefit of man. These views are usually grounded on a view of nature as *designed*, or *created*, by an intelligent designer or creator, who, in most such opinions, is also thought of as perfect and beneficent in his making. Sometimes it is even implied that a conscious and intelligent purposive being is *always* actively at work in the world, suitably arranging and ordering all these means and ends. Extrapolating from the peculiarly human purposive activities (e.g., intending, planning, choosing, making),

holders of these views recognize and attribute purpose and purposiveness only where there is active intelligence at work: if purpose,
then mind; or, no mind, no purpose.

But these notions of external teleology, and of design and conscious purposive planning, are all secondary notions. The primary
home of teleological thought is the internal and immanent purposiveness of individual organisms, in their generation, their structure, their
activities. It was almost certainly his viewing the activities of living
things in and for themselves that called forth Aristotle's view of nature as directed toward ends, and led him to consider the "that for the
sake of which" as an indispensable part of an understanding of natural
phenomena. Let us review the evidence.

First, consider the *generation of living things.* Each organism
comes to be not at random, but in an orderly manner, starting from
some relatively undifferentiated but nevertheless specific seed or zygote produced by parents of the same species, and developing, unfolding, and informing itself from within in successive stages that tend
toward and reach a limit, itself, the fully formed organism. There is a
natural end to the process of development that defines the previous
motion, through its various stages, *as* a development, and the *specific*
character of the end determines whether it is the development of a
dolphin or a daffodil. Each stage depends on the preceding one out of
which it emerges, and prepares for the succeeding one into which it
passes. The emergence of the differentiating parts of the whole is coordinated, each part being related always to every other part but also,
prospectively, to the mature form of itself and of the whole. The adult
that emerges from the process of self-development and growth is no
mere outcome, but a completion, an end, a whole. To use Aristotle's
terminology, living beings come to be by a process that has a natural
end or *telos;* the *telos* of coming to be is the mature *eidos* (i.e., species
or "looks") of the particular organism in question. To use a more up-
to-date formulation—which adds only to the *how* and not to the *what*
of generation—a chicken comes to be as the result of numerous biochemical reactions, coded for, specified, regulated, directed, and ordered by the blueprint for making a chicken, which is "in" the chicken
DNA contained in the nucleus of a fertilized chicken egg.*

Second, consider any *fully formed mature organism or adult.*

*This is the crucial sense, usually overlooked by molecular biologists, of the common
claim that DNA is the bearer of *in-form-ation.* Though the genetic code and the machinery for its translation are, for the most part, uniform and universal, the ordered
expression of genes and the overall organization of the genetic messages in each organism bears the unmistakable stamp of the species-specific pattern or "form." Chicken
DNA and human DNA orchestrate different melodies using the very same notes and
instruments.

1. It is an organic whole, an articulated whole, composed of parts. It is a structure, not a heap, and to be a structure means to have a function; to understand a structure is to understand *what it is* in terms of *what it does*. And function, I need not remind you, is a kind of teleological notion; function is not a material or a mechanism, nor does it have extension; rather, it is the end of the extended material structure and its mechanism. (We note, in passing, that machines, too, are structures, not heaps; thus, one cannot *fully* understand even a machine mechanistically [i.e., by knowing only the how of its workings]. To understand a machine *as* a machine is to understand its work, to know it as a functioning whole. The Lilliputians might have been taught to take apart and reassemble Gulliver's watch, and might have learned everything about the workings of its gears and springs, without ever learning that it was an instrument for telling time rather than the God he worshipped.) Usually, the search for a mechanism begins only *after* one has identified the function, the activity, the work—usually one asks "*how* does it work" only after one knows the "what" and the "what for" of the "it."

2. The parts of an organism have specific functions, which define their nature as parts: the bone marrow for making red blood cells; the lungs for exchange of oxygen and carbon dioxide; the heart for pumping the blood. Even at a biochemical level, every molecule can be characterized in terms of its function: hemoglobin is a protein that binds and carries oxygen; hexokinase is a protein that phosphorylates glucose; DNA polymerase is a protein that copies the existing DNA molecules, faithfully reproducing the sequence of the bases so that the encoded messages are not scrambled when the cells multiply and divide. Organisms also have enzymes that produce genetic novelty or mutations; even the production of so-called accidental variations is to some extent an organic (and purposive) function. The parts, both macroscopic and microscopic, contribute to the maintenance and functioning of the other parts, and make possible the maintenance and functioning of the whole.

3. A mature organism shows itself as a whole, maintains itself as a whole, and functions as a whole in characteristic ways *above and beyond merely maintaining itself.* The mockingbird delights in its own imitative sounds, the adult cardinal hops from ground to low branch and squawks for the young bird to follow suit, a coyote howls at the moon, the otter turns identical underwater somersault after somersault for hours on end, the beaver cuts down trees, the dog sniffs the ground for traces of his pals, the young deer engages in ceremonial duels, the lizard sunbathes on a rock, the penguin struts and parades, the peacock shows off its plumage. These looks and ways serve to de-

fine the animal; its activities, taken together, are most of all what the animal *is* and what the animal *is for.*

4. In addition to the self-producing, self-organizing, self-maintaining, self-preserving, and self-fulfilling characteristics of organisms, there is that most remarkable power of self-healing. A plant cutting will regrow the missing roots, each half of a divided planarian will regenerate the missing half, and more generally, nearly all living things heal wounds or breaks and tend to restore wholeness. This tendency toward wholeness also encompasses, in some cases, the rejection of additions to the whole, as in the immune system of higher animals, which combats the entry of alien elements, whether infectious agents, tumors, or grafted tissue.

Third, living things display a *directedness,* an inner "striving" toward a goal, both in their coming to be and in many of their activities.[7] These activities point forward in time, as what is here and now prepares itself (often) for the over there, (always) for the not yet. A seedling sprouting beneath a large rock will bend and grow around the rock to reach the light. A young bird will continue to struggle to coordinate wing and tail motions until it finally learns to fly. A beaver will make many trips to build a dam, or a bird a nest, or a spider a web.* And for many animals there is an elaborate pattern of behavior leading up to mating. *In none of these cases is the activity planned or conscious or intended, yet it is just the same a directed and inwardly determined activity to an end for a purpose.*

Fourth, many organisms function not only as individuals but also as *members of some larger group* toward ends that reside with the group: The specialized contributions of the various kinds of honeybees to the work of the hive is only the best-known example. Even the so-called unsocial or solitary animals are social animals in respect of reproduction. Much of their appearance and activity is for the sake of attracting a mate, and this for the sake of reproducing their kind. Among the truly social mammals (deer, wolves, primates), the goals of social life may go beyond preservation and reproduction, to include expression and communication, play and display—regarded to some

---

*The products themselves—dam, nest, web—are also evidence of natural purposiveness. They clearly serve a useful function for the animal (therefore, they are "for the sake of something"), and they are of entirely natural origins. So artful, well-ordered, and obviously useful are the products of some of these animal artists that people once suspected that it was by intellect that they did their work. Yet, as Aristotle already pointed out, "it is absurd to suppose that something does not come-to-be for the sake of something unless they see the moving thing deliberating. . . . This is most clear whenever someone heals himself; for nature is like that."[8] Nature (i.e., their specific nature) is this unreasoning yet reasonable cause that leads animals to produce things useful to themselves.

extent as ends in themselves (see Chapter 13, "Looking Good: Nature and Nobility").

Before proceeding further, let us take stock of the ingredients of the notion of internal teleology developed to this point: (1) orderly coming to be, reaching an internally determined end or completion; (2) an organic and active whole, a unity of structure and function, with parts contributing to the maintenance and functioning of the whole; (3) directedness, activity pointing toward a goal; (4) serving and preserving one's kind. Implied in all of these notions is a fifth element, the notion of "good" or "well" or "fit" or "successful" or "perfect." Was the animal perfectly formed? Did the part function well? Was the end successfully reached? Was the organism fit for reproduction (i.e., fertile)? When considering the functioning of an animal, we ask not only whether it functions in a characteristic way, but whether it functions well or normally as such and such an animal. For example, we regard a child born blind or with a congenital heart deformity as imperfect, not fully whole, not completely formed. These notions of "good" and "well" are, in a way, implied by the notion of "complete," "whole," and "functioning"; on the other hand, it may be useful to distinguish "having a function" and "performing it well." This holds for machines, no less than for organisms, indeed for anything that works toward an end. *The end is a standard as well as a goal.* Teleological analysis will be concerned both to identify the end and to evaluate how well or badly it is achieved.

The questions of completeness and of the well- or ill-functioning of a particular animal relative to a standard for that kind of animal invite a further question about completeness and well-functioning simply. If there is a completeness or fitness or excellence of moss, of barnacles, of grasshoppers, and of chimpanzees, is there no completeness or fitness or excellence overall? I shall return to this question toward the end of the chapter (pages 270–275).

What I have presented in this section are some of the rather obvious reasons why we must regard living things as purposive beings, as beings that cannot even be looked at, much less properly described or fully understood, without teleological notions. I am well aware that many scientists believe this view is vitalistic—which it certainly is not—or unscientific, and that some philosophers of science claim that all teleological descriptions can be reformulated in nonteleological terms without loss of meaning. I know that, drunk on the truly amazing successes of modern biology—including the discoveries of Mendelian genetics; the chemical structure of the gene; the detailed biochemical mechanisms of gene replication, transcription, and translation; the achieved synthesis of proteins, genes, and even a self-replicating virus—many biologists believe that physics and chemistry are adequate to the full explanation of living things, or, in other words, that the complex phenomena of living things—including their *apparent* purposiveness, and including awareness, memory, desire, and thought—are in principle fully explicable in terms of molecular structures and biochemical reactions. Let me, for now, appeal only to the phenomena and simply assert that they are all mistaken.* In the study of organisms, teleological notions are so far from being undermined by or reducible to mechanical and materialist notions as to be the indispensable principles both of formulating a mechanistic inquiry and of rendering its findings intelligible. As we shall see, it is *history* and *not* mechanism that presents the real challenge to teleology.

The Explicit Teaching of Darwinism: Nature Has No End

The more challenging and problematic question about plants and animals is whether their teleological character reveals anything about the natural backdrop before which they purposively come and work and play. Is nature as a whole purposive? Are plants and animals, in some sense, also *purposes* of nature? Why are living things purposive? How did they and how do they come to be that way?

Before Darwin, there were several grand alternatives competing

---

*A more complete defense of this assertion would examine the following questions, among others: (a) *On explanation:* What do scientists and philosophers of science mean by "explanation"? Is "to know" or "to understand" identical with "to explain"? If explanation is to be understood in opposition to "description," is it true that description is no part of understanding? (b) *On being and becoming:* What or how much can we learn about *what* something *is* from learning about its genesis? Can one ever deduce "the what" from "the out of what" or "the from what"? For example, knowing only that Hector is "son of Priam," can we know or predict that he is also "breaker of horses"? Can one, in principle, know what a chicken is from knowing the full base sequence of chicken DNA? Or does the *what* of a chicken—its looks and its power—require an account of its own? (c) *How to put the questions about aliveness and wholeness:* Is not the mechanist-vitalist quarrel misconceived and inappropriate to the phenomena of life? Vitalism can only be understood in relation to mechanism, to which the vitalists themselves adhere in dealing with nonliving things. But here both sides err, for both believe that material and efficient causes are sufficient to fully understand, say, machines. Therefore, the vitalists are obliged to try to account for living things in terms of some mysterious, thing-like, addition to a machine (see, e.g., the speculative writings of Hans Driesch). But is not the question about purposiveness of living things (and machines) better formulated in terms of part and whole, or material and form, or system and activity? (d) *On the compatibility of mechanism and teleology:* Are these not complementary rather than antithetical modes of presenting the "why" of something? For example, when asked to give an account of the energetic late-autumnal activity of squirrels, must we not say *both* that they are storing up food for the winter, and that they are behaving as a result of a genetically determined, "instinctive" pattern programmed into their nervous system? For a full treatment of this question, see Kant's *Critique of (Teleological) Judgment,* especially §§ 69–78, "Dialectic of the Teleological Judgment." (Kant's exploration of teleology is especially useful to us because it comes with full knowledge and appreciation of the ways of modern natural science.)

as an answer. Some held that the cause was naturally and eternally resident in living things, whereas others held that it had been implanted in them by an intelligent being or designer. Some looked to a cause largely within the material of a living thing, while others turned to the organic form of the thing itself. Some views credited combinations of matter and form, and of nature and design. Perhaps the two grandest and most influential alternatives are these: The biblical view of a teleological and created world with its various forms specially created after God's plan, and the Aristotelian view of a teleological but eternal nature with its various forms kept in being, generation after generation, by the immanent workings of eternal species (*eidē*).*

Neither the teleology of special creation and design nor the teleology of the eternal *eidē* could survive unchanged once Darwin's theory of evolution joined with them in the struggle for existence. However, and this I want to emphasize again, the view of living beings as purposive, as self-producing, self-maintaining, and self-reproducing, is in no way undermined by Darwin's theory. In fact, in certain respects, the teleological emphasis is accentuated.

Darwin himself, in addition to seeking an account of the origin of species, was mainly interested in explaining their teleological character, *which he took for granted.* This is the task he set for himself as recorded in his Introduction to *The Origin of Species:*

> In considering the Origin of Species, it is quite conceivable that a naturalist, reflecting on the mutual affinities of organic beings, on their embryological relations, their geographical distribution, geological succession, and other such facts, might come to the conclusion that each species had not been independently created, but had descended, like varieties, from other species. Nevertheless, such a conclusion, even if well founded, would be unsatisfactory, until it could be shown *how* the innumerable species inhabiting this world have been modified, so as to *acquire that perfection of structure and co-adaptation which most justly excites our admiration.*[10]

Darwin was interested in accounting not only for the internal purposiveness of plants and animals, but also and especially for the perfec-

---

* It is difficult to translate the Aristotelian notion of *eidos* (plural, *eidē*). The usual translations, "form" or "idea," are bound to be misleading, given the long history of these English words as technical philosophical terms. Though clumsy in English, "looks" would be a more appropriate translation. It would preserve Aristotle's appreciation that the distinctive look of an animal reveals what it is, that the "inside" shines forth and makes itself known in its presentation on and to the "outside"—in its appearance and in its visible activities. *Eidos* refers both to the *what* of the particular animal, as distinguished from its *hyle*, or material, *out of which* it is composed, and thus also to the *kind* or *species* of animal it is. Each dog is what it is because it *is* dog.[9] See Chapter 13, "Looking Good: Nature and Nobility," for a more extensive treatment of the meaning of animal looks.

tion of their structure and for the perfection of their usefulness to each other (i.e., for their relative purposiveness or *externally* teleological character). He wanted to account for why everything was so perfectly ordered, for why everything *appeared* to be designed.

What is the Darwinian account? The core of Darwin's theory, as presented in *The Origin of Species,* has two major theses: the transformation of species and natural selection. The first holds that species are neither eternal nor separately created, but have come into being by descent with modification from preexisting species. The second holds that natural selection has been the major *means* of modification. The argument of *The Origin* proceeds in the reverse order. In the first part of the book, Darwin puts forth his theory of natural selection; the argument draws much of its rhetorical force from the ingenious opening chapter in which human artfulness in animal breeding is used as the mystery-solving analogy for the "craftsmanship" of nature. In the latter part of the book, Darwin convincingly shows that a theory of descent with modification—which he now assumes, but no longer argues, is due to natural selection—is superior to a theory of separate creation in accounting for numerous mysteries in the geological record, in geographical distribution, in the classification and morphology of organisms, and in embryology.

The theory of natural selection, Darwin's most novel proposal, comprises four elements: variation, inheritance, the struggle for existence, and natural selection proper. Though offspring by and large resemble their parents, variations occur frequently, and in all parts of the organism. These variations can themselves be transmitted to the progeny of the variant (i.e., they can be inherited). Some of these variations may be "beneficial" (i.e., they may enhance the individual organism's chance for survival under the conditions in which it finds itself); others may be "injurious." Since all species tend to reproduce themselves in geometrical proportions, more offspring are produced than can be supported by the available resources; death of large numbers of each species is a necessary occurrence. Variations that give to a plant or animal some advantage in this struggle to survive—say, an increased ability to escape the notice of predators, or a greater resistance to disease, or a greater efficiency in securing food—will give those individuals a greater chance of surviving, and those surviving individuals will transmit the advantageous traits to their offspring. This preservation of favorable heritable variations, and the rejection of injurious variations, Darwin named "natural selection."

In concluding this brief summary of Darwin's theory, let me emphasize that there are two aspects to change through natural selection: the *origin* of novelty and the *preservation* of novelty. Variation (or mutation, as we now generally call it) is responsible for the *ap-*

pearance of novelty (both the "beneficial" and the more common "injurious" sorts), while natural selection proper (i.e., the selective elimination of the less fit) is responsible for *preservation*. The second step is dependent on the first; as Darwin put it,

> Unless profitable variations do occur, natural selection can do nothing.[11]

This fact led Samuel Butler to complain that Darwin overrated natural selection, that "The 'Origin of Variation,' whatever it is, is the only true 'Origin of Species.' "[12] Of this, more later.

What are the implications of Darwin's theory for teleology? To a first approximation, with the theory of natural selection, Darwin succeeds in providing a nonteleological account of the origin of and basis for the teleological character of organisms. Purposive and co-adapted beings and races rise and change and fall by processes that are themselves *not* purposive, out of a nature that is seemingly purposeless and aimless.

The following two qualifications should, however, be noted: First, the teleological character of organisms is (at least tacitly) reaffirmed. But in Darwin's hands, it is reduced to the single dimension of self-preservation, with the consequence that one looks for and only at the *survival* value of all aspects of structure and behavior. This overzealous effort at a retrospective "teleological" explanation later bothered Darwin himself, as he records in this passage from *The Descent of Man:*

> I now admit . . . that in the earlier editions of my 'Origin of Species' I perhaps attributed too much to the action of natural selection or the survival of the fittest. . . . I did not formerly consider sufficiently the existence of structures, which, as far as we can at present judge, are neither beneficial nor injurious; and this I believe to be one of the greatest oversights as yet detected in my work. I may be permitted to say, as some excuse, that I had two distinct objects in view; firstly, to shew that species had not been separately created, and secondly, that natural selection had been the chief agent of change. . . . I was not, however, able to annul the influence of my former belief, then almost universal, *that each species had been purposely created; and this led to my tacit assumption that every detail of structure, excepting rudiments, was of some special, though unrecognized, service.* Any one with this assumption in his mind would naturally extend too far the action of natural selection. . . . [13]

Second, Darwin's nonteleological explanation—variation, inheritance, struggle for existence—not only assumes but even depends upon the immanent teleological character of organisms. The desire or tendency of living things to stay alive and their endeavor to increase their numbers, which are among the minimal conditions of the theory, are taken for granted and are unexplained. It would only be part of an explanation to say that those beings with no tendency to maintain and reproduce themselves have died out. Why are the other ones, the self-maintaining and reproducing beings, here at all? They are not teleological *because* they have survived; on the contrary, they survived (in part) because they were teleological. Why should a nonteleological nature generate and sustain teleological beings?

Yet, despite these qualifications, Darwin's theory of natural selection is not a theory of nature acting to an end, but of nature acting of necessity and through chance coincidences of its necessitated workings and effects.

Let us then make explicit the implications of Darwin's theory for the grand alternatives of teleology due to design and teleology due to self-actualization of immanent and eternal *eidē*. The doctrine of special creation and design was the view Darwin explicitly tried to overthrow, and to a large extent he succeeded. Yet the argument from design has not and need not become extinct, but survives by retreat and modification: the retreat is to the first beginnings, with design postulated only at the origin of the world; the modification is to consider specially created not any particular beings or kinds but only the special "laws of nature" impressed upon matter, which both necessarily and designedly thereafter gave rise to the world as we know it, with its profusion of living forms. Darwin himself was sometimes at least partially disposed to this opinion:

> Another source of conviction in the existence of God, connected with the reason and not with the feelings, impresses me as having much more weight. This follows from the extreme difficulty or rather impossibility of conceiving this immense and wonderful universe, including man with his capacity of looking far backwards and far into futurity, as the result of blind chance or necessity. When thus reflecting, I feel compelled to look to a First Cause having an intelligent mind in some degree analogous to that of man; and I deserve to be called a Theist. This conclusion was strong in my mind about the time, as far as I can remember, when I wrote the *Origin of Species*. . . .*[14]

*Consider also the following similar passage, from Darwin's correspondences with Gray:[15]

> . . . I cannot see as plainly as others do, and as I should wish to do, evidence of design and beneficence on all sides of us. There seems to me too much misery in the world. I cannot persuade myself that a beneficent and omnipotent God would have designedly created the Ichneumonidae with the express intention of their feeding within the living bodies of caterpillars, or that a cat should play with mice. Not believing this, I see no necessity in the belief that the eye was expressly designed. On the other hand, I cannot anyhow be contented to view this wonderful universe, and especially the nature of man, and to conclude that everything is the result of brute force. I am inclined to look at everything as resulting from designed laws, with the details, whether good or bad, left to the working out of what we may call chance. Not that this notion *at all* satisfies me. I feel most deeply that the whole subject is too profound for the human intellect. A dog might as well speculate on the mind of Newton. Let each man hope and believe what he can.

And this is certainly what Thomas Huxley had in mind when he praised Darwin, as did Asa Gray, for his reconciliation of teleology and morphology:

> But perhaps the most remarkable service to the Philosophy of Biology rendered by Mr. Darwin is the reconciliation of Teleology and Morphology, and the explanation of the facts of both, which his views offer. The teleology which supposes that the eye, such as we see it in man, or one of the higher vertebrata, was made with the precise structure it exhibits, for the purpose of enabling the animal which possesses it to see, has undoubtedly received its death-blow. Nevertheless, it is necessary to remember that there is a wider teleology which is not touched by the doctrine of Evolution, but is actually based upon the fundamental proposition of Evolution. This proposition is that the whole world, living and not living, is the result of the mutual interaction, according to definite laws, of the forces possessed by the molecules of which the primitive nebulosity of the universe was composed. . . .
>
> . . . The teleological and the mechanical views of nature are not, necessarily, mutually exclusive. On the contrary, the more purely a mechanist the speculator is, the more firmly does he assume a primordial molecular arrangement of which all the phenomena of the universe are the consequences, and the more completely is he thereby at the mercy of the teleologist, who can always defy him to disprove that this primordial molecular arrangement was not intended to evolve the phenomena of the universe.[16]

This theory of "design on the installment plan," as John Dewey ridiculed it, is perhaps only a vestigal rudiment of the full-blown account, but it remains undefeatable.

As for Aristotelian teleology, at least insofar as it rests on the eternity of the species, it has suffered perhaps even more. True, it is not disturbed by the rejection of special creation; nature and the natural kinds, for Aristotle, are not *creatures,* nor are they the products of intellect or divinity. True, if one considers the world of our experience—and the world since the beginning of civilization—and ignores the immense lapse of time that preceded, the *stability* and *permanence* of the *eidē* in generation is by far the most impressive phenomenon, much more so than the appearance of significant novelty. Nevertheless, if Darwin is right about descent with modification—and, no doubt, he is right about that—then the *eidē* are not eternal. And if he is right about natural selection, the *eidē* may be mere outcomes of a nonteleological nature, even though they still can be the principles (*archai*) or the ends (*telē*) of change that rule over the generation, growth, and reproduction of individual organisms. Species come and go—albeit very, very, very slowly. In the beginning, there was only one or perhaps a few species, and out of these have come millions. And if Darwin is right about natural selection, the process is not yet finished; evolution continues and will continue. It is a process without a clear directedness, with no definite term or end or goal, and *a fortiori*, not a process for the sake of its completion.

Let me emphasize that each of the two parts of Darwin's theory has significant yet distinctive implications for teleology. Although any showing of the transformation of species would be fatal to notions both of the eternity of species and their special creation as fixed species,* it is Darwin's special theory, natural selection, that—at least on first, and also on second, glance—seems to rule out even an ordered and goal-directed unfolding of created nature. Lamarck and others had put forth theories of evolution, of unfolding, but the changes that appeared in Lamarck's account, even those in response to changes in the environment, were the result of nature working from *within* organisms. For Lamarck, organic change was purposive, directed either for the sake of adaptation to changed conditions, or toward the goal of achieving greater complexity. For Darwin, descent with modification was not an evolution, an unfolding of a prearranged plan, and, in fact, the term "evolution" does not occur in *The Origin of Species*.† The changes that occur, or more precisely, the changes that are preserved, are not due to causes *within* organisms, but rather are the result of competition *between* and *among* organisms. There is no necessary, let alone goal-directed, connection between the variations that nature sportively tosses up and the circumstances that deem some of them "advantageous" and lead to their preservation.

Thus, to sum up, though plants and animals and man are purposive beings, they are not nature's purposes, nor anyone else's. Moreover, they are part of a process that began by accident, proceeded by necessity, and has no end. Nature is blind and dumb, aimlessly but persistently going nowhere.

This, at least, is what the theory of natural selection tells us. But is it true? Are all aspects of evolution really without a direction and really open-ended? We should have another look at *The Origin of Species* and also at the evidence.

---

*Would not Aristotle have been greatly perplexed by the very title of Darwin's work, *peri eidōn geneseōs*, the genesis of the *eidē*? (Though *genesis* is not a proper translation of "origin," Darwin's book is really about "origin" understood as *process*—coming to be—rather than as *source*. *The Genesis of Species* would have a more precise title.)

†The word "evolved" occurs once; it is the very last word in the book: "There is grandeur in this view of life, with its several powers, having been originally breathed into a few forms or into one; and that, whilst this planet has gone cycling on according to the fixed law of gravity, from so simple a beginning endless forms most beautiful and most wonderful have been, and are being, evolved."[17] One can only suspect that Darwin's theological waverings were here leaning toward belief; in later editions he inserted "by the Creator" after "originally breathed." This may account for the novel and unprepared-for use of the term "evolved."

The Origin of Species Reconsidered: The Directions of Evolution

The Origin of Species is replete with teleological terms and passages, both explicit and implicit, not only about the functioning of individual animals but also about the overall course of evolution. Terms such as "useful," "important," "purpose," "adapted," "fit," "the good of each being," "profitable," "harmful," "beneficial," "injurious," "advantageous," "good," "tendency," "success," "welfare," "improvement," "perfection," "low" and "high" in the "scale of nature," and "absolute perfection" occur frequently, almost on every page. Most of these terms are treated as self-evident and apt; Darwin apparently felt no need to define or discuss most of them nor to defend their use. One might say they flowed naturally into his account.

Consider some selected passages that illustrate the range of these teleological references. (1) On *adaptation* and its *excellence:* "How have all those exquisite adaptations of one part of the organisation to another part, and to the conditions of life, and of one distinct organic being to another being, been perfected?"[18] (2) On *directedness:* "Every organic being is constantly endeavoring to increase in numbers."[19] (3) On *absolute perfection:* "Natural selection will not necessarily produce absolute perfection; nor, as far as we can judge by our limited faculties, can absolute perfection be everywhere found;"[20] and again, "The wonder indeed is, on the theory of natural selection, that more cases of the want of absolute perfection have not been observed."[21] (4) On *purpose* and its *alterability:* "The swimbladder in fishes . . . shows us clearly the highly important fact that an organ originally constructed for one purpose, namely flotation, may be converted into one for a wholly different purpose, namely respiration."[22] (5) On the *tendencies* of natural selection: "There will be a constant tendency in natural selection to preserve the most divergent offspring of any one species;"[23] and again, "Hence we may look with some confidence to a secure future of equally inappreciable length. And as natural selection works solely by and for the good of each being, all corporeal and mental endowments will tend to progress towards perfection."[24]

Even the subtitle of the book invites teleological consideration: "The preservation of favoured races in the struggle for life." Did Darwin really mean to say the "preservation of *favoured* races," rather than "the preservation of preserved races" or simply, "the preservation of races?" What is *favored* about the "favoured races"?

Many of these teleological terms and passages are difficult to understand, especially in the light of the theory Darwin is propounding. Most of Darwin's followers have purged these notions from the theory. The current neo-Darwinian theory, which defines natural selection as differential reproduction (i.e., leaving more offspring) rather than differential survival, and which focuses on gene pools and gene frequencies rather than on plants and animals, makes these notions seem even more problematic. Nevertheless, the terms and passages appear in Darwin's book in abundance. We must, like good scientists, ask why. Did Darwin intentionally include these teleological references? Were they deliberate attempts to coat the bitterness of his teaching, either for himself or for his readers? Were they rhetorical devices to win adherents? Was Darwin simply sloppy of speech, confused, or undecided? Was he merely the victim of a flowery nineteenth-century English prose? Or was Darwin, the devoted naturalist and lover of nature, simply recording his impressions of the appearances, of a nature whose beings indeed shine forth "that perfection of structure and co-adaptation which most justly excites our admiration"?

We cannot readily answer this question, at least not in general. I suspect that in some cases, Darwin was indeed merely whistling in the dark, for example, in the surprising last sentence of the chapter on the "Struggle for Existence":

> When we reflect on this struggle, we may console ourselves with the full belief, that the war of nature is not incessant, that no fear is felt, that death is generally prompt, and that the vigorous, the healthy, and the happy survive and multiply.[25]

In other cases, it is clear that Darwin is undecided, while in yet others, he seems to use certain terms against his own strictures (e.g., a memorandum in his copy of *The Vestiges of Creation* says "Never use the word 'higher' or 'lower' ").

But it may be more interesting to consider the meaning of these teleological passages than to continue to speculate about how they came to be in Darwin's book; knowing their origins cannot teach us whether or not they are true. It may turn out that we should be grateful to Darwin for recording and preserving his intuitions, his sentiments, his uncertain musings, or even only his fond hopes, which a more scrupulous devotion to his theory might have caused him to delete. I propose to examine two such passages, one in which I will argue Darwin was teleological without adequate justification, a second in which he was not teleological enough. The first passage deals with the small workings of natural selection at the level of individual animals; the second deals with the full sweep of evolution.

Darwin says in many places that "natural selection can act only through and for the good of each being."[26]

> As man can produce and certainly has produced a great result by his methodical and unconscious means of selection, what may not nature effect? Man can act only on external and visible characters: nature cares nothing for appearances, except in so far as they may be useful to any

being. She can act on every internal organ, on every shade of constitutional difference, on the whole machinery of life. Man selects only for his own good; *Nature only for that of the being which she tends.*[27]

Let us leave aside what Darwin means by "Nature," and also the vexed question of whether natural selection is indeed an *agent* of change (i.e., whether natural selection is the name of a cause or of a result), and let us consider the meaning of the phrase "the *good* of *each being.*" There are two problems: What is *good,* and *whose* good is it?

The chief good that appears to rule over natural selection is *survival.* But this is a good that natural selection can but very imperfectly give to any living being, since every living being dies. Moreover, those variations that give it advantage over its competitors arise not by natural selection but by mutation; *if* these variations are indeed beneficial, and *if* they are inherited, natural selection means that they will be preserved and conferred as "benefits" *on the offspring.* Thus, each individual may be said to be the beneficiary of *previous* natural selection, but no *living* being is tended or served by the *current* action of natural selection.

Then perhaps one should say that natural selection acts for the good of each *kind* of being (i.e., for the good of the *species*). Temporarily, even for long periods of time, this may be so. But for this improvement to *remain* an improvement depends upon the stability of the environment: Introduce climatic changes or new plants or animals, and yesterday's gift of natural selection may today turn out to be a kiss of death. Moreover, in the long run, the species itself is not a permanent or stable entity; this is the whole point of the story. Species, like individuals, come and go. Enough natural selection will produce not an improved species, but a different one. "Life" continues, but beings *and* kinds equally perish. If survival is the good, it cannot be *their* good. Is there possibly some other "being" whose good it could be?

Our difficulty in identifying *whose* good is survival must lead us to reconsider whether in fact mere survival can be the good, or the only good, of living things. If permanence and stability are the end, why should life have appeared and persisted at all? According to Whitehead, this is the true puzzle in evolution:

> In fact life itself is comparatively deficient in survival value. The art of persistence is to be dead. Only inorganic things persist for great lengths of time. A rock survives for eight hundred million years; whereas the limit for a tree is about a thousand years, for a man or an elephant about fifty or one hundred years, for a dog about twelve years, for an insect about one year. The problem set by the doctrine of evolution is to explain how complex organisms with such deficient survival power ever evolved.[28]

The same point is picked up and enlarged by Hans Jonas. Life, he says,

> is essentially precarious and corruptible being, an adventure in mortality, and in no possible form as assured of enduring as an inorganic body can be. Not duration as such, but "duration of what?" is the question. This is to say that such "means" of survival as perception and emotion are never to be judged as means merely, but also as qualities of the life to be preserved and therefore as aspects of the end. It is one of the paradoxes of life that it employs means which modify the end and themselves become part of it. The feeling animal strives to preserve itself as a feeling, not just a metabolizing entity, i.e., it strives to continue the very activity of feeling: the perceiving animal strives to preserve itself as a perceiving entity—and so on. Without these faculties there would be much less to preserve, and this *less* of what is to be preserved is the same as the *less* wherewith it is preserved.[29]

If natural selection aims at some good, if evolution has direction, it must go beyond mere survival.

Perhaps we can understand better how evolution works in the small, if we can see where it is headed in the large. Though Darwin speaks little of the overall directions of evolution, and occasionally cautions us against trying to generalize, we can extract from Darwin's own words that there are at least three prominent tendencies in the evolution of life, which I shall name "Diversity," "Plenitude," and "Ascent": Diversity, the tendency to greater and greater variety ("Nature is prodigal in variety, though niggard in innovation."[30]); Plenitude, the tendency to more life; Ascent, for now, a mystery. Diversity and Plenitude are related. Natural selection tends to produce what Darwin called "the divergence of character"—more and more widely differing species—and divergence of character leads toward plenitude, that is, more organisms:

> The advantage of diversification in the inhabitants of the same region is, in fact, the same as that of the physiological division of labour in the organs of the same individual body. . . . In the general economy of any land, the more widely and perfectly the animals and plants are diversified for different habits of life, so will a greater number of individuals be capable of there supporting themselves.[31]

More animals, more species, tending to complete and fill the democratic polity of nature. Life, it seems, is given to overflowing: into new forms, new varieties, and in an ever-growing number of organisms. Though not his intent, Darwin makes us wonder whether the principle of survival is sufficient to account for nature's prodigality. Has Darwinism adequately identified, much less explained, the immanent native "strivings" of organic life? Might appearance, show, diversity, even beauty be independent and additional "concerns" or "principles" of nature? (See Chapter 13, "Looking Good: Nature and Nobility.")

But Darwin speaks not only of the range and breadth of animal profusion. Without proper justification from his theory, and not without some reticence, Darwin also records for us the notion that some living beings are *higher* than others:

> Thus, from the war of nature, from famine and death, the most exalted object which we are capable of conceiving, *namely the production of the higher animals, directly follows.*[32]

What does Darwin mean by "higher"?

It is hard to be sure what he means, generally and in this passage, by "the higher animals." In some passages, the term seems to imply a ranking of structure and activity. For example:

> The embryo in the course of development generally rises in organisation: I use this expression, though I am aware that it is hardly possible to define clearly what is meant by the organisation being higher or lower. But no one probably will dispute that the butterfly is higher than the caterpillar.[33]

and again,

> But in some genera [of cirripedes] the larvae become developed either into hermaphrodites having the ordinary structure, or into what I have called complemental males: and in the latter, the development has assuredly been retrograde; for the male is a mere sack, which lives for a short time, and is destitute of mouth, stomach, or other organ of importance, excepting for reproduction.[34]

But in other cases, "higher" seems merely to mean "the latest and the surviving":

> There has been much discussion whether recent forms are more highly developed than ancient. I will not here enter on this subject, for naturalists have not yet defined to each other's satisfaction what is meant by high and low forms. But in one particular sense the more recent forms must, on my theory, be higher than the more ancient; for each new species is formed by having had some advantage in the struggle for life over other and preceding forms. If under a nearly similar climate, the eocene inhabitants of one quarter of the world were put into competition with the existing inhabitants of the same or some other quarter, the eocene fauna or flora would certainly be beaten and exterminated; as would a secondary fauna by an eocene, and a palaeozoic fauna by a secondary fauna.[35]

Yet despite Darwin's hesitancy when he explicitly considers the meaning of the term "higher," and despite the fact that his theory only clearly permits the application of terms "higher" and "lower" on the basis of success in the struggle for life, Darwin himself appears to be sensitive to, and, at the very least, unwittingly points us toward, a higher view of "higher." We are led to remember our common sense view that some survivors are higher (i.e., more fully developed and more fully alive) than others, that there is a hierarchy among living beings, that moss, barnacles, grasshoppers, and chimpanzees, though all equally prospering and successful, are nevertheless unequal in some decisive respects. Moreover, if one looks at the appearance of organic beings over the span of evolution, it is clear that there has been, in general, an upward trend, and this despite the fact that the life of the higher forms is often more precarious than the lower. To be sure, "low" forms persist and flourish, but higher and higher types have successively appeared.

Higher types in what sense? From here on, I shall present, in a crude and sketchy fashion, some free speculations, which, though not foreclosed by Darwin's theory, go beyond it in some important respects. To repeat, "Higher types in what sense? *Higher in terms of soul.* Somehow, as thousands upon thousands of species came and saw and went under, the processes of evolution not only produced more and more organisms, new species and new forms—or, in other words, produced both more soul and more forms for the expression of soul—but also *higher grades* of soul.

The reader may well be puzzled by the sudden appearance, not to say resurrection, of the notion of soul, a notion long deemed either meaningless or useless in biology (though preserved, albeit with shrunken meaning—consciousness or mind—in psychiatry and common discourse, as *psyche*). Neither has any philosopher of the first rank, since Descartes abandoned the practice, used *anima* or soul as a principle of *life* (rather than as a principle only of mind or consciousness). By soul I mean nothing mystical or religious, not a disembodied spirit or person, not one of Homer's shades or anything else that departs the body intact, not a ghost in a machine or a pilot in a ship, not, indeed, a "thing" at all. Rather, following Aristotle's reflections in *Parts of Animals, Book I,* and in *On the Soul, Book II,* I understand by soul the integrated vital powers of a naturally organic body, always possessed by such a body while it is alive ("animated," "ensouled"), even when such powers are not actively at work (e.g., in sleep or before maturation). Thus understood, soul is not a possession of man alone, as it seems to have been for Descartes. All living things have soul. Precisely Darwinism itself, which regards the life of man as continuous with the rest of living nature, invites one to reconsider whether a notion of soul might be necessary in order to understand the aliveness of all living things, down to the very simplest.

Though the first beginning of life lies in darkness, we may surmise that the first organisms could do little more than maintain and reproduce themselves; though possessing the rudiments of irritability, theirs was primarily the life of nutrition and growth. After this, there were only two or perhaps three more *big* "surprises" in the hierarchical history of soul: (1) the emergence of sensitivity and awareness (i.e., of the sensitive soul); (2) the emergence of locomotion, and with it, desire (i.e., of the full-fledged animal soul); and finally, (3) the emergence of speech and intellect (i.e., of the rational soul).

It is not enough to call these new and distinct powers merely more complex or more organized forms of life: They represent and make possible new and essentially different ways of life. Sensing is not just a more complicated form of nourishing; thinking is not just a complicated form of sensing or moving. Though these higher kinds of soul may have emerged, and probably did emerge, gradually by cumulative differences of degree, the difference of degree eventually became a difference in kind. A certain critical threshold in development and organization was reached and crossed, and a new activity appeared. A ready analogy for such a gradual emergence of new levels of soul is available in the development of a human being from a single cell—the fertilized egg—which, as far as we know, neither senses nor desires nor thinks.

The ascent of soul has meant the possibility both of an ever-greater awareness of and openness *to* the world, and an ever-greater freedom *in* the world. The growth of soul has produced a hierarchy among living things in their powers both to be affected by and to affect the world. The hierarchy of soul is a hierarchy of openness and purposiveness. Touching, smelling, tasting, hearing, seeing, moving toward, exploring, desiring, remembering, imagining, loving, judging, speaking, naming, thinking, and self-consciousness—all these forms of *openness* to things, to others, and to ourselves, are the pride of the human soul. They make possible a pronounced and inwardly felt sense of self, set over against the othernesses of a diversely articulated world, with many of whose members it can enter into genuine relations and even communication. They also greatly enhance our possibilities for action on the world. The freedom from immediate preoccupation with necessity provided by human reason and its arts, and the inner awareness of this freedom and of our purposiveness, permit us the opportunity to set *our own purposes,* to have goals, plans, dreams, decisions, projects, and activities that soar far beyond mere survival. Yet to claim these things for man is not to deny some or many of them to animals or even plants. Some openness and purposiveness is seen all along the rise, even at the bottom.

The End of the Climb: Beyond Chance and Necessity?

If we have correctly identified soul as at least one "subject" or "being" of evolution, and if we have correctly identified, historically, the emergence of ever higher forms of soul in evolution, we can now ask some teleological questions about this overall process. Has evolution itself an end, a completion? Is this end a cause, or an effect of some other cause, or a mere outcome of some accidental process? Are openness and purposiveness in some sense purposes of nature? (These questions, though prompted by Darwin, are foreign to his thinking, though not to Lamarck's.)

First, is there a natural term to evolution at least in its tendency of ascent? The answer would appear to be man; at least it is hard to see how it could be anything else. Man is the peak, both in possessing the *highest*, and also in possessing the *complete range* of, faculties of soul.* Even looking to the future, what could be higher than man? To be sure, many more species will come and go. Man himself will no doubt undergo many changes. Certain changes in the human constitution have been occurring fairly rapidly over the past century, albeit more due to human interventions than to natural selection in a strict sense— the average height has risen; the average age of menarche has decreased. The frequency of genes for myopia and diabetes in the human population continues to rise, as eyeglasses and insulin have compensated for the disadvantage of these genetic defects. In the future, modern science may find a way to graft chlorophyll-containing tissue onto our foreheads so that we may make our own food, or, alternatively, to endow us with paper-digesting enzymes so that we may more fully digest the great books we so much enjoy. But even these somewhat farfetched possibilities would represent mere wrinkles of shape and form, compared to the great powers of soul. Like air-conditioning and power steering, they would add nothing essential to the Model-T. Granted, we could be, and might in time become, more intelligent, more

---

*This old-fashioned conclusion, supported in the Western tradition largely by biblical religion, seems now to be unfashionable, not least because of the challenge Darwinism seems to pose to a literal reading of Genesis, and because biology can find no way on the basis of its present principles to speak about genuine hierarchy—and also because of charges that such a conclusion is anthropocentric and, therefore, a cause or a justification of the terrible harms done by men to other living things. My argument is strictly philosophical, not moral, and rests on what I take to be the biopsychological evidence. I intend no support for anthropocentrism or the exploitation of animals. The reader is asked to consider only whether the argument is true.[36]

alert, less forgetful, more energetic, etc., but can we imagine for our-selves or for soul anything really new? What great innovations can we imagine?*

That we, or at least I, can't imagine any startling new faculties of soul may indeed point to a deficiency in our imaginations that the next great emergence will rectify. And one should never be comfortable pre-dicting the end of change or history—though I can be bold with im-punity, knowing that I will not be proved wrong in my lifetime. But, in fact, there *are* a few things we can imagine, but some are not *de-cisively* novel, and others strike me as out of reach. To the former class belong new faculties of sensation, responsive to X-rays, radio waves, television, gravity, and magnetism, and to whatever it is that those people with extrasensory perception now perceive, if they do indeed perceive. To the latter class belong bodily immortality and the ability to be in two places at the same time. And while both of these fantastic possibilites would radically change the nature of things, they might not greatly add to the kinds of activities we might pursue forever or in two places at once.† For these reasons, I suggest that with nutri-tion, reproduction, sensation, appetite, imagination, emotion, and in-tellect, the story of the ascent of soul may already be complete, though improvements can surely be made, and though evolution and its ten-dencies to diversity and plenitude may continue without bound.

What is responsible for this upward trend in evolution? This re-mains an important and fascinating question, even if I am roundly mistaken in suggesting that the ascent has reached the summit. Let us recall the distinction between the origin and the preservation of novelty, between the first appearance of a "higher" form of life—say the first sensitive being, the first animal with the rudiments of touch—and the preservation of that higher form. Let us provisionally say that the origin of sensitivity was due to chance. But is it due to chance that sensitivity was preserved or that it began to flourish? Could it not be said that, if it is a sensible world, sensing beings would be at home in it, or to put it in Darwinian terms, sensing beings would, other things being equal, have an advantage over their insensitive fore-bears? Ought we to be surprised, should we regard it as an accident, that, in a visible, odorous, and sounding world, the powers of sight, or smell, or hearing once they appeared should have been preserved, magnified, perfected? Likewise with intellect. However accidentally

---

*Please note: I am speaking not about the particular details of the human genome, but only of the large powers of the human soul. Also, I do not deny the great influence of changing cultural conditions on the way human beings live. Nor do I mean to rule out great future changes in social organization, including improved social and technical means for harnessing and developing native human capacities.

†See Chapter 12, "Mortality and Morality: The Virtues of Finitude," for a discussion of the implications of an indefinitely prolonged human life span.

intellect first appeared, is it surprising that it should have been preserved in a world of cause and effect, past and future, means and ends,
all of which can be brought to consciousness and used to advantage
in a being endowed with memory, a sense of time, self-awareness, and
the ability to order means to ends in securing the future? If it is an
intelligible world, is it surprising that an intelligent being, once one
appears, will be at home in it or, to put it in the less complete terms
of survival, will be likely to survive and flourish? Is it not only not
surprising; is it perhaps even necessary that such a world would be
hospitable for the maturation of soul (i.e., for evolutionary ascent)?

Having gone this far, let me try out a most speculative speculation: Let me suggest that the kinds and levels of soul complement and
answer to the kind of a world this is, and that by evolving to complement the things that are, soul *completes* the things that are. Are not
the looks and beauty of a flower incomplete until there exists a seeing
being open to and aware of this beauty? Are the laws of nature in their
intelligibility fully intelligible, and hence *fully* themselves, before they
are discovered by an intelligent being? And is the unfolding and emergence of soul complete until there exists a being who discovers the
emergence of soul, and what is more important, understands soul, independent of whether and how it emerged?

Now one may ask: What have the sense of beauty or knowledge
of nature, or for that matter, philosophy or music or mathematics or
morals—what have all these things to do with natural selection?
Surely, these activities of human beings, which we admire and which
are dependent on human rationality, go well beyond the uses of reason
that may have enabled intelligent beings to survive. Even if we grant
that reason was not accidentally preserved—that intelligence, in the
form of arts and cunning and planning, naturally helped its possessors
to survive—is it not an *accident* that these powers of intelligence were
turned to other uses? Or is it possible, or even likely, that faculties
which are first preserved for the sake of living already have or soon
acquire, so to speak, a life of their own, and continue to flourish for
the sake of living well?

Let us turn, in some closing speculations, from preservation and
flourishing to the first appearance of life and of each level in the ascent
of soul. Is it true, as we provisionally asserted, that the beginnings
were due to chance? What *is* due to chance is, perhaps, the time and
circumstance in which life first arose and also the beings in which the
first steps of further ascent were taken. But what emerged on these
occasions were possibilities already present as possibilities in the lower
forms of life, and ultimately, in matter itself. Everything present to
the higher animals, including man, must have been "present" *potentially* in preanimate matter. Matter, if not actually alive, was poten-

tially alive; given the right circumstances, it came alive *on its own.* The accident of circumstance merely *released* the *inherent capacities* of matter to become organic, to form structures that produce and maintain and reproduce themselves. Analogies may be found in the releasing of developmental capacities by the chance union of egg and sperm, or in the bursting into flame of wood brought to the kindling temperature, which flame maintains itself as long as further material is available for combustion, or in the releasing of the potentiality to form water caused by the chance joining of hydrogen and oxygen gases. Just how this releasing took place, just what circumstances made possible the emergence of life, we may never know. But at least formally, some account like this one must be correct. Matter must not be thought of simply as inert and passive, as resistance to acceleration, as little billiard balls, as waves or particles, as understandable only externally and through laws of relation. Matter has character; matter has possibility; matter is prefiguratively alive. For all his errors on this subject, Aristotle was perhaps closer to the truth even about matter than are the modern physicists:

> Animals and their parts exist by nature, and so do plants and *the simple bodies*, for example, earth, fire, air, and water; for we say that these and other such exist by nature: . . . All things existing by nature have *in themselves* a principle of motion [or "change"; *hinosis*] and of rest. . . .[37]

Whether or not Aristotle correctly identified these principles of motion and rest, and whether or not he correctly identified the material elements, the insight that the elements no less than the plants and animals were self-moving—each in its own way—strikes me as fundamentally sound, and useful in thinking about the connectedness of nature lifeless and nature alive.

To be sure, preanimate matter probably had numerous potentialities including many that were never realized—although if my earlier account is correct, the majority of the most interesting possibilities may have been perfected. One probably cannot infer, strictly speaking, that the emergence of soul and of the human soul was *necessary* simply from the fact that it happened. One probably cannot even say that life or intelligent life would come to be, if nothing obstructed. One comes closer to the truth, perhaps, if one says that life, then sentient life, then intelligent life came to be and would most likely come to be if circumstances were not unfavorable to their release. Given enough time, given a durable deck of cards, given a tireless dealer and shuffler, sooner or later the cards will be dealt out in the right order, ace-to-king, spades, hearts, diamonds, and clubs—if not always, at least in most such protracted dealings. In and from the beginning, the ascent of soul was, as they say, in the cards.

CHAPTER ELEVEN

Thinking About the Body

And, though all other animals
are prone, and fix their gaze
upon the earth, he gave to
man an uplifted face
and bade him stand
erect and turn his eyes
to heaven.
—Ovid, *Metamorphoses* (I, 84–86)

My poor gentlemanlike carcass.
—Ben Jonson, *Everyman in His Humour*
(Act IV, Scene 5)

Man is to himself the most wonderful
object in nature; for he cannot
conceive what the body is, still
less what the mind is, and
least of all how a body
should be united to a mind.
This is the consummation of his
difficulties, and yet it is his
very being.
—Blaise Pascal, *Pensées* (II,72)

What is the relation between a human being and his body? Never a simple question, it is today even more puzzling, thanks, in part, to new surgical and technological developments that also give it great practical importance. On one side, we have a living body apparently devoid of all human activity in the permanently unconscious young woman who still manages to breathe spontaneously on her own for several years. On the other side, we have a human being alienated from his living body in the man who believes he is really a woman trapped inside a man's body and who undergoes surgery for "gender reassignment." In between, an increasing number of people walk around bearing other people's blood, corneas, kidneys, hearts, and livers; successful transplantation even of brain cells is currently proceeding in animals. To meet the shortage of organs for transplantation, some people have proposed that we allow the buying and selling of such human "spare parts," transferable both before and after death. Implantable and attachable mechanical organs add to our possible confusion, as do the more prevalent but less spectacular phenomena of wigs, tattoos, silicone injections, and various forms of body-building and remodeling.

If practice turns to theory for clarification and assistance, it finds there nearly equal disorder. Philosophers should certainly not be faulted for failure to "solve" the mind-body "problem"—though they are perhaps to be blamed for how it is defined and for presenting it as a "problem" to be "solved." But certain dominant fashions of thought do not even face up to the difficulty. On one side are the corporealists, for whom there *is* nothing but body and who aspire to explain all activities of life, including thought and feeling, in terms of the motions of inorganic particles. On the other side, say especially in ethics, are the theorists of personhood, consciousness, and autonomy, who treat the essential human being as pure will and reason, as if bodily life counted for nothing, or did not even exist. The former seeks to capture man for dumb and mindless nature; the latter treats man in isolation, even from his own nature. At the bottom of the trouble, I suspect, is the hegemony of modern natural science, to whose view of nature even the partisans of personhood and subjectivity adhere, given that their attempt to locate human dignity in consciousness and mind presupposes that the subconscious living body, not to speak of nature in general, is utterly without dignity or meaning of its own. These prejudices of theory do not accord well with our experience.

Several times during the past few years, I have led discussions with freshman medical students immediately after their first experience with the human cadaver in the gross anatomy laboratory. My purpose has been to explore with them their responses—their thoughts and feelings—to what is for most of them their first encounter with a dead body. As one might expect, responses varied considerably, from the student who had become physically ill to the matter-of-fact fellow who could not understand what the fuss was all about. Several students surprised themselves with their own reactions. Though all understood the necessity of anatomical dissection for their own proper training, many found the activity repulsive. "What if the relatives were to walk in? I feel as if I am abusing the family." "Did this guy knowingly consent to be dissected? Did he really know what we are going to do to him?" "I would never let this happen to my father—or to myself." Some commented on the youth and beauty, or on the decrepitude and ugliness, of their particular specimens; many expressed curiosity about the individual lives that once were led by these bodies.

Someone expressed gratitude for the gift of the body and the invitation to study it, but for someone else, this reminder of missing personhood made working on the cadaver only more difficult. Many said they could not bear to look upon the face; I myself had seen that, in almost all cases, the face was the first part to be covered and wrapped. Some could not look at the genitalia; others could not touch the hands. Reservations were expressed about performing unavoidable "invasions of privacy," "objectification," and "reduction to nothing" through dissection. Almost everyone who spoke acknowledged expressly or tacitly the need, and his own desire, to respect the mortal remains of a human being, but those who were most troubled somehow intuited the impossibility of doing so. They understood and felt that they were engaged in something fundamentally disrespectful—albeit in a good cause.

Such responses have, of course, been noted by others; they are the subject of a careful sociopsychological study by Renée Fox.[1] But what has not been sufficiently observed, in my view, is the fact that all these responses—perfectly natural ones to a layman—are entirely inappropriate and unreasonable, unreasonable, that is, on the scientific view of the body that our medical students are taught and to which they adhere. Their science—*our* science—regards the living body in terms of nonliving matter in motion. Extended matter in necessary and purposeless motion, organized by necessity on an inherited plan and functioning as pure mechanism, the body even in life is, on this scientific view, no object for shame, awe, or respect. And in death, it is a gradually decaying, inoperative, worthless heap of finally homogeneous stuff. What, as the true corporealist said, is all the fuss about?

Soon the class would be hard at work, digging away in dead earnest, and reaching uniform agreement on the names, locations, relations, and functions of each of the separate organs, tissues, nerves, and blood vessels. Soon no one would think of the cadaver as a whole, never mind in relation to a person. The powerful scientific way of analysis would in fact and in thought dissolve the whole, and with it those original "unscientific"—indeed natural—repugnances. And yet, those initial reactions and thoughts strike me as sound. For the body—even the dead body, the mortal remains of a singular human being—is more than our present science can say. And what of the living body? Does modern medicine, grounded in modern reductive and mechanistic science, have an adequate account of the living body—as an organic whole; as lively and self-moving; as a personal center of awareness, felt need, and self-concern; as a vehicle of individuated self-presentation and communication? Is there a biology, an anthropology, that does justice to the being and meaning of the body and of bodily life—as we live it?

A second story cuts in a different direction. Roughly eleven years ago, I went to the local hospital to visit an extraordinary man I had come to know. Though I knew him only in his waning years, he displayed even then the most amazing mind I had ever encountered; no offense to my readers, his was a mind whose power, learning, and understanding were virtually off the scale occupied by the rest of us. He had written a shelf of luminous books and had as many left in him, if only he could remain fit long enough to get them written down. I had attended some of his seminars. Flattered that a medical doctor would find his humanistic teaching interesting, he honored me with the opportunity for occasional private conversation—though this sometimes turned to medical problems, not surprisingly, his own. He was weak, frail, sickly—and even in health had a body that promised nothing of the wonders of his mind. The first things he read in the newspaper were the obituaries: "To see if I died," he would explain. I had seen him well on the previous day, as his official physician pronounced him fit to travel for a lecture. He became ill en route and was brought back to hospital. Looking to cheer him up, I walked briskly to the hospital room where I had visited him during several previous hospitalizations. I asked the nurse who was just leaving his room how he was doing. "Don't you know," she replied, "Mr. _____ expired an hour ago." I entered the room, thunderstruck. There he lay, peacefully, a frail figure in a large bed, half-smiling, as if in a pleasant dream. Dreaming, I would have thought had I not met the nurse. Moments later, I found myself on my knees at the foot of the bed, full of awe and horror. Over and over, I asked myself, "Where *is* he? Where did he go? Where is that mind, that learning and understanding, those unwritten books that no one will now write?"* There he lay or seemed to lie, but lay not; there he was or seemed to be, but was not. The body, the still warm and undisfigured body, identical in looks to what I had seen the day before, mocked me with its unintentional dissembling and camouflage of extinction. Here, there was vastly *less* than meets the eye. The dead body may be more than what our science teaches, but it is also less than what it appears to us to be. The body may be more than stuff, but the man seems to be more than his body.

*Any death raises acutely the question of the relation of the human being to his body; the death reported in this story is special only because of the magnitude of what has vanished without trace. Even in my medical days, well before I acquired philosophical interests in these matters, I found the disappearance of a human life from a human body to be a simply incomprehensible occurrence. For this reason, I always disliked the autopsy room, where confident pathologists gave anatomical or physiological explanations, adequate to their limited purpose, that only increased my bewilderment regarding the questions that most troubled me: What *happened* to *my patient*? What was responsible for his extinction? (For a sensitive treatment of this topic, see Richard Selzer, M.D., "An Absence of Windows," in his *Confessions of a Knife,* and "The Corpse," in *Mortal Lessons: Notes on the Art of Surgery.*²)

These two stories drive one in two contradictory directions: first, to suspect that a man is self-identical with his body; second, to suspect that the best part of a man is somehow not corporeal. The first invites us to return from our abstract scientific notions again to treat seriously ordinary appearances; the second requires us also to look beyond the appearances (though not in the way of modern science) to something utterly invisible and intangible. Neither of these conclusions or directions is comfortably at home in current thought. The first, by identifying man and body, might seem to agree with the scientist's or corporealist's view that man *is* nothing but "body," but in fact this "body" points not to "matter" but to notions of bodily wholeness, individuation, and active form—in short, to a very different idea of *body*. The second invites speculation about an incorporeal soul or mind, a notion still present in religious but long absent from philosophical or biological discourse.

Once such perplexities are raised, there is no alternative but to think about the body. In keeping with this search for a more natural science, I will pursue mainly the first line of inquiry in an effort to make manifest some plain truths about the body, which a proper biology and anthropology would not ignore. I do so without any hope—or even desire—of dispelling or denying the mystery of the body's nature and being. On the contrary, I seek to recover and reaffirm it.

There are, of course, obstacles to thinking about the body. First of all, the body—or, to avoid begging the question, most of it—is mute. True enough, each of us has experience of his or her own body, but that experience is entirely subrational (i.e., inarticulate and speechless) and probably even largely unconscious. The materials for thought are available, but the handles are not ready made. In fact—a second obstacle—it seems that there may be no *naturally* or universally appropriate way to think about the body and no universally valid "plain truths" about the body, since different cultures vary widely in their assessments of the nature and worth of what we call "the body." Questions about the body are tied to questions about life, death, and soul; the whole cosmic picture is soon at issue, and about such matters, we are well aware, cultures differ. Some believe in the transmigration of souls, others believe there is no soul; some are panpsychists, others pancorporealists, still others dualists. In our tolerant age, we are reluctant to declare another culture's beliefs less worthy or true than our own. Indeed, we are generally quick to criticize as ethnocentric any of our own passionately held beliefs—except, of course, for cultural relativism itself, a belief, we generally forget, that is itself culture bound.

This is not a new difficulty. One of our oldest texts on the subject of the power and relativity of law and custom, in fact, deals with the treatment of the dead body. It is found in Book III of Herodotus's *Inquiries:*

> ... If one were to offer men to choose out of all the customs in the world such as seemed to them the best, they would examine the whole number, and end by preferring their own; so convinced are they that their own usages far surpass those of all others. Unless, therefore, a man was mad, it is not likely that he would make sport of such matters. That people have this feeling about their laws may be seen by many proofs: among others, by the following. Darius, after he had got the kingdom, called into his presence certain Greeks who were at hand, and asked what he should pay them to eat the bodies of their fathers when they died. To which they answered, that there was no sum that would tempt them to do such a thing. He then sent for certain Indians, of the race called Callatians, men who eat their fathers, and asked them, while the Greeks stood by, and knew by the help of an interpreter all that was said, what he should give them to burn the bodies of their fathers at their decease. The Indians exclaimed aloud, and bade him forbear such language. Such is men's custom; and Pindar was right in my judgment, when he said "Law [or "custom" or "convention" or "mores": *nomos*] is king over all."[3]

Men's customs regarding dead bodies, like customs in general, are both powerful and powerfully different. But not all-powerful or altogether different. Those who attend carefully may learn from the story that custom may be king over almost all, but not over all. Against its own explicit relativistic conclusion, the story presents at least two universal and related facts. First, everybody—Greeks, Indians, even Persians—dies. Everybody, sooner or later, becomes a body. Second, everybody does *something* with the dead bodies of the deceased ancestors. Human beings everywhere recognize human mortality; human beings everywhere feel a sense of responsibility to the deceased, elicited by ties of kinship. These samenesses seem to me at least as significant as the differences in funeral practice. Beneath and beyond the different ways human beings think or feel or act, there do seem to be at least a few universal truths about the body and its human meaning. I take this prospect as my license to try to think—and not merely ventilate the prejudices of my culture—about *the* body—and not merely some body, say, bodies of twentieth-century Americans. As it turns out, the results of the inquiry will cast doubt on certain reigning American opinions and (implicitly) practices.

The Body in Speech and Experience

Let us begin with the word "body." In its original Old English usage, it meant "the physical or material frame of a man or any animal; the whole material organism (as in 'to keep body and soul together')," or, in another sense, the "main portion of the animal frame," that is, the trunk ("all head and very little body"). In both senses, body has a correlative term: in the first case, soul; in the second case, head (or limbs). It is thus doubtful whether body at first denominated the whole living organism, or whether the whole was regarded as the body and then some. What is clear is that body meant primarily *living* body; only later (in Middle English) was the same term, as short (or euphemistic) for "dead body," applied to the corpse.* By extension, it came later to be used to refer to the person or the individual being. Not until the sixteenth century was the term transferred from the material part of man to material things generally; the first known use of "body" to mean "matter" occurs in 1586 ("A bodie is a masse or lump, which as much as lieth in it, resisteth touching, and occupieth a place").

In its prime usage, body is always body *of:* body of *an* animal or human being. *The* body is an abstraction. *The* body is always *some*-body, somebody's body, some body in particular. To seek greater clarity about the being and status of the body, let us consider our own bodies, how we speak about them in ordinary speech, how we experience them.

Sometimes we say "he pinched my arm," sometimes "he pinched me." The former seems to imply that I am distinct from the bodily parts, which parts are *mine*—my equipment or tools; the latter identifies me and my body. One might say that the former expression gives a more objective or cognitive account—I localize the act of pinching as it can be outwardly seen—while the latter gives a more subjective or pathic account, in which the pain and affront caused by the act to my person are central, and identification of me and my arm makes sense. But our usages are not so consistent. (Consider, in this regard: "Sticks and stones will break *my* bones, but names will never hurt *me.*") True, our speech in self-reference reflects a certain self-division, in which we are linguistically both subject and object, the viewer and the viewed. Indeed, the fact of such self-consciousness is largely responsible for the difficulty in understanding our relation to our own bodies. Yet, there is no fixed or constant biopsychological referent for the grammatical subject "I." The I who speaks of "my body" is not the same "I" who speaks of "losing my mind." Sometimes "I" denotes my totality, sometimes only my conscious part; sometimes my whole soul or psyche, sometimes only the rational principle of my will-

---

*Similar changes occurred also with the Greek *soma* and the Latin *corpus:* both these languages used the same word to apply first to the living body and later on to the corpse. Both are contrasted with soul (*psychē* and *anima*) and mind (*nous* and *animus* or *mens*). Even ancient corporealists, like Lucretius, preserve the distinction between body and soul or mind, notwithstanding that they claim that soul is in fact a very rarefied kind of body (i.e., matter). German, however, has a distinct word for the living human body, *leib.*

ing or thinking. Such confusion is, I suspect, not merely linguistic or superficial. Though we have, as it were, inside knowledge that we are somehow a "one," a whole, a psychophysical unity, though we sense ourselves immediately as both feeling and embodied beings, we are also in various moments and kinds of self-consciousness more or less aware that we are also and at the same time two (or more); indeed, as I now try to think about my body, I am aware that my thinking, though it is *my* thinking—*of* as well as *about* my body—is not related to my body in the same way as, say, my pain or hunger or cough or disease. True, *thinking* is unmistakably done by me in a body, but in a body set aside, at ease and unobtrusive. In fact, were I fully absorbed in my thinking—oblivious to my need to come before the reader—I would not even identify my thinking as *mine*. Thought entails self-forgetting, even self-overcoming, as the thinking one becomes one with the things thought.*

Here, a word or two more on the curious usage of the possessive pronoun to identify *my* body. The locution makes sense in several ways: Often we regard our bodies as tools (literally, organs) of our souls or wills. Our organism is organized: for whose use?—why, for our own. My rake is mine, so is my arm with which I rake. The "myness" of my body also acknowledges the privacy and unshareability of my body. Sometimes we assert possession against threats of unwelcome invasion, as in the song, "My Body's Nobody's Body But Mine," which reaches for metaphysics in order to teach children to resist potential molesters. My body may or may not be mine or God's, but as between you and me, it is clearly mine. And yet I wonder. What kind of *property* is my body? Is it mine or is it *me*? Can it be alienated, like my other property, like my car or even my dog? And on what basis do I claim property *rights* in my body? Have I labored to produce it? Less than did my mother, and yet it is not hers. Do I claim it on merit? Doubtful: I had it even before I could be said to be deserving. Do I hold it as a gift—whether or not there be a giver? How does one possess and use a gift? Is it mine to dispose of as I wish—especially if I do not know the answer to these questions? The bearing of this on organ donation is clear; so, too, on our loose talk about an absolute right to do whatever one pleases with or to one's body.

---

*In this way, the thinker can "think himself" far outside of his body, here and now, to encompass even the cosmic whole and times long past. It is this openness and transcendent possibility of thought that once made philosophers doubt the self-identity of the human being and his body. In the Platonic dialogues, Socrates's arguments for the immortality of the soul are connected with soul understood as the principle of knowing, not with soul as the principle of life. And Aristotle argued that, although soul (*psyche*) was generally the form of the organic body, intellect (*nous*) could have no bodily organ or permanent form of its own; otherwise, it would not be possible for *nous* to be open to all things, and some things would be unthinkable. I return briefly to this matter later.

Experience of ourselves as embodied provides no greater clarity. We can only give confusing answers to the curious question of where in this whole corpus we think we truly live. Science tells us the brain, but we do not experience the brain, and no one would naturally give such an answer. Much of the time, I think, we feel ourselves concentrated just behind the eyes; when someone says "Look at me," we look at his face—usually at the eyes, expecting there to encounter the person or at least his clearest self-manifestation.* But *where are we* when we are exhausted or suddenly terrified? When we hit the baseball or make love? When one looks at gymnasts or dancers, are not legs and trunk as important as face? How indeed can we know the dancer from the dance?

We are, as we live no less than as we speak, deeply unsure of who or what we are most of all. Happily, in most of what we do, we feel no need to decide this question. We go about our business, usually with immediacy and without apparent self-division. Moral philosophers may busy themselves elaborating a theory of personhood based on autonomy of will, devoid of all references to the body; Descartes can declare—not seriously, in my view—that he *is* only a thinking thing;† biologists and behavioristic psychologists may advance their global corporeal or deterministic explanations, denying the causal independence or even the reality of feeling or thought or the existence of free will—we don't care: We both know and don't know who we are, and appear none the worse for our ignorance. The way we live gives the lie to all this theorizing, and implies that the truth is both much more mysterious and complicated than is dreamt of in our philosophizing, and, at the same time, largely irrelevant to getting along in life.

Largely, but not completely, irrelevant. Sometimes we must choose. One of the most unsettling—yet, for the thoughtful man, also interesting—things about confronting cadavers, dead bodies, or the question of organ transplantation is that we are by practice *forced* to decide who or what we think we are, really, and most of all. How to treat dead bodies may seem to be a trivial moral question, compared with all the seemingly vital problems that confront the living. But, from a theoretical point of view, few are as illuminating of our self-conception and self-understanding. I return to this point at the end.

---

*Hence our natural repugnance when we find instead silver-mirrored reflecting sunglasses. For more on the face, self-presentation, and the meaning of our looks, see Chapter 13, "Looking Good: Nature and Nobility."

†This is an obviously fallacious inference from "I think, therefore I am"; for from "I *know* only that I am a thinking thing," it in no way follows that "I *am* only a thinking thing."

Looking Up to the Body

Our insider's view of our relation to our own bodies is certainly useful, especially in relation to fancy and abstract theories that purport to know better. In his essay, "Is God a Mathematician? The Meaning of Metabolism,"[4] Hans Jonas makes elegant use of this "immediate testimony of our bodies" decisively to refute the claims that organic life—even the lowly life of an amoeba—can be fully understood, without remainder, on the basis of the principles of mathematical physics, so successful in dealing with inorganic body. Still, though "inwardness" is a fact, its precise character is hard to describe and its relation to our bodies elusive. In search for something more easily grasped, we turn from our speech about our bodies and our experience of ourselves as embodied to look directly at the human body itself. Most of what I will talk about is evident on the surface—though I do not think therefore only superficial. Our science distrusts the surface and finds the clearest and most certain truths buried within. To be sure, *how* the body works can only be learned by mechanistic analysis. But what the body *is* and especially what it *means* can be grasped, if at all, only by looking on it whole, as we encounter it.

Perhaps the first thing that strikes us in looking at the body—any living body—is that it *is* a whole, a unity, a one. It has a boundary, a surface, that clearly delimits it from everything that it is not. It is solid but shapely, corporeal but articulated, enmattered but most definitely formed. The forms are distinctive—each one, though individuated, is always one of a *kind,* with a distinctive shape, attitude, look, and way of moving. When the living body moves, it moves as a whole; if we are able to observe it growing, we see that it grows as a whole. It is capable, when injured, of making itself whole, through remarkable powers of self-healing, and it generates other wholes formed like itself. And, in the higher animals, the form and patterns of the body acquire a plasticity useful for communicating to other wholes of the same species, expressing in look or in gesture something of the state of the life within.*

In his study of *Animal Forms and Patterns,*[5] Adolf Portmann explores the meaning of bodily form and demonstrates its revelatory character. He observes, for example, that with ascent up the mammalian line comes a marked accentuation of two poles of the animal body, one the center of awareness and expression, the other the center of reproductive activity. At the head pole, the head is progressively

*See also Chapter 13, "Looking Good: Nature and Nobility."

demarcated off from the rest of the body, and a marked and mobile face is eventually formed in the higher mammals, receiving and communicating meaningful looks; the genital pole is also progressively distinguished, by the descent of the testes, special patterns of hair or coloration, and other ornamentation. (We note in passing that the activities here centered provide the two major ways mammals transcend their particularity: in reproduction, and in communication and awareness.) These developments reach a certain peak in man—though the head and tail poles are no longer poles, due to the fact that man has acquired an upright posture, which places his head high above his groin. I turn next to the human body and consider the meaning of our peculiar way of standing-in-the-world, keeping this Portmannesque observation in mind. In a second reflection, I shall speak also about our sexuality.

Nearly all of what I have to say about the upright posture comes from an essay, "The Upright Posture,"[6] by the late German-American neurologist-psychologist, Erwin Straus, an essay one can hardly praise too highly and whose riches I barely begin to tap. Straus seeks to articulate a biologically oriented psychology that interprets human experience not as a train of percepts, thoughts, or volitions occurring in a sequestered mind or consciousness, but as a manifestation of man's position in the world, directed toward it, acting and suffering. He shows the close correspondence between human physique and certain basic traits of human experience and behavior—and ultimately connects our rationality with our bodily uprightness. "While all parts contribute to the upright posture, upright posture in turn permits the development of the forelimbs into human shoulders, arms, and hands and of the skull into the human skull and face."[7] I summarize but a few points about (1) standing, (2) the arms and hands, (3) eyes and mouth, and (4) the direction of our motion.

Though upright posture characterizes the human species, each of us must struggle to attain it. "Before reflection or self-reflection start, but as if they were a prelude to it, work makes its appearance within the realm of the elemental biological functions of man. In getting up, in reaching the upright posture, man must oppose the forces of gravity. It seems to be his nature to oppose nature in its impersonal, fundamental aspects with natural means."[8] Moreover, automatic regulation does not suffice; staying up takes continuous attention and activity. Awakeness is necessary for uprightness; uprightness is necessary for survival. Yet our standing in the world is always precarious; we are always in danger of falling. Our natural stance is, therefore, one of "resistance," or "withstanding," of becoming constant, stable.

This instituted and oppositional but precarious posture introduces an ambivalence into all human behavior. "Upright posture removes us from the ground, keeps us away from things, and holds us aloof from our fellow-man. All of these three distances can be experienced either as gain or as loss."[9] We enjoy the freedom of motion that comes with getting up, but we miss and often sink back to enjoy the voluptuous pleasures of reclining and relaxing. We miss the immediate commerce with things given to animals and crawling infants, but enjoy instead the pleasures of confronting a true and distant horizon, as interested seeing becomes detached beholding. As upright, we enjoy our dignity and bearing and the opportunity to encounter one another "face-to-face," yet this very rectitude makes us distant and aloof— verticals that never meet. To meet, we must bend or incline toward one another, or express our intentions to one another in some departure from strict verticality.

In upright posture, the upper extremities, no longer needed to support and carry the body, are free to acquire new tasks. Much has been made of the significance of the opposable thumb and the prehensile hand. But this is a small part of the story. The free swinging of the arms is crucial to the psychological experience of what Straus calls "action space," not the neutral homogeneous space of objective Cartesian science, but lived space, my space, a sphere of *my* action, which somehow both belongs to and gives rise to my sense of myself and to which I am related through body, limbs, and hands. In relation to action space, the hands develop into a true sense organ—a tool—of "gnostic touching," ranking with the eye and ear in powers of discrimination. The hand also functions, in cooperation with eye and ear and mind, to form new kinds of world-relations. Among its many new functions is pointing:

> In pointing, also, man's reach exceeds his grasp. Upright posture enables us to see things in their distance without any intention of incorporating them. In the totality of this panorama that unfolds in front of us, the pointing finger singles out one detail. The arm constitutes intervening space as a medium which separates and connects. The pointing arm, hand, and finger share with the intervening space the dynamic functions of separating and connecting. The pointing hand directs the sight of another one to whom I show something, for pointing is a social gesture. I do not point for myself; I indicate something to someone else. To distant things, within the visible horizon, we are related by common experience. As observers, we are directed, although through different perspectives, to one and the same thing, to one and the same world. Distance creates new forms of communication.[10]

Pointing points ultimately to both friendship and philosophy.

With upright posture come major changes in the head and face, and a reordering of the relation of the senses. Sight replaces smell as

the dominant sense, and in so doing is itself transformed, finally coming into its own as the sense of forms and wholes:

> In every species, eye and ear respond to stimuli from remote objects, but
> the interest of animals is limited to the proximate. Their attention is
> caught by that which is within the confines of reaching or approaching.
> The relation of sight and bite distinguishes the human face from those of
> lower animals. Animal jaws, snoot, trunk, and beak—all of them organs
> acting in the direct contact of grasping and gripping—are placed in the
> "visor line" of the eyes. With upright posture, with the development of
> the arm, the mouth is no longer needed for catching and carrying or for
> attacking and defending. It sinks down from the "visor line" of the eyes,
> which now can be turned directly in a piercing, open look toward distant
> things and rest fully upon them, viewing them with the detached interest
> of wondering. Bite has become subordinated to sight.[11]

Whereas smell, like taste with which it is intimately connected, is
a chemical sense indifferent to the forms of things, sight—especially
in higher forms—brings awareness of wholes. Thus, when sight is liberated from subordination to the mouth, it is open to become interested in forms *as such*, apart from the utility of such perception for
feeding and defense.

> Eyes that lead jaws and fangs to the prey are always charmed and spell-
> bound by nearness. To eyes looking straight forward—to the gaze of up-
> right posture—things reveal themselves in their own nature. Sight
> penetrates depth; sight becomes insight.[12]

Though man remains a nourishing being, his being-in-the-world is
not oriented solely or even primarily as eater.

> Animals move in the direction of their digestive axis. Their bodies are
> expanded between mouth and anus as between an entrance and an exit,
> a beginning and an ending. The spatial orientation of the human body is
> different throughout. The mouth is still an inlet but no longer a begin-
> ning, the anus, an outlet but no longer the tail end. Man in upright pos-
> ture, his feet on the ground and his head uplifted, does not move in the
> line of his digestive axis; he moves in the direction of his vision. He is
> surrounded by a world panorama, by a space divided into world regions
> joined together in the totality of the universe. Around him, the horizons
> retreat in an ever growing radius. Galaxy and diluvium, the infinite and
> the eternal, enter into the orbit of human interests.[13]

These prospects for wonder and thought are supported also by
striking changes in the mouth itself. Animal jaws, previously equipped
to grasp and crush, are extensively remodeled, as are the snout, teeth,
tongue, and muscles of the face. The human mouth—still the organ of
ingestion, taste, and mastication—has acquired the flexibility and
subtle mobility to serve the expression of emotions and especially the

articulation of speech. Where sight once served the mouth, now the mouth gives utterance to what mind through eyes has seen. The mouth not only homogenizes form to capture its matter; it helps now to preserve and communicate perceived and intelligible form through articulate speech. What enters the mouth nourishes the body; what departs the mouth nourishes the mind.

The dumb human body, rightly attended to, shows all the marks of, and creates all the conditions for, our rationality and our special way of being-in-the-world. Our bodies demonstrate, albeit silently, that we are more than just a complex version of our animal ancestors, and, conversely, that we are also more than an enlarged brain, a consciousness somehow grafted onto or trapped within a blind mechanism that knows only survival. The body-form *as a whole* impresses on us its inner powers of thought and action. Mind and hand, gait and gaze, breath* and tongue, foot and mouth—all are part of a single package, suffused with the presence of intelligence. We are *rational* (i.e., *thinking*) animals, down to and up from the very tips of our toes.† No wonder, then, that even a corpse still shows the marks of our humanity.

Looking Down on the Body

We can, it seems, be justly proud of our upright posture and the other bodily marks of the rational existence for which we are natured.‡ But

---

*Human respiratory patterns undergo marked changes during speech, and without conscious effort or awareness. More air is inhaled, the time of inspiration is shortened and the time of expiration is prolonged, the number of breaths per minute decreases markedly, chest and abdominal muscular activity differs, and so on. We tolerate these modifications of breathing for almost unlimited lengths of time, without suffering respiratory distress. Special inherited anatomical and physiological adaptations in breathing enable us to talk for hours. They constitute a crucial part of the composite package that is "rational animal," the animal having *logos* or thoughtful speech. See Curtis A. Wilson, "*Homo Loquens* from a Biological Standpoint," *The St. John's Review*, Summer 1983. See also E. H. Lennenberg, *The Biological Foundations for Language*, New York: John Wiley & Sons, 1967.

†I leave out of the present account the way in which the *gestalt* of our entire form shows forth what and who we are. I also neglect the various bodily marks of our individuality, from obvious things like face and gait and gesture to fingerprints, and, even on the cellular level, the unique cell-surface antigen patterns, which are responsible for such things as our unique blood type and our immunological rejection of alien material, including organ transplants. See Chapter 13, "Looking Good: Nature and Nobility." See also Adolf Portmann, *op.cit.*

‡A careless reader may think that I am here suggesting that man evolved (or was created) *for the purpose* of rationality, and that the entire argument about upright posture depends on such a teleological view. But these questions of human *origins* and their *causes* are beside the present point. The point is that we are *naturally* prepared, not just in mind but also in body—better, as a unified composite of both—for a life everywhere colored by thought. That such *is* our natural endowment may invite teleological speculations, but it neither requires nor presupposes them. It is an evident fact.

pride goeth before a fall. Our bodies are not only organized and self-organizing wholes, independent centers of awareness, thought and desire, and sources of purposive motions; we are not only self-maintaining and self-healing beings, individuated, well-defined, and discrete; not only upright, well-proportioned, and dignified in carriage; not only clever and dextrous, separate but in face-to-face communication with our fellows, through pointings, gestures, and articulate speech. Our bodies are also isolated, finally unshareable—yes, even in sexual union—and privatizing; vulnerable and weak; often mute and opaque; and frequently concealing rather than revealing of the soul within. Though highly touted as compliant tools, they all too often are an impediment and obstacle to our wills that refuse to do what we want them to—have you, too, perhaps, recently tried to slide into second base? And our bodies are sometimes ugly and misshapen, and very frequently ridiculous: in short, a positive embarrassment to anyone with pride. Such is the ancient discovery of our race when, its pride newly aroused, it first began to think about the body, which seems to have occurred when it first began to think at all. The body, after all, first comes to light as naked.

With the help of our tradition's most famous text on this subject, let us take a more sobering and less celebratory look at natural or original man—upright, and, as Milton said, "of far nobler shape, erect and tall, God-like erect, with native honor clad. . . ." Though we know from his later development that he was even then a being potentially possessed of reason, and hence of choice, man was, to begin with, guided by nature, instinctively seeking the things needful for life, which, his needs being simple, nature adequately provided. A prescient and benevolent God, solicitous of the man's well-being, might have sought to preserve him in this condition and to keep him from trying to guide his life by his own lights, exercised on the things of his experience, from which he would form for himself autonomous—self-prescribed—knowledge of good and bad, which is to say, knowledge of how to live. This tempting but dangerous prospect of autonomy, of choice, of independence, of the aspiration to full self-command, lay always at the center of human life, for to reason is to choose and to choose for oneself (even to choose to obey) is not-to-obey, neither God nor instinct nor anything else.

When the voice of reason awoke, and simple obedience was questioned (and hence no longer possible), the desires of the man began to grow. Though he did not know what he meant exactly, he imagined that his eyes would be opened and he would be as a god—that is, self-sufficing, autonomous, independent, knowing, perhaps immortal, and free at last. Such did the serpent promise—the voice that asked the world's first question and so disturbed its peace of mind forever.

Yet the rise of man to choice and knowledge brought none of these divine attributes—indeed, quite the contrary. The serpent had said, "Your eyes shall be opened and ye shall be as God, knowing good and evil." But, as the biblical author points out, with irony, "the eyes of them both were opened and they knew that they were naked; and they sewed fig-leaves together and made themselves girdles."[14] The first human knowledge relevant to life is knowledge of our nakedness and knowledge that nakedness is shameful and bad.

What is the meaning of nakedness? Why is the awareness of one's nakedness shameful? To be naked means, of course, to be defenseless, unguarded, exposed—a sign of our vulnerability before the elements and the beasts. But the text makes us attend, as did our ancient forebears, to our sexuality. In looking, as it were, for the first time upon our bodies as sexual beings, we discover how far we are from anything divine. As a sexual being, none of us is complete or whole, either within or without. We have need for and are dependent on a complementary other, even to realize our own bodily nature. We are halves, not wholes, and we do not command the missing complementary half. Moreover, we are not internally whole, but divided. We are possessed by an unruly or rebellious "autonomous" sexual nature within—one that does not heed our commands (any more than we heeded God's); we, too, face within an ungovernable and disobedient element, which embarrasses our claim to self-command. (The punishment fits the crime: The rebel is given rebellion.) We are compelled to submit to the mastering desire within and to the wiles of its objects without; and in surrender, we lay down our pretense of upright lordliness, as we lie down with necessity. On further reflection, we note that the genitalia are also a sign of our perishability, in that they provide for those who will replace us. Finally, all this noticing is itself problematic. For in turning our attention to our own insufficiency, dependence, perishability, and lack of self-command, we manifest a further difficulty, the difficulty of self-consciousness itself. For a doubleness is now present in the soul, through which we scrutinize ourselves, seeing ourselves as others see us, no longer assured of the spontaneous, immediate, unself-conscious participation in life—no longer enjoying what Rousseau longingly referred to as "the sentiment of existence," experienced with a whole heart and soul undivided against itself. Self-scrutiny, self-absorption, attention to ourselves being seen by others, vanity, and that perhaps greatest evil which is self-loathing—all these possible ills of thinking are coincident with self-consciousness; and self-consciousness is coincident with learning of our nakedness—our incompleteness, insufficiency, dependence, mortality, and the lack of self-command. Reason's first and painful discovery was of its own poor carcass. Rational we may be, but abidingly animal.

What are we to think of this double-ness imprinted on our bodies and essential to our being: on the one hand, our uprightness, our dignity, our capacity though we are only a part, here and now, to stand up before and to the world, to contemplate the whole and to think the eternal; and, on the other hand, our being weighted down, self-divided, naked, needy, and alone? We have, as it were, been demonstrating a possible and proper answer. Necessity may be a mark of our lowliness, but recognizing and owning up to our relation to necessity is not itself lowly. On the contrary, it is a mark of our dignity. Indeed, since most of dignity consists in our thoughtful response to necessity, we must even be grateful for it, just as we are indebted to gravity for the dignity of our posture, which, though exercised against gravity, depends absolutely upon gravity's power to bind us down. The rise of man may be ambiguous, but it is nonetheless a rise.

The animals, too, are naked, but they know no shame. They, too, experience necessity, but they neither *know* it nor know it *as necessary*. Thinking about the body may sober the thinker, and dispel his delusions of autonomy, but it does not cripple him. For one thing, the discovery of nakedness, however humbling, is a genuine discovery; our eyes are indeed opened. The so-called fall of man is identical to his mental awakening. Moreover, the discovery of his insufficiency becomes his spur to rise. "And they sewed fig-leaves together and made themselves girdles." Man does not take his shame lying down. Aroused from dormant potentiality, human ingenuity and manual dexterity give birth to the arts, at first glance, to cover our shame, but in truth to elevate and humanize the otherwise degradingly necessary. For in awareness of our need, we are capable not only of succumbing to it, but of meeting it in a knowing and dignified way: The story about our nakedness addresses us not only as naked, but as lovers of stories. In fact, the capacity for shame means also the capacity for the beautiful and the aspiration to the noble (see Chapter 13, "Looking Good: Nature and Nobility"). And, finally, in—and *only* in—the discovery of our own lack of divinity comes the first real openness to the divine. *Immediately* after making themselves girdles, reports the biblical author, "they heard the Lord God walking in the Garden," the first explicit mention that man attended to or even noticed the divine presence.

The significance of this stage of anthropological self-development has been marvelously summarized by Kant, in his "Conjectural Beginning of Human History," which is, in effect, largely a commentary on the Garden of Eden story:

> In the case of animals, sexual attraction is merely a matter of transient, mostly periodic impulse. But man soon discovered that for him this attraction can be prolonged and even increased by means of the imagination—a power which carries on its business, to be sure, the more

moderately, but at once also the more constantly and uniformly, the more its object is removed from the senses. By means of the imagination, he discovered, the surfeit was avoided which goes with the satisfaction of mere animal desire. The fig leaf (3:7), then, was a far greater manifestation of reason than that shown in the earlier stage of development. For the one [i.e., desiring the forbidden fruit] shows merely a power to choose the extent to which to serve impulse; but the other—rendering an inclination more inward and constant by removing its object from the senses—already reflects consciousness of a certain degree of mastery of reason over impulse. *Refusal* was the feat which brought about the passage from merely sensual to spiritual attractions, from merely animal desire gradually to love, and along with this from the feeling of the merely agreeable to a taste for beauty, at first only for beauty in man but at length for beauty in nature as well. In addition, there came a first hint at the development of man as a moral creature. This came from the sense of decency, which is an inclination to inspire others to respect by proper manners, i.e., by concealing all that which might arouse low esteem. Here, incidentally, lies the real basis of true sociability.

This may be a small beginning. But if it gives a wholly new direction to thought, such a beginning is epoch-making. It is then more important than the whole immeasurable series of expansions of culture which subsequently spring from it.[15]

Crucial to the development of genuine sociability and culture is the perception of one's place in the line of generations. Those who aspire to autonomy and self-sufficiency are prone to forget—indeed eager to forget—that the world did not and does not begin with them. Civilization is altogether a monument to ancestors biological and cultural, to those who came before, in whose debt one always lives, like it or not. We can pay this debt, if at all, only by our transmission of life and teachings to those who come after. Mind, freely wandering, in speculation or fantasy, can forget time and relation, but a mind that thinks on the body will be less likely to do so. In the navel are one's forebears, in the genitalia our descendants. These reminders of perishability are also reminders of perpetuation; if we understand their meaning, we are even able to transform the necessary and shameful into the free and noble. For even in yielding to our sexual natures—I must add, only heterosexually—we implicitly say yes to our own mortality, making of our perishable bodies the instruments of ever-renewable human life and possibility. Embodiment is a curse only for those who believe they deserve to be gods.

Where do we now stand regarding the body? What has our thinking about the body thus far revealed? The body bears throughout the marks both of human dignity and human abjection. It points us beyond itself, even to the heavenly and divine, and permits us to see and think and scheme; but it reminds us, too, of our debt and our duties to those who have gone before, that we are not our own source, neither in body nor in mind. Our dignity consists not in denying but in thoughtfully acknowledging and elevating the necessity of our embodiment, rightly regarding it as a gift to be cherished and respected. Through ceremonious treatment of mortal remains and through respectful attention to our living body and its inherent worth, we stand rightly when we stand reverently before the body, both living and dead.

But thinking about the body is revealing not only about body; on reflection, it sheds light also on thought and its puzzling relation to the being that thinks. Consider your present thoughtful activity. You are seated, solidly enough, and holding book in hand. Your eyes scan the visible symbols on the tangible pages. Yet without your conscious effort, your body has silenced awareness of thigh touching chair, hand grasping book, eye crossing page. You have, for the time being, suppressed or suspended all other concerns, somatic and psychic, concentrating your attention and energy to receive the ordered units of intelligibility, themselves incorporeal, which are in some mysterious way "linked to" or borne by these visible symbols on these tangible pages (or, in oral speech, by audible sounds), units of intelligibility once somehow "associated" with me, now "at work" on you. My thought is, at least in principle, present to you, even in my absence. One is forced to wonder: Can thought be corporeal?

The living body of the thinker has extension—length, breadth, width—and place; his thoughts have neither. He is here and now; they can be anywhere and of any time—in the best case timeless and enduring. Necessarily embodied, the thinking man is mortal, yet his thought—thought as such—may live on, especially as it is revivified in other and later minds. However much our minding depends on the proper organization and function of our bodily parts, we cannot but suspect that thought and mind are not corporeal. And, in any case, the thinker and his body are not simply of one mind. The body, even in upright posture, has its own subrational needs and aspirations— not to speak of pains and disorders—that get in the way of thinking: an empty stomach or a full bladder make thinking difficult; the aphrodisiac pleasure makes it impossible.* Can we equate the human being

*For these reasons, among others, philosophers have sometimes railed against their bodies and wished instead for disembodiment, for existence as pure minds, undistracted and unencumbered. The philosopher, it was said long ago, lives as close to death as possible, turning his back on things all too human, aspiring to live in accordance with that small but most divine element within him, intellect. Such men, it seems, exaggerated, else they knew not for what they were asking. Pure minds could be neither ours nor us; it makes no sense to wish something for ourselves whose attainment depends absolutely on our own disappearance. Even if thoughts (and minds) are incorporeal, the

as thinker with his body, even with his living, breathing, and moving body? How exactly can organic body think—or feel or desire or wonder or know? Thinking about the body of thinkers returns us to that mystery of mysteries which is its own ground: the being of an embodied mind or a thoughtful body. This is not a problem to be solved, but a question and perplexity to be faced, I suspect, permanently. We can here do little more than acknowledge it.

Thinking about the body is both exhilarating and sobering for the thinker: exhilarating because it shows the possibility of a more integrated account of his own psychosomatic being—against the prejudices of corporealists, subjectivists, and dualists—by showing the way in which his body prepares him (or, shall I say, itself) for the active life of thought and communication; sobering because it teaches him his vulnerability, dependence, and connectedness, exploding his illusions of and pretensions to autonomy. Thinking about the body is also constraining and liberating for the thinker: constraining because it shows him the limits on the power of thought to free him from embodiment, setting limits on thought understood as a tool for mastery; liberating because it therefore frees him to wonder about the irreducibly mysterious union and concretion of mind and body that we both are and live.

Thinking about the body ought also to be useful for thinking about and evaluating practices that deal with the body, in life and in death, though one should not expect to derive rules of conduct from such philosophical reflection. Although a proper examination of our practices lies beyond the scope of the present inquiry, I close with some thoughts about the treatment of the dead body prompted by the foregoing discussions.

Looking Rightly on the Body: Funereal Practices Reconsidered

Let us return to the story about the Greeks and the Indians. The story on its surface establishes the fact that each people thinks its own cus-

---

possibility of *our* having them depends absolutely on our being embodied, on our being a concretion—in fact, *this* concretion—of mind (soul) and body. Eyes and ears, mouths and tongues are the scouts and servants of thought; the visible cosmos provides to thought most of its nourishment; and some other bodies help it to feed and flourish. True enough, our body is sometimes a pain in the neck. True enough, our self-division often prevents us from self-fulfillment. Yet other bodies, harboring other minds, are a spur to thought (and other activities); the "doubleness" of friends helps heal the self-division of each. Shared thoughts and speeches, the highest activity of friends, draw us out of ourselves. In the best case, the one and the other, in truth the same mind but diversely incarnate, transcend particularity and are opened to the universal and the eternal. But here, too, embodiment aids in its own transcendence.

toms and mores are best; moreover, people generally do not believe
that their own customs are merely *customary*, but think them inherently—naturally—right or good or best. *Nomos*—law or custom or
convention—(and not nature or God or reason) is powerful and
authoritative (kingly, not tyrannical) over all: over all people who live
under it; over all aspects of human life, including especially what they
think is right. Yet we who have learned of this power of convention
are to that extent liberated from its rule. Contained in the discovery
of the conventionality of convention is the simultaneous discovery of
nature, of that which is everywhere the same and independent of human agreement. We are free to look to nature, to seek the underlying
and universal, which each culture then rules over differently. And we
are even free to ask whether all customs and beliefs about human life—
about body and soul—are equally true to nature or especially good for
a flourishing human life. Law may be king over all; but as there may
be better and worse, wiser and more foolish kings, so, too, with law.
Indeed, a more careful reading of the story bears out our suspicion. It
also suggests that different cultural attitudes and practices toward
the dead body may be emblematic of fundamental differences in ways
of life, some better, some worse.

The Greeks, men who, one infers, burn the bodies of their fathers,
answer Darius's question, declaring themselves absolutely unwilling
to adopt the practice of eating their dead. The Indians, men who eat
their fathers, refuse to answer the question as put, instead exclaiming
aloud to protest the very speech about burning the dead. The Indians,
more pious, or if you prefer, more superstitious, conflate deed and
speech; the undoable is also unthinkable, or at least unspeakable. Their
customs completely dominate their thought. They will never attain to
the insight that *nomos* is king over all, thereby discovering the difference between convention and nature; they will never discover a
world beyond their confines, never think freely, which is to say, never
really think. The Greeks, though closed in their practice, are open in
thought. They stand by and calmly listen, through an interpreter, to
the exchange with the Indians. The Greeks are pious, but mindfully
so. They are sober and attached to their own, but they know the difference between the love of their own and the truth. The Greeks, far
more than the Indians, behave as rational animals.

Is there any connection between this difference in behavior and
the difference in funereal practice? One cannot be sure, because one
cannot know what the people involved thought they were doing. Still,
the following suggestion seems at least plausible. The Indians in-
gested the bodies of their ancestors, thus preserving literally their connection with their own past, perhaps even in the belief that their
ancestors lived on inside them. Theirs was a death-defying and even

death-denying act, inasmuch as they swallowed much of the evidence of its occurrence. The Indians of this story probably made no distinction between soul and body, in identifying their fathers with the corpses of their fathers. Soul and especially mind had no independent origins or being, or if they did, their worth was subordinate to bodily life.

The Greeks, in contrast, knew the difference between the father and his mortal remains. Not by silent ingestion and incorporation, but in memory, through acts of loving speech and symbolic deed, do they remain respectfully mindful of their ancestor. Though its identity and integrity were respected, the dead body was not allowed to pretend to be the man it no longer was. It was dispatched, cleanly and purely, through ceremonious firing,* in full view of those who mourned. Greek ancestral piety is compatible with the independence of mind.

Though little noted, our story features a third people: the Persians, in the person of their king, Darius. Darius is presented as the man who has seen through the mere conventionality of conventions. Indeed, he revels publicly in his discovery. He compels people to look upon ways that are not their own, to confront what must be seen from his detached and enlightened view as the simple arbitrariness of their own way. Having transcended the limits of law—especially those tied to ancestral piety—he makes sport at the expense of the pious. Strict rationality is the Persian way: "The most disgraceful thing in the world, they think, is to tell a lie."[16] We learn elsewhere in Herodotus that the Persians looked to nature as divine—but only to the aloof, remote, permanent, and regularly moving bodies of the heavens (sun, moon, and stars), beings so unrelated and indifferent to human affairs that they might for all practical purposes just as well be absent. (In practical terms, the Persians were indistinguishable from atheists— and their practices show it.) Their funereal practice is what you might expect:

> There is another custom which is spoken of with reserve, and not openly, concerning their dead. It is said that the body of a male Persian is never buried, until it has been torn either by a dog or a bird of prey.[17]

The Persians also practice mutilations of their own living bodies, in the service of shaping themselves according to their will and taste.

The Greeks, it seems, are a mean between the superstitious Indians and the autonomous Persians, reverent rather than fanatical or impious, reasonable rather than either irrational or hyperrational. In honoring the bodies of their ancestors, they acknowledge their own

*In this regard, the Greeks may be said to have somewhat less regard for the body as mortal remains than those people—like the Jews—who condemn cremation, bury the body whole, and, in orthodoxy, oppose all operations on a corpse, including autopsy.

gratitude for the unrepayable gift of embodied life. Yet they make their peace with mortality by facing up to it and, through such representatives as Pindar and Herodotus himself, seek the enduring through memories, poems, and inquiries into the naked truth of things.

We, on the other hand, with our dissection of cadavers, organ transplantation, cosmetic surgery, body shops, laboratory fertilization, surrogate wombs, gender-change surgery, "wanted" children, "rights over our bodies," sexual liberation, and other practices and beliefs that insist on our independence and autonomy, live more and more wholly for the here and now, subjugating everything we can to the exercise of our wills, with little respect for the nature and meaning of bodily life. We expend enormous energy and vast sums of money to preserve and prolong bodily life, but, ironically, in the process, bodily life is stripped of its gravity and much of its dignity. Rational but without wonder, willful but without reverence, we are on our way to becoming Persians.

CHAPTER TWELVE

Mortality and Morality
The Virtues of Finitude

So teach us to number our days that we may get a heart of wisdom.
—Psalms (90:12)

". . . for you are dust and to dust you shall return." The man called his wife Eve because she was the mother of all living.
—Genesis (3:19-20)

Why should we die? Why should we, the flower of the living kingdom, lose our youthful bloom and go to seed? Why should we grow old in body and in mind, losing our various powers—first gradually, then altogether in death? Until now, the answer has been simple: We should because we must. Aging, decay, and death have been inevitable, as necessary for us as for other animals and plants, from whom we are distinguished in this matter only by our awareness of this necessity. We *know* that we are, as the poet says, like the leaves, the leaves that the wind scatters to the ground.

Recently, this necessity seems to have become something of a question, thanks to research into the phenomena of aging. Senescence, decay, and even our species-specific life span are now thought to be the result of biological processes that are, at least in part, genetically controlled, open to investigation, and in principle subject to human intervention and possible control. Slowing the processes of aging could yield powers to retard senescence, to preserve youthfulness, and to prolong life greatly, perhaps indefinitely. Should these powers become available, Whether to wither? and Why? will become questions of the utmost seriousness.

I think they should be serious questions even now, for several reasons. First, the project to control biological aging is already underway, and is part of the mission of the new National Institute on Aging. Whether and how vigorously to pursue this project is thus already a matter of public policy, and demands most thoughtful deliberation.

Second, the consequences of any success in the campaign against aging are likely to be vast and far-reaching, affecting all our important social institutions and our fundamental beliefs and practices. No other area of present biomedical research promises such profound alterations of our way of life, not to say of our condition.

But there is a more far-reaching reason for looking at the project to control aging, inasmuch as its objectives are, in many respects, continuous with the aspirations of modern medicine for longer life and better health. Indeed, prolongation of healthy and vigorous life—and, ultimately, a victory over mortality—is perhaps the central goal and meaning of the modern scientific project, associated in its founding with such men as Bacon and Descartes. Bacon it was who first called mankind to "the conquest of nature for the relief of man's estate," and there is ample suggestion in Bacon's writings that he regarded mortality itself as that part of man's estate from which he most needs relief. Bacon himself engaged in immortality research and may well have been its first martyr, sacrificing his life on the altar of longevity: He apparently contracted his fatal illness while performing freezing experiments on a chicken.

Descartes, in a famous passage in Part VI of the *Discourse on the Method*, rejects the speculative philosophy of his predecessors in favor of a new practical philosophy that would "render ourselves as masters and posessors of nature." He continues: "This is not merely to be desired with a view to the invention of an infinity of arts and crafts which enable us to enjoy without any trouble the fruits of the earth and all the good things which are to be found there, but also principally because it brings about the preservation of health, which is without doubt the chief blessing and the foundation of all other blessings in this life." Descartes prophesied "that we could be free of an infinitude of maladies both of body and mind, and even also possibly of the infirmities of age, if we had sufficient knowledge of their causes, and of all the remedies with which nature has provided us."[1]

Examining the campaign against aging might therefore shed some light on our entire scientific and technological project—its promise and its danger, its benefits and its costs. Thought about this future—albeit somewhat futuristic—prospect may along the way illuminate current practice and belief. At a minimum it will cause us to reexamine some of the basic assumptions on which we have been proceeding—for example, that everything possible should be done to make us healthier and more vigorous, that life should be prolonged and death postponed as long as possible, and that the ultimate goal of medical research is to help us live in health and vigor, indefinitely. Most important, we might become more thoughtful about the meaning of mortality and its implications for how to live.

Prospects for Retarding Aging

But allow me first to set out some preliminary observations and assumptions.

1. By *aging*, I mean the biological processes, distinct from disease, that makes the body progressively less able to maintain itself and to perform its various functions. Aging entails a gradual decline in vigor, a gradual degeneration of bodily parts and functions, an increasing susceptibility to disease, and an increasing likelihood of death. These changes are thought to be governed by a built-in biological clock or clocks, whose rate is species-specific and genetically determined.

2. By *life span*, I mean the biologically determined upper limit on longevity, different for different species, between ninety to one hundred years for human beings. This would be the life span of most of us in the absence of specific mortal diseases and fatal accidents. Thus, this specific age represents a biological "wall" against further increases in longevity by further improvements in medicine or our habits of life.

3. The biological clock and its midnight hour are probably linked; slowing the rate of aging could very well lead to a longer life span. Most knowledgeable people agree that the rate of aging probably can be slowed, but how much slowing or lengthening of life span is theoretically possible or technically feasible is anybody's guess.

4. The processes of aging are extremely complex and variable. Very little is known about their causes or about how to retard them. Still, the many theories that have been advanced now stimulate a growing amount of research. Other researchers believe that methods to slow aging can be discovered empirically, in advance of a full understanding of the causes. Such research is also currently being pursued.

5. The primary biological effects of age-slowing technologies could vary considerably—from increases in vigor with no gain in longevity, to a longer life span with all stages prolonged, to longer life with a prolonged period of decline or with partial or uncoordinated increases in vigor (e.g., stronger joints but weaker memory)—and cannot now be predicted. I shall here assume what is held to be the most attractive prospect, an increase in life span with parallel increases in vigor, for ten to twenty years, but perhaps longer. I shall also assume an anti-aging technology that is easy to administer, inexpensive, and not burdensome or distasteful to the users—a technology that will be widely demanded and used.

All these observations and assumptions warrant critical examination of the evidence and much further discussion.[2] Yet they suffice
to provide a plausible and concrete basis from which to approach our
more fundamental questions. And though one should not spend too
much time deliberating about the impossible, one should not unduly
encumber discussions of desirability with details of technique. Besides, it is ends, not means, that I wish to consider here.

Social Consequences

Aging research is pursued and supported by those who aspire to longer
life for man, recognizing as they do that medicine's contributions to
longevity have nearly reached their natural limit. As more fatal diseases and other causes of death are brought under control, more and
more people are living out the natural human life span. But aging research is also pursued and encouraged by many more who hope that
it will help to prevent or treat the infirmities, degenerations, and general loss of vigor that afflict the growing number of old people. These
ailments are, in large part, the hitherto necessary price for the gift of
longevity, a gift made possible by previous advances in hygiene, san-
itation, medicine, and general living conditions. The benefits of success for individuals are obvious—who would not like to avoid or
minimize for himself or his loved ones the burdens of weakness, im-
mobility, memory loss, and progressive blindness, deafness, and de-
mentia? These burdens to individuals are also costly for the society:
There is loss of productivity and expensive medical and social services. By reducing these losses and these costs, the community, too,
would presumably benefit from alleviating the handicaps and dependencies of the aged.

     Yet this is but a narrow view of the social implications of retarding
aging and contains a rather shrunken view of the old. The elderly are
related to us not only as nonproducing objects of care and expendi-
ture. They are, it should go without saying, in the first instance human
beings—now our ancestors, soon ourselves—most of whom do not
think of themselves as belonging to a separated class, insultingly called
"senior citizens." Especially as they are fit and able, they participate
as individuals in the complex network of functions, institutions, customs, and rituals that bind us all together. Yet for some purposes it
is useful to recognize what each of the age groups has in common and
to notice as well the interdependence of these groups. It should then
be clear that one cannot change the lot of one segment of the population without affecting the entire network of relations.

     To begin with, if life were extended ten to twenty years, what
would be the effects on the size and distribution of the population?
The percentages and number of people over age sixty-five continues

to increase; in 1900 they were 4 percent, today more than 11 percent of our population; in 1900 roughly 3 million, today roughly 26 million. How would still further increases in these numbers and percentages, or the growing numbers of nonagenarians and centenarians, affect work opportunities, retirement plans, new hiring and promotion, social security, housing patterns, cultural and social attitudes and beliefs, the status of traditions, the rate and acceptability of social change, the structure of family life, relations between the generations, or the locus of rule and authority in government, business, and the professions? Clearly these are very complex issues, affected not only by changing demographic patterns, but also by social attitudes and practices relating these various matters to perceived stages of the life cycle, and also by our ability to anticipate and plan for, or at least to respond flexibly to, dislocations and strains. Still, even the most cursory examination of any of these matters suggests that the cumulative effect of the result of aggregated individual decisions for longer and more vigorous life could be highly disruptive and undesirable, even to the point that many individuals would be *worse off* through most of their lives, and worse off enough to offset the benefits of better health afforded them near the end of life. Indeed, several people have predicted that retardation of aging will present a classic instance of the Tragedy of the Commons, in which genuine and sought-for gains to individuals are nullified or worse, owing to the social consequences of granting them to everyone.

Let me illustrate with one example. Consider employment. How will the large numbers of seventy- and eighty- and ninety-year-olds occupy themselves? Less infirm, more vigorous, they will be less likely to accept being cut off from power, work, money, and a place in society, and it would seem, at first glance, to be even more reprehensible than it now is to push them out of the way. New opportunities and patterns for work or leisure would appear to be needed. Mandatory retirement could be delayed, permitting the old to remain active and permitting society to gain from the continued use of their accumulated skills. But what about the numerous tedious, unrewarding, or degrading jobs? Would delaying retirement be desirable or attractive? Also, would not delayed retirement clog the promotional ladders and block opportunities for young people just starting out, raising obstacles to the ambitions and hopes of all—save for longer job security for those who have made it aboard?

The planned undertaking of second and third careers could provide alternatives to later retirement, but with few exceptions such opportunities would require reeducation, during mid-career, especially now that knowledge and skills needed for work are increasingly sophisticated and require more and more specialized education. These same educational requirements render difficult the development of new and rewarding uses of postretirement leisure, and it is far from clear that leisure is most fruitfully used when stacked up at the end of a life in which work is regarded as the main source of dignity. And, in any case, if the old are to be at leisure, the middle aged will have to pay—a task they are unlikely to want to undertake, strapped as they are by the mounting costs of caring for their young. A basic question we are already struggling with, and not very well, is how to accommodate our growing elderly population in a society whose young people are greatly troubled by feelings of powerlessness, frustration, and alienation. If people lived healthily to 100 or 120, if institutions were altered to meet their needs, we would likely have traded our problems of the aged for problems of youth. Retardation of aging could really mean prolongation of functional immaturity. Consider the young: isolated not only from the top of the ladders of power, but even from some of their lower rungs; supported by or even living with parents into their thirties or beyond; kept in a protracted sexually mature adolescence; frustrated; disaffected; some rebellious, the rest apathetic—the picture is not difficult to complete.

Clearly, to avoid such strains and disasters, great changes in social patterns and institutions would probably be needed, changes unlikely to occur except through strong centralized planning. The coming of such centralized planning will have consequences of its own, not all of them attractive or desirable, to say the least.

I have but scratched the surface of only one of the myriad areas of concern. The implications for society will be immense, and it boggles the mind to think of identifying them, much less to evaluate whether or not, on balance, we shall be better off. Some take a very gloomy view. One scientist colleague advised me to think of society as an organism, its individual members as cells. In this metaphor, unlimited prolongation of individual life would appear as a cancer, eating away at the body politic, preventing new life and new growth—a matter to which I shall return. Still, one should stress that questions about consequences are always in large part empirical and cannot be assessed in advance—though it is none too early to begin to formulate the questions worth asking.[3] One thing is clear: The stakes are very high and the issues very complex—enough to make us suspect utopian promises, projected from shallow glances through narrow lenses.

Against all these concerns about social consequences, it will be argued that we will soon enough adjust to a world of longevity. We will figure out a way. This confidence rests on what seems to be good evidence: We have always adjusted in the past. Let us grant this point. Let us for now overlook the fact that adjustment does not necessarily yield a more desirable state of affairs, and that not all change is progress. Let us not try to show that this technologically induced change may produce unprecedented changes, for which the history of past adjustments to novelty is an irrelevant source of optimism. Let us accept the optimist's view: longer life for individuals is an unqualified good; we will, in due time, figure out a way to cope with the social consequences.

The Proper Life Span

Conceding all this, how *much* longer life is an unqualified good for an individual? Ignoring now the possible harms flowing back to individuals from adverse social consequences, let us consider only the question, How much more life is good for us as individuals, other things being equal? How much more life do we want, assuming it to be healthy and vigorous? Assuming that it were up to us to set the human life span, where would or should we set the limit and why?

The simple answer is that no limit should be set. Life is good, and death is bad. Therefore, the more life the better, provided, of course, that we remain fit and our friends do, too.*

This answer has the virtues of clarity and honesty. But most public advocates of prolonging life through slowing aging deny such greediness. Immortality, or rather indefinite prolongation, is not their goal—it is, they say, out of the question (one wonders whether this is only because they deem it impossible). They hope instead for something reasonable; just a few more years.

How many years is reasonably few? Let us start with ten. Which of us would find unreasonable or unwelcome the addition of ten healthy and vigorous years to his or her life, years like those between ages thirty and forty? We could learn more, earn more, see more, do more. Maybe we should ask for five additional years? Or ten more? Why not fifteen, or twenty, or more?†

---

*These qualifications are, of course, crucial. Jonathan Swift's satirization of the wish for immortality depends on the fact that his immortal Struldbrugs (*Gulliver's Travels*, Voyage to Laputa) outlived all their contemporaries and became senile and decrepit. He thus failed to present the best case for immortality.

†One is reminded of the perhaps apocryphal story about Samuel Gompers, who once gave testimony at a trial involving alleged Communist infiltration into the labor movement. The Communist defense attorney cross-examined Mr. Gompers:

"Mr. Gompers, can you tell the Court, what is the goal of the American labor movement?"

"Very simply," said Gompers, "in one word: More."

"And when you have achieved more, what will be your goal then, Mr. Gompers?"

"More."

"And after that, then what?"

"More."

At which point the attorney addressed the bench:

"You see, your honor, the American labor movement is far more radical than the Communist party. We only want everything."

If we can't immediately land on the reasonable number of added years, perhaps we can locate the principle. What is the principle of reasonableness? Time needed for our plans and projects yet to be completed? Some multiple of the age of a generation, say, that we might live to see great-grandchildren fully grown? Some notion—traditional, natural, revealed—of the proper life span for a being such as man? We have no answer to this question. We do not even know how to choose among the principles for setting our new life span. The number of years chosen will have to be arbitrary, barring some revelation or discovery.

Under such circumstances, lacking a standard of reasonableness, we fall back on our wants and desires. Under liberal democracy, this means on the desires of the majority. Though what we desire is an empirical question, I suspect we know the answer: The attachment to life—or the fear of death—knows no limits, certainly not for most human beings. It turns out that the simple answer is the best: We want to live and live, and not to wither and not to die. For most of us, especially under modern secular conditions in which more and more people believe that this is the only life they have, the desire to prolong the life span (even modestly) must be seen as expressing a desire *never* to grow old and die. However naive their counsel, those who propose immortality deserve credit: They honestly and shamelessly expose this desire.[4]

Some, of course, eschew any desire for longer life. They profess still more modest aims: not adding years to life, but life to years. No increased life span, but only increased health, increased vigor, no decay. For them, the ideal life span would be our natural fourscore and ten, or if by reason of strength, fivescore, lived with full powers to the end, which end would come rather suddenly, painlessly, at the maximal age.

This has much to recommend it. Who would not want to avoid senility, crippling arthritis, the need for hearing aids and dentures, and the degrading dependencies of old age? Yet leaving aside whether such goals are attainable without simultaneously pushing far back the midnight hour, one must wonder whether, in the absence of these degenerations, we could remain content to spurn longer life, whether we would not become still more disinclined to exit. Would not death become even more of an affront? Would not the fear and loathing of death increase, in the absence of its antecedent harbingers? We could no longer comfort the widow by pointing out that her husband was delivered from his suffering. Death would always be untimely, unprepared for, shocking.

Montaigne saw it clearly:

I notice that in proportion as I sink into sickness, I naturally enter into a certain disdain for life. I find that I have much more trouble digesting this resolution when I am in health than when I have a fever. Inasmuch as I no longer cling so hard to the good things of life when I begin to lose the use and pleasure of them, I come to view death with much less frightened eyes. This makes me hope that the farther I get from life and the nearer to death, the more easily I shall accept the exchange. . . . If we fell into such a change [decrepitude] suddenly, I don't think we could endure it. But when we are led by Nature's hand down a gentle and virtually imperceptible slope, bit by bit, one step at a time, she rolls us into this wretched state and makes us familiar with it; so that we find no shock when youth dies within us, which in essence and in truth is a harder death than the complete death of a languishing life or the death of old age; inasmuch as the leap is not so cruel from a painful life as from a sweet and flourishing life to a grievous and painful one.[5]

Withering is nature's preparation for death, for the one who dies and for those who look upon him. We may wish to flee from it, perhaps, or seek to cover it over, but we must be cognizant of the costs of doing so.

By the way, it is well worth pausing to ask, Of *what* will we die in that golden age of prolonged vigor? Perhaps there will be a new spate of diseases, as yet unknown. More likely, the unnatural or violent causes will get us, as they increasingly do: some by auto; some by pistol; some by fire and some by drowning; some by lightning and some by bombing; some through anger and some through mercy; and some by poison from their own hand. Should we wish to avoid spilling blood, or desire a clean technological solution, we could require that our drink from the fountain of youth be accompanied by the implantation into our midbrains of an automatic self-destruction device, preset to go off at an unknown time some eighty to one hundred years hence. The control of natural decay might intensify the fear of violent death.

But to return from these macabre speculations to the main point: It is highly likely that either a modest prolongation of life with vigor or even only a preservation of youthfulness with no increase in longevity would make death even less acceptable, and would exacerbate the desire to keep pushing it further away . . . unless, for some reason, such life should also prove to be less satisfying.

Could longer, healthier life be less satisfying? How could it be, if life is good and death is bad? Perhaps the simple view is in error. Perhaps mortality is not simply an evil, perhaps it is even a blessing—not only for the welfare of the community, but even for us as individuals. How could this be?

The Virtues of Mortality

It goes without saying that there is no virtue in the death of a child or a young adult, or the untimely or premature death of anyone, before they had attained to the measure of man's days. I do not mean to imply that there is virtue in the particular *event* of death for anyone. Nor am I suggesting that separation through death is ever anything but pain for the survivors, those for whom the deceased was an integral part of their lives. Nor have I forgotten that, at whatever age, the process of dying can be painful and degrading, smelly and mean—though we now have powerful means to reduce much of, at least, the physical agony. Instead, my question concerns the fact of our finitude, the fact of our mortality—the fact *that we must die,* the fact that a full life for human beings has a biological, built-in limit, one that has evolved as part of our nature. Does this fact also have value? Is our finitude good for us—as individuals? (I intend this question entirely in the realm of natural reason and apart from any question about a life after death.)

To praise mortality must seem to be madness. If mortality is a blessing it surely is not widely regarded as such. Life seeks to live, and rightly suspects all counsels of finitude. "Better to be a slave on earth than the king over all the dead," says Achilles in Hades to the visiting Odysseus, in apparent regret for his prior choice of the short but glorious life.[6] Moreover, though some cultures—such as the Eskimo—can instruct and moderate somewhat the lust for life, ours gives it free rein, beginning with a political philosophy founded on the fear of violent death and on the mastery of nature for the relief of man's estate, and reaching to our current cults of youth and novelty, the cosmetic replastering of the wrinkles of age, and the widespread, and not wholly irrational, anxiety about disease and survival. Finally, the virtues of finitude—if there are any—may never be widely appreciated in any age or culture, if appreciation depends on a certain wisdom, if wisdom requires a certain detachment from the love of oneself and one's own, and if the possibility of such detachment is given only to the few. Still, if it is wisdom, the rest of us should harken, for we may learn something of value for ourselves.

It is, I recognize, awkward, and perhaps improper, for a relatively young man—I am forty-five—to praise mortality, especially before his elders. Doubtless, there are people reading this essay who are close to death, who may indeed know that they or a loved one is dying, and my remarks may give offense or may appear insensitive. More important, because of the apparent remoteness of my own end of days, I may simply not know what I am talking about. If wisdom comes through suffering, perhaps only among the old can there be wisdom about mortality. I am acutely aware of these possibilities, but I persist, offering as my excuse that, if I am off the mark, time will teach me my lessons, and that, in any case, whether my answer be right or wrong, the question is certainly worth thinking about.

Let us consider, then, the problem of *boredom* and *tedium*. If the life span were increased—say by twenty years—would the pleasures of life increase proportionately? Would professional tennis players really enjoy playing 25 percent more games of tennis? Would the Don Juans of our world feel better for having seduced 1,250 women rather than 1,000? Having experienced the joys and tribulations of bringing up a family until the last left for college, how many parents would like to extend the experience by another ten years? Similarly, those who derive their satisfaction from progressing up the career ladder might well ask what there would be to do for fifteen years after one had become president of General Motors or been chairman of the House Ways and Means Committee for a quarter of a century. Even less clear are the additions to personal happiness from more of the same of the less pleasant and less fulfilling activities in which so many of us are engaged so much of our time. It seems to be as the poet says: "We move and ever spend our lives amid the same things, and not by any length of life is any new pleasure hammered out."[7]

The problem of boredom is worse for us than it once might have been because of how we have come to understand it. For us, with our self-centered views, the fear of boredom is the fear that sooner or later the world and its objects will fail *us*. For the medievals, boredom meant that *we* will fail the world. They regarded boredom as a defect within oneself. It is an aspect of sloth—one of the seven deadly sins—according to Thomas Aquinas, a sin against the Sabbath, against the created order, not, as we might think, against the work week.*

The question of boredom leads directly to the second and more serious question, the question of *seriousness*. Could life be serious or meaningful without the limit of mortality? Is not the limit on our time the ground of our taking life seriously and living it passionately? To know and to feel that one goes around only once, and that the deadline is not out of sight, is for many people the necessary spur to the pursuit of something worthwhile. To number our days is the condition for making them count, to treasure and appreciate all that life brings. Homer's immortals, for all their eternal beauty and youthfulness, live shallow and rather frivolous lives, their passions only transiently engaged, in first this and then that. They live as spectators of the mortals, who by comparison have depth, aspiration, genuine feeling, and

---

*I am indebted to William F. May for these observations.

hence a real center to their lives. Mortality makes life matter—not only in the chemist's sense.

There may be some activities, especially in some human beings, that do not require finitude as a spur. A powerful desire for understanding can do without external proddings, let alone one related to our mortality; and as there is never too much time to learn and to understand, longer, vigorous life might be simply a boon. The best sorts of friendship, too, seem capable of indefinite growth, especially when growth is somehow tied to learning—though whether real friendship doesn't depend somehow on the shared perceptions of a common fate is a good question. But, in any case, I suspect these are among the rare exceptions. For most activities, and for most of us, I think it is crucial that we recognize and feel the force of not having world enough and time.

A third matter: *Beauty*. Death, says the poet, is the mother of beauty.[8] What he means is not easy to say. Perhaps he means that only a mortal being, aware of his mortality and the transience and vulnerability of all natural things, is moved to make beautiful artifacts, objects that will last, objects whose order will be immune to decay as their maker is not, beautiful objects that will bespeak and beautify a world that needs beautification, beautiful objects for other mortal beings who can appreciate what they themselves cannot make because of a taste for the beautiful, a taste perhaps connected to awareness of the ugliness of decay.

Perhaps the poet means to speak of natural beauty as well, which beauty—unlike that of objects of art—depends on its *im*permanence. Does the beauty of flowers depend on the fact that they will soon wither? Does the beauty of spring warblers depend upon the fall drabness that precedes and follows? What about the fading, late afternoon winter light or the spreading sunset? In general, is change necessary to the beautiful? Is the beautiful necessarily fleeting, a peak that cannot be sustained? Or does the poet perhaps mean not that the beautiful is beautiful because mortal, but that our appreciation of its beauty depends on our appreciation of mortality—in us and in the beautiful? Does not love swell before the beautiful precisely on recognition that it (and we) will not always be? It seems too much to say that mortality is the cause of beauty and the worth of things, but not at all much to suggest that it may be the cause of our enhanced appreciation of the beautiful and the worthy and of our treasuring and loving them.

Finally, there is the matter of that peculiarly human beauty: the beauty of *character*, of *virtue*, of *moral excellence*. To be mortal means that it is possible to give one's life, not only in one moment, say, on the field of battle—though that excellence is nowadays improperly despised—but also in the many other ways in which we are able in action to rise above attachment to survival. Through moral courage, endurance, greatness of soul, generosity, devotion to justice—in acts great and small—we rise up above our mere creatureliness, for the sake of the noble and the good. We free ourselves from fear, from bodily pleasures, or from attachments to wealth—all largely connected with survival—and in doing virtuous deeds overcome the weight of our neediness; yet for this nobility, vulnerability and mortality are the necessary conditions. The immortals cannot be noble.

Of this, too, the poets teach. Odysseus, long suffering, has already heard Achilles's testimony in praise of life when he is offered immortality by the nymph Calypso. She is a beautiful goddess, attractive, kind, yielding; she sings sweetly and weaves on a golden loom; her island is well-ordered and lovely, free of hardships and suffering. Says the poet, "Even a god who came into that place would have admired what he saw, the heart delighted within him." Yet Odysseus turns down the offer to be lord of her household and immortal:

> Goddess and queen, do not be angry with me. I myself know that all you say is true and that circumspect Penelope can never match the impression you make for beauty and stature. She is mortal after all, and you are immortal and ageless. But even so, what I want and all my days I pine for is to go back to my house and see the day of my homecoming. And if some god batters me far out on the wine-blue water, I will endure it, keeping a stubborn spirit inside me, for already I have suffered much and done much hard work on the waves and in the fighting.[9]

To suffer, to endure, to trouble oneself for the sake of home, family, and genuine friendship, is truly to live, and is the clear choice of this exemplary mortal. This choice is both the mark of his excellence and the basis for the visible display of his excellence in deeds both noble and just. Immortality is a kind of oblivion—like death itself.*

Longings for Immortality

In arguing the case for mortality I have tried to show that necessity is the mother of virtue, but some might argue that I am rather trying to make a virtue of necessity, and soon, not such a necessary necessity. Perhaps if we lived indefinitely, we would have no need of engagement, seriousness, beauty, or virtue. For we would be altogether different beings, perhaps capable of other satisfactions and achievements—though God only knows what they would be. And if mortality were such a blessing, why do so few cultures recognize it as such? Why do so many teach the promise of life after death, of something eternal, of something imperishable? We must face this challenge, for it leads

*The name Calypso means "one who hides or conceals or covers over."

us to the very heart of the question about mortality and the way we think about it.

What is the meaning of this concern with immortality? We are interested here not in the theological question but in the anthropological one: Why do human beings seek immortality? Why do we want to live longer or forever? Is it really first and most because we do not want to die, because we do not want to leave this embodied life on earth or give up our earthly pastimes, because we want to see more and do more? I do not think so. This may be what we say, but it is not what we mean. Mortality as such is *not* our defect, nor bodily immortality our goal. Rather, mortality is at most a pointer, a derivative manifestation, or an accompaniment of some deeper deficiency. That so many cultures speak of a promise of immortality and eternity suggests, first of all, a certain truth about the human soul: The human soul yearns for, longs for, aspires to some condition, some state, some goal toward which our earthly activities are directed but which cannot be attained during earthly life. Our soul's reach exceeds our grasp; it seeks more than continuance; it reaches for something beyond us, something that for the most part eludes us. True happiness, a genuine fulfillment of these deepest longings of our soul, is not in our power and cannot be fully attained, much less commanded. Our distress with mortality is the derivative manifestation of the conflict between the transcendent longings of the soul and the all-too-finite powers and fleshy concerns of the body.

What is it that we lack and long for? Notwithstanding their differences, many of our poets and philosophers have tried to tell us. One possibility is completion in another person. In Plato's *Symposium*, the comic poet Aristophanes speaks of the tragedy of human love and its unfulfillable aspiration. You may recall how we are said to spend our lives searching for our own complement, our other half, from whom we have been separated since Zeus cleaved our original nature in half:

> When one of them—whether he be a boy-lover or a lover of any other sort—happens on his own particular half, the two of them are wondrously thrilled with friendship and intimacy and love, and are hardly to be induced, as it is said, to leave each other's side for a single moment. These are they who continue together throughout life, though they could not even say what they would have of one another. No one could imagine this to be the mere sexual connexion, or that such alone could be the reason why each rejoices in the other's company with so eager a zest: obviously *the soul of each is wishing for something else that it cannot express, only divining and darkly hinting what it wishes.* Suppose that, as they lay together, Hephaestus should come and stand over them, and showing his implements should ask: "What is it, good mortals, that you would have of one another?"—and suppose that in their perplexity he asked them again: "Do you desire to be joined in the closest possible union, so that

you shall not be divided by night or by day? If that is your craving, I am ready to fuse and weld you together in a single piece, that from being two you may be made one, that so long as you live, the pair of you, being as one, may share a single life; and that when you die you also in Hades yonder *be one* instead of two, having shared a single death. Bethink yourselves if this is your heart's desire and if you will be quite contented with this lot." Not one on hearing this, we are sure, would demur to it or would be found wishing for anything else: Each would unreservedly deem that he had been offered just what he was yearning for all the time, namely to be so joined and fused with his beloved that the two might be made one. For this is the cause, that our ancient nature was this way and we were wholes: to the desire and pursuit of the whole, then, we give the name *eros.*[10]

Plato's Socrates both agrees and disagrees with Aristophanes. He agrees that we long for wholeness, completeness, but not in bodily or psychic union with a unique beloved. Rather, *eros* is the soul's longing for the noetic vision—for the sight of the beautiful truth about the whole: our soul aspires most to be completed by knowledge, by understanding, by wisdom; for only by possessing such wisdom about the whole could we truly come to ourselves, could we be truly happy. Yet Plato too strongly hints that wisdom is not given to human beings, at least not in this life; *philosophia,* yes, the love and pursuit of wisdom, yes, but its possession, no.

The Bible also teaches of human aspiration. Once we dwelled in the presence of God, the source of all goodness and righteousness; now we are estranged. That separation from God's presence occurs as the immediate result of eating of the tree of knowledge of good and evil, itself an act of autonomy (since all choice is nonobedience) and hence of separation. The serpent promised "and your eyes shall be opened and you shall be as God," but "their eyes were opened and they saw that they were naked." No, we are not as God; we are naked, weak, not-self-sufficient, possessed by powerful and rebellious desires that we can neither master nor satisfy alone. We are ashamed before ourselves, and we hide from God, even before we are caught and punished, and well before we are blocked from the possibility of tasting of the tree of life. The expulsion from the Garden merely ratifies our estrangement from God and testifies to our insufficiency, of which our accompanying mortality is but a visible sign—or perhaps even God's gift to put an end to our sad awareness and deficiency.

The decisive facts about all these—and many other—accounts of human aspiration, notwithstanding their differences, are the following:

1. Man longs not so much for deathlessness as for wholeness, wisdom, goodness.
2. This longing cannot be satisfied fully in our embodied earthly

life—the only life, by natural reason, we know we have. Hence the attractiveness of any prospect or promise of a different and thereby fulfilling life hereafter. We are in principle unfulfilled and unfulfillable in earthly life, though human happiness—that semblance of complete happiness of which we are capable—lies in pursuing that completion to the full extent of our powers.

3. Death itself, mortality, is not the defect, but a mark of that defect.

From these facts, the decisive inference is this:

This longing—any of these longings—cannot be answered by prolonging earthly life. No amount of "more of the same" will satisfy our own deepest aspirations.

Even the Christian promise of the end of days, which includes a resurrection of the body, is not to be understood vulgarly as the beginning of a never-ending and greatly eased earthly life of the sort we have, an uninterrupted gala of wining and dining, of winters in the Bahamas and summers on the Riviera, of disco dancing in golden slippers and Super Bowls on the heavenly turf, of listening to Elvis Presley or Caruso, of playing ball with Babe Ruth or making love to Marilyn Monroe. The kingdom of heaven is a promise of redemption, of purity, of wholeness in the presence of love and holiness.

If this is correct, then the proper meaning of the taste for immortality, for the imperishable and eternal, is not a taste that the conquest of aging would satisfy: we would still be incomplete; we would still lack wisdom; we would still lack God's presence; we would still lack purity. Mere continuance will not buy happiness. Worse, its pursuit threatens human happiness by distracting us from the goal(s) toward which our souls naturally point. By diverting our aim, by misdirecting so much individual and social energy toward the goal of bodily immortality, we may seriously undermine our chances for living as well as we can and for satisfying to some extent, however incompletely, our deepest longings for what is best. The implication for human life is hardly nihilistic: Once we acknowledge and accept our finitude, we can concern ourselves with living well, and care first and most for the *well-being* of our souls, and not so much for their mere existence.

Perpetuation

But perhaps this is all a mistake. Perhaps there is no such longing of the soul. Perhaps there is no soul. Certainly modern biology doesn't speak about the soul; neither does medicine or even our healers of the soul, our *psychi*atrists. Perhaps we are just animals, complex ones to be sure, but animals nonetheless, content just to be here, frightened in the face of danger, avoiding pain, seeking pleasure.

Curiously, however, biology has its own view of our nature and its inclinations. Biology also teaches about transcendence, though it eschews talk about the soul. For self-preservation is one thing, reproduction quite another; in bearing and caring for their young, many animals risk and even sacrifice their own lives. Indeed, in all higher animals, to reproduce *as such* implies both acceptance of the death of self and participation in its transcendence: The salmon, willingly swimming upstream to spawn and die, makes vivid this universal truth.

But man is natured for more than spawning. Much as it acknowledges and delineates our capacities and instincts for self-preservation and our remarkable powers to restore and maintain our wholeness, human biology, too, teaches us how our life points beyond itself—to our offspring, to our community, to our species. Man, like the other animals, is built for reproduction. Man, more than other animals, is also built for sociality. And, man, alone among the animals, is built for culture—not only through capacities to transmit and receive skills and techniques, but also through capacities for shared beliefs, opinions, rituals, traditions. The origins of these powers for culture and their significance are matters of dispute, but their existence is not.

Many have called attention to man's remarkable biological characteristics that prepare him for culture, including the following: (1) the prolonged period of neonatal yet still embryonic dependence and development, called by Portmann the period in the social womb, during which the child learns to speak and stand and begins to perform voluntary actions; (2) the upright posture, which permits a beholding of the world (in turn eliciting our curiosity), which exposes things at a distance and at the same time frees the hands to fashion means for overcoming distance, which brings us face-to-face with our fellows, opposed but in communication; (3) our capacity for speech, requiring special laryngeal and respiratory, as well as cerebral, development, and a relation to others with whom that capacity is actualized through a learned language; (4) a sense of time and powers of imagination and forethought for the future; (5) special social passions, such as friendliness, shame, pity and respect, which permit and are cultivated in community (see Chapter 13, "Looking Good: Nature and Nobility"); and (6) special ethical powers, including a capacity for acquiring a sense of responsibility, of fairness, and of concern for posterity, which culture requires but also nurtures.

To be sure, the present orthodoxy in sociobiology treats our sociality as but a fancy mechanism geared to the sole end of the survival of the human gene pool. A richer sociobiology might come to understand that it is not just *survival*, but survival of *what*, which matters. It might again remember that sociality and culture, admittedly part of the means of preservation, are also part of the end for which we seek to preserve ourselves, and that only in community and through culture do we come into our own as that most special animal. But however this may be, biology does teach that we must see ourselves as community- and species-directed, and not merely self-directed. We are built with leanings toward and capacities for perpetuation.

Is it not possible that aging and mortality are part of this construction, and that the rate of aging and life span have been selected for their usefulness to the task of perpetuation? Could not overturning the process of aging place a great strain on our nature, jeopardizing our project and depriving us of success? For interestingly, perpetuation is a goal that *is* attainable. Here is transcendence of self that is largely realizable. Here is a form of participation in the enduring that is open to us, without qualification—provided, that is, that we remain open to it.

Biological consequences aside, simply to covet a prolonged life span for ourselves is both a sign and a cause of our failure to open ourselves to this—or any higher—purpose. It is probably no accident that it is a generation whose intelligensia proclaim the meaninglessness of life that embarks on its indefinite prolongation and that seeks to cure the emptiness of life by extending it. For the desire to prolong youthfulness is not only a childish desire to eat one's life and keep it; it is also an expression of a childish and narcissistic wish incompatible with devotion to posterity. It seeks an endless present, isolated from anything truly eternal, and severed from any true continuity with past and future. It is in principle hostile to children, because children, those who come after, are those who will take one's place; they are life's answer to mortality, and their presence in one's house is a constant reminder that one no longer belongs to the frontier generation. One cannot pursue youthfulness for oneself and remain faithful to the spirit and meaning of perpetuation.

In perpetuation, we send forth not just the seed of our bodies, but also a bearer of our hopes, our truths, and those of our tradition. If our children are to flower, we need to sow them well and nurture them, cultivate them in rich and wholesome soil, clothe them in fine and decent opinions and mores, and direct them toward the highest light, to stand straight and tall—that they may take our place as we took that of those who planted us and who made way for us, so that in time they, too, may make way and plant. But if they are truly to flower, we must go to seed; we must wither and give ground.

To be fair, I must confess that to seek immortality through one's children can be a snare and a delusion, perhaps today more than ever. Continuity of lineage, and, more important, of mores and beliefs, is in no way assured, not least because our ethos has become less hospitable to the concern for transmission, in our effort to push back our own deaths and ensure our private rights to the endless pursuit of happiness, understood as end-less pursuit. But there *is* something that we can certainly preserve and perpetuate, and only through sowing fresh seed. To see this, we need to look again at the nature of growing old.

Those who look primarily at the aging of the body and those who look upon the social and cultural aspects of aging forget a crucial third aspect: the psychological effects simply of the passage of time—of experiencing and learning about the way things are. After a while, no matter how healthy we are, no matter how respected and well-placed we are socially, most of us cease to look upon the world with fresh eyes. Little surprises us, nothing shocks us, righteous indignation at injustice dies out. We have seen it all already, seen it all. We have often been deceived, we have made many mistakes of our own. Many of us become small-souled, having been humbled not by bodily decline and not by "the system" but by life itself. So our ambition also begins to flag, or at least our noblest ambitions. As we grow older, we "aspire to nothing great and exalted and crave the mere necessities and comforts of existence."[11] At some point, most of us turn and say to our intimates, Is this all there is? We settle, we accept our situation—if we are lucky enough to have been prepared to accept it. In many ways, perhaps in the most profound ways, most of us go to sleep long before our deaths. In the young, aspiration, hope, freshness, boldness, openness spring anew—even if and when it takes the form of overturning our monuments. Immortality for oneself through children may be a delusion, but participating in the natural and eternal renewal of human possibility through children is not—not even in today's world.

For it still stands as it did when Homer made Glaukos say to Diomedes:

> As is the generation of leaves, so is that of humanity. The wind scatters the leaves to the ground, but the live timber burgeons with leaves again in the season of spring returning. So one generation of man will grow while another dies.[12]

And yet it also still stands, as this very insight of Homer's itself reveals, that human beings are in another and decisive respect unlike the leaves; that the eternal renewal of human beings embraces also the eternally human possibility of learning and self-awareness; that we, too, here and now may participate with Homer, with Plato, with the Bible, yes with Descartes and Bacon, in catching at least some glimpse of the enduring truths about nature and human affairs; and that we, too, may hand down and perpetuate this pursuit of wisdom and goodness to succeeding generations for all time to come.

CHAPTER THIRTEEN

Looking Good
Nature and Nobility

O Chestnut-tree, great-rooted blossomer,
Are you the leaf, the blossom or the bole?
O body swayed to music, O brightening glance,
How can we know the dancer from the dance?
> —W. B. Yeats, "Among School Children"

Student: Look, professor, at this beautiful feather I found.
Biologist: What do you take me for, a milliner?
> —Anonymous

Wisdom does not inspect, but behold.
We must look a long time before we can see.
> —Henry David Thoreau, *Natural History*
> *of Massachusetts*

Purposive yet perishable, needy yet aspiring, living things seek their own good. Over against felt necessity, life seeks to be, to persist and to perpetuate itself. Living things seek to transcend their own isolation, particularity, and impermanence, through awareness, communication, and, above all, through reproduction. Necessity, the mother of aspiration, provokes in man, because he alone *knows* and not only feels necessity, the search for the object of life's—and his own—aspiration. *We wonder*, even quite concretely: To what do we aspire? Is it to mere presence or continuity, a fleeing of the perishable? Or is it also to specific positive kinds of flourishing—to participation in what is beautiful or good or true? Moreover, are these objects of human aspiration purely matters of our own human contrivance? Or does nature herself have something good to teach, something beautiful that beckons? If necessity is aspiration's mother, has it no natural father?

Can knowledge of nature, or of human nature, show us anything about how better to live? Conversely, can any of our moral and political notions find support in the nature of things? Such questions have been agitated over and over in the Western tradition, at least since classical antiquity. "The noble and the just things," says Aristotle at the start of his ethical inquiry, "possess so much variety and confusion that they are thought to be only by convention (*nomos*) and not by nature (*physis*)";[1] his *Nicomachean Ethics* may be said to be one long exploration—and partial refutation—of this "conventionalist" thesis. Yet neither Aristotle's nor anyone else's teaching of a naturally supported ethics has ever been allowed to rule with ease. Skeptics in these matters frequently controverted the alleged ethical teachings of nature by pointing out that dumb nature had not voice to speak: Nature was, of course, mute about better and worse because nature was silent altogether.

Skepticism about any connection between nature and good is now more than ever in the saddle, and anyone who breaks silence to reopen these questions must be prepared to ride against the pack. In fact, skepticism has given way to a confident new dogmatism: that nature, if it could only speak, would say nothing at all attractive to human ears, about the good, the beautiful, or even the absolutely true. The beautiful and the good are seen not as ideas that the human mind may discover and behold, but as purely human creations, constructs made by the human mind to give order to a chaotic and inhospitable world. And not, mind you, by *the* human mind, but by a plethora of human minds, each relative to a particular time, place, culture, and individual point of view. What few nonrelativists survive look no more to nature, but instead either to some transcendental reason or to history, seen as the progress of human freedom in its overcoming of nature.

All these modern beliefs—the historicisms, cultural relativisms, and existentialisms—are intellectually rooted in the rise of modern natural science. The most powerful obstacle to finding humanly meaningful clues from nature is modern science's approach to and assumptions about the nature of nature. Whether we look to science's mathematization of nature and its break with ordinary experience and common sense, or to the denial of natural teleology and its dissolution into chance and necessity, or to the analytic and reductive modes of understanding wholes in terms of parts (e.g., animals in terms of DNA), or to the insistence on the absolute distinction between facts and values, or to the demand for an artificial objectivity that removes the knower from amidst the things known—in all these aspects, modern science teaches the uselessness of natural knowledge for ethics. Moreover, certain substantive teachings of science—say, corporealism, mechanism, or Darwinism—will likely incline anyone interested in ethics not only to ignore but even to *flee* the teachings of cosmic nature, because they show cosmic nature as, to say the least, indifferent to human longing and aspiration, and in the end, godless. The knowledge of nature that science finds has been, for ethics, at best only negatively helpful—or if you prefer, subversive—insofar as it embarrasses the ethical claims of traditional philosophical or religious teachings, by contradicting what they say about nature, man, and the whole.

If one suspects, as I do, that the questions about nature, man, and the human good are not settled once and for all—indeed, many of the previous chapters implicitly and even explicitly challenge the adequacy of our beliefs, say, about life, health, teleology, evolution, the body, or the place of man—then one must seek some way to have them reopened. If the problem lies with our view of nature, perhaps we need to take another look. If science errs in deliberately abstracting from the world of ordinary experience, sacrificing the vivid for the mathematically precise, and forsakes knowledge of beings for knowledge of laws permitting prediction and control of events, perhaps we need a fresh look at what is immediately and manifestly evident. Looking toward a more natural science, one more truthful *to* the phenomena of life, as ordinarily lived and encountered, we can do no better than to look *at* the phenomena, literally, at the things that appear. We will look to the superficial in the hope that it may turn out to be profound.

Though we are, of course, looking for clues for human life, we should begin not with man but with his animal relations. We do so not because we think man and the animals are alike on all fours; indeed, we have argued (in Chapter 10 and elsewhere) that though continuous in origin with the rest of nature, man's difference makes him at the same time also a thing apart. Rather, as we are looking for something strictly natural—something uncontaminated by contributions of human consciousness or culture—that might nevertheless point toward the human good, we begin by looking afresh at the looks of animals.

Our interest is both in the looking and the looks. The better view of nature that we seek requires relearning how to view. Indeed, although we are flooded with sophisticated visual images, through advertising, television, photography, and film, we are not adept at looking. Moreover, our culture tends more generally to share science's depreciation of surfaces and distrust of appearances—in a way, rightly so, for we know that much of what we put before one another's eyes is either trivial or false. We believe that the truly real lies always buried beneath the surface—in unconscious motives, hidden agendas, underlying causes, latent tendencies, and deeper reasons. We need to retrain and concentrate our visual powers, and open our minds to the possibility that there is more than we think in what meets the eye. Nature's surfaces, at least, are not only apparent but real, and, as we shall see, important. Accordingly, in the spirit of looking afresh, we offer more a showing or display, less a proof. We mean to provoke questions, not to solve problems. We aim to stir the mind to wonder by means of sightings, not to quiet it by means of logical deduction.

The Looks of Animals

What do I mean by the looks of animals? I mean, in the first instance, its visible form, its surface appearance. The look is more than shape and more than a mere aggregate of parts. It is a unified whole on display, what in German might be called *gestalt*. The look is, at least in speech, distinct from the materials that carry it, as the representation of a painting is other than the oils and canvas. The looks of animals differ, each more or less after its kind; "looks" might be a literal translation of the Latin *species*. Unlike the looks of artifacts, the looks of animals are mobile, as mobile as the animals that bear them. Yet even the motions of the looks are true to form.

What has modern biology to say about animal looks? For the most part, very little. The prevailing principles and methods push in other directions. The way of analysis treats the whole organism, looks included, in terms of smaller and smaller parts, the positions and motions of which are said to account for the whole. The complementary genetic way explains the whole in terms of its origins. Our special theory of origins, evolution by means of natural selection, regards organisms largely as survival machines, contributing to the survival of the population, or perhaps only its genes. On this view, animal looks are understood as contributing to individual and species survival, by making the animal less visible or more impervious to predators and more attractive to the opposite sex. Further, the superficial *gestalt* holds little interest for an experimental biology whose chief aim is to discover how to predict and control phenomena by figuring out *how* things work.

All of these legitimate and valuable ways and notions have, however, a homogenizing tendency. They search mainly for universal "laws" of life, whether it be the universal genetic code and its translation or the uniform mechanisms of natural selection and speciation. Biology, with the notable but only partial exceptions of ethology and functional and evolutionary morphology, is now largely the study of vital phenomena in general, not primarily the study of plants and animals in their particularity, manyness, and distinctiveness. Little if any effort is made to understand the looks of animals, or, by means of the looks, the natures of the animals that display them.

Yet living nature cares a great deal for surfaces. Unbeknownst to most of us, nature continues in its profusion to put forth and sustain things in natural kinds—not eternal or imperishable kinds to be sure, but kinds none the less. And the significance of the looks is not a matter of indifference to nature, or at least to the respective kinds. A rabbit is more impressed with the difference between the looks of a rabbit and the looks of a fox than it is with the fact that rabbits and foxes both share a common ancestor and the same genetic code.

A rarity among contemporary biologists, Adolf Portmann, the late Swiss zoologist, was deeply interested in animal *gestalt*, and made it an object of lifelong attention and careful research.* Portmann understood—and can help us to see—the profundity of the superficial:

> What more than anything else urges and indeed compels us to take an interest in these animal forms is the impression, conveyed by their appearance, that *their life is related to our own* and *possesses an inwardness revealed through the animal's form and its independent behavior.* To find some way of understanding this inwardness of theirs is one of the strongest incentives of biological research.[3]

If one begins to attend to the surfaces of things, and looks without prejudice at the myriad animal species, one cannot fail to be impressed by the astonishing variety and elaborateness of animal looks. Nature's exuberance in surface display is simply staggering. This lavishness alone might render doubtful the view that the surface patterns and colorations are but more or less accidental by-products of unrelated, vitally important developmental processes, which may have incidentally produced certain rhythmic patterns on the surface. But Portmann shows further that an animal's look, especially among the higher animals, is the result of a genetically determined plan directed solely toward the production of the look. Patterns of the adult are formed from independent primordia in the embryo, located far apart from one another, but coordinated for later fusion under strict genetic control. For example, individual bird feathers form, within separate horny sheaths, relatively independent patterns, but when they later come together a "picture" is formed (Fig. 1.)[†] In other cases, the adult pattern appears quite belatedly, only after several moultings of the skin. The same peacock feather germ, for example (Fig. 2), will produce three different but rather drab feathers before it produces the brilliant eye-containing tail feather after the third moult. Judging from the preci-

---

*This part of the chapter draws heavily on two of Portmann's books, *Animal Forms and Patterns* and *Animals as Social Beings.*[2] My purpose is less to persuade the reader of the correctness of his conclusions and more to commend his questions and the importance of his concern.

†Figs. 1 to 8 (including legends) have been taken from Adolf Portmann, *Animal Forms and Patterns* (translated by Hella Czech, illustrated by Sabine Baur), Faber and Faber, London, 1964 (paperback edition, Schocken Books, New York, 1967). Reproduced by permission of the publisher.

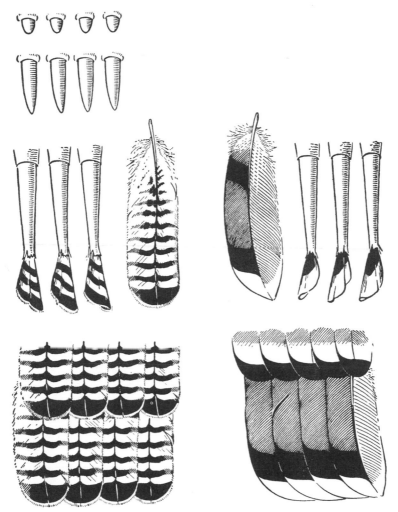

Fig. 1. We see a series of feather germs, at an early stage of growth when they are within their horny sheaths. In the upper pictures each feather has developed a relatively independent pattern; below, on the other hand, a 'picture' is produced by several feathers taken together.

Reprinted by permission of Faber and Faber Ltd from ANIMAL FORMS AND PATTERNS by Adolf Portmann, Fig. 3, p. 21.

sion and strictness of the genetic controls, one might say that the look was every bit as important as the so-called vital organs (e.g., heart, liver, kidneys).

Important for what? Portmann argues that the importance or meaning of the look goes far beyond its useful contributions to survival or sexual selection. True, the animal's skin is an organ for pro-

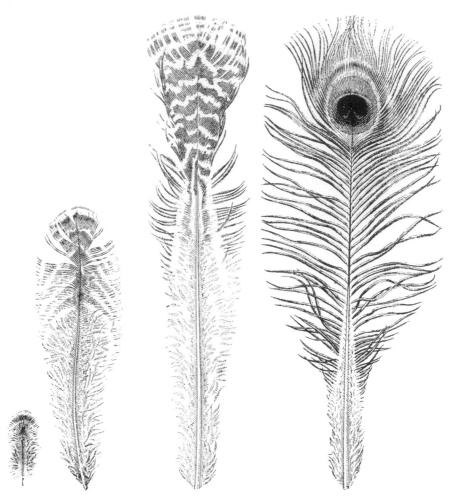

Fig. 2. The primordium of a single feather from the long train of a peacock is able to form quite different kinds of feathers during the course of juvenile development. The illustration shows the four feathers successively produced from the same feather germ. They have not only become gradually larger, but have also changed as regards the formation of their pattern. What transformation must take place in the embryonic tissue of the skin to allow the beautiful structure of the peacock's eye to develop after the third moult! As yet we know next to nothing about these hidden processes.

Reprinted by permission of Faber and Faber Ltd from ANIMAL FORMS AND PATTERNS by Adolf Portmann, Fig. 92, a and b, pp. 162–163.

tection and heat exchange. True, among animals possessed of sight, the look of an animal is useful in attracting a mate or warding off a predator. Yet the patterns seen, their intricacy, their ornateness, and their exuberance of detail go far beyond what is needed for these utilitarian functions. Moreover, the specific details of the pattern cannot be accounted for by these functions. Consider the peacock fan. Studies have shown that the female responds to the small kernel-shaped objects seen in some of the feathers, and also to the liveliness of the dance. But isn't the elaborateness of the pattern far in excess of what is needed for these goals? Moreover, what is one to make of the elaborate and colorful patterns displayed by animals that cannot see? Or the patterns of those that cannot be seen, such as the microscopic, richly structured one-celled animals known as radiolarians, never seen by anyone until the microscope was invented (Fig. 3)? What selective advantage can these patterns confer?

Or consider the progressive descent of the testes in mammals, from their well-protected intra-abdominal location to their exposed position in thin-walled scrotal sacs. Some try to attribute this to the fact that sperm production in mammals proceeds vastly better at the lower temperature of the scrotum than at the higher temperatures within the body cavity. But this is more likely a secondary adaptation to, and not a cause of, the descent. Birds have body temperatures higher than ours and yet their intra-abdominal testes function perfectly well. What selective advantage could there be in this precarious display?

We are invited to consider the possibility—Portmann deems it a likelihood—that the surface has a meaning beyond any direct contribution to species survival:

> It is our aim to arrive at as comprehensive an understanding as possible of animal form. For many people the view is still often obscured by a one-sided functional method of approach. They are only prepared to see horns as weapons or as a sexual character. Both these explanations are true; but they forget that this view does not enable us to grasp fully the peculiar shape and position of these structures. The slender limbs of many hoofed mammals are viewed merely as instruments for running, for speedy flight in forest or steppe. This is undoubtedly correct—but in addition to that, *they are also part of a higher grade of differentiation.* This need not be a functionally more efficient one; when it becomes further developed it may finally contribute to the extinction of the species.[4]

We note, in passing, that a higher level of organization does not necessarily imply better prospects for survival, for the higher ranked are also exposed to greater dangers and risks. For example, such animals face greater dangers of starvation, unknown to lower forms, which are able to avoid death by adopting a condition of inactivity. The chase does not always yield the prey.

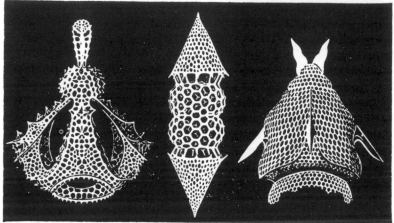

Fig. 3. Skeletons of Radiolarians (after E. Haeckel). These unicellular animals, which all live in the deeper zones of the sea, are visible only under the microscope.

Reprinted by permission of Faber and Faber Ltd from ANIMAL FORMS AND PATTERNS by Adolf Portmann, Fig. 58, p. 106.

What is a higher grade of differentiation? What generalizations emerge if one looks up the evolutionary scale? The more highly organized animals, those capable of receiving and processing richer and more complex impressions of their surroundings and capable of greater "freedom" of motion in their environment, all show advanced development of the brain, and within the brain, of the higher centers, of the cerebrum. This increasing cerebralization, the condition of higher awareness and richer experience of the world, correlates well with increasing overall differentiation, and is widely regarded as a mark of higher animal rank.

Looking to the outside, one observes coordinate signs of rank. Internal developments of the brain are roughly paralleled by progressive differentiation of a head region. Lower molluscs, like the clam, and the somewhat higher Nautilus lack any clearly demarcated head; the more advanced members, like the cuttlefish and octopus, have fully developed heads, combining in one locus the activities of feeding, sentience, and direction of locomotion (Fig. 4). Among the vertebrates, in fish

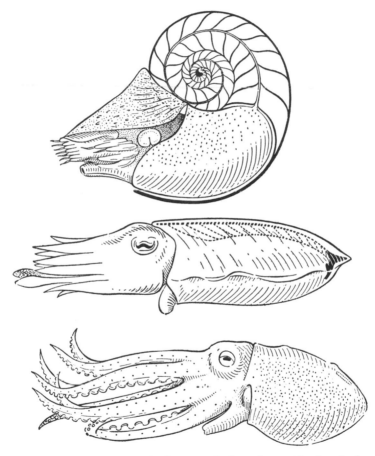

Fig. 4. Reduction of the shell in cephalopods as the brain becomes increasingly differentiated. *Nautilus, Sepia, Octopus.* At the same time the tentacles increase in size as compared with the rest of the body, whereas their number is reduced.

Reprinted by permission of Faber and Faber Ltd from ANIMAL FORMS AND PATTERNS by Adolf Portmann, Fig. 56, p. 104.

the head is continuous with the body, whereas birds and mammals have a distinctive, marked off head. In higher mammals, the distinctive head is further accentuated by color, hair, and pattern, to produce a face, as can be seen in comparing the head of a lower mammal, such as the dormouse, with that of the tiger or chimpanzee (Fig. 5).

The correlations between internal structure and outward appearance are striking. First, an increased power to see is paralleled by an increased complexity and subtlety of what is to be seen. The looking and the looks grow richer together. Second, the outward appearance, because it parallels inward structure, comes to be more and more revealing. In higher mammals, the received look conveys more of life's self-manifestation. Not only does the head present a face, but the en-

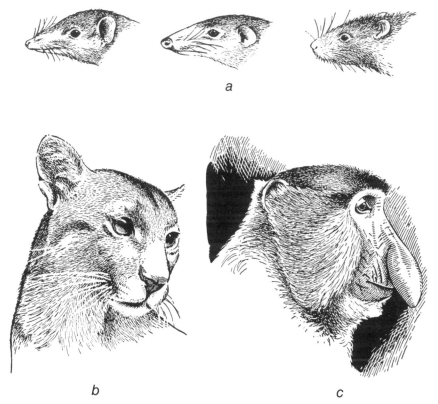

Fig. 5a. Pouched mouse (Australia); tree shrew (Eastern Asia); rat.
Fig. 5b. Puma (America).
Fig. 5c. Proboscis monkey (Borneo). Only the male animal has the nasal pouch which serves as an organ of resonance.

Reprinted by permission of Faber and Faber Ltd from ANIMAL FORMS AND PATTERNS by Adolf Portmann, Figs. 28a to 28c, p. 71.

tire body expresses a unity of form that reveals not only the inner structure but also something of the "inner" life. Behold the look of the tiger (Fig. 6):

> How elaborately is the tiger's body divided up by the arrangement of stripes! The transverse markings around the limbs and body have their various directions unified by cleverly arranged gussets where the legs leave the trunk: In the magnificent pattern of the head, the most beautiful figuring is reserved for the purpose of bringing eyes and ears into relief. For all its free rhythm, the pattern nevertheless still follows strictly the arrangement of the particular parts. It emphasizes the leading pole, it frames and accentuates the important sense organs. This correlation between internal structure and outward appearance creates a unity of the visible form which appeals to us clearly in its powerful impression. It is as if the higher animal becomes, in some peculiar way, transparent, since the importance of the special parts of its form, and the play of its limbs, is stressed by the accentuations in the pattern. How extraordinary it is that in the higher animals significant places should be made to stand out![5]

If the look of an animal is toward the eye of the beholder, and if that look can reveal something of life within, the surface becomes an organ of communication. In the higher animals, the look can communicate stages of youth or maturity, or certain seasonal and cyclical stages such as readiness for mating. Expressions of the mood of the moment are projected to friend and foe: The facies and postures of fear, annoyance, tense attention, or rage are well known to the students of the cat family or monkeys. Not all expressions of inwardness

Fig. 6. The tiger as an example of the pattern on one of the higher mammals.

Reprinted by permission of Faber and Faber Ltd from ANIMAL FORMS AND PATTERNS by Adolf Portmann, Fig. 33, p. 75.

are necessarily addressed to others. Nevertheless, many of them *can* be read and properly interpreted by members of same and different species, for the look is more than an disinterested view correlated to the beholding eye as a television transmitter is to the attuned receiver. In many social animals, the looks communicate and arouse the socially relevant emotions that make possible a rich social life. Chemicals and other simple sensory stimuli can keep animals together in swarms or mere crowds. But a *social* group requires mutual expression of moods "so that the being together is raised to a richer relationship, a relationship that goes beyond use and power, to a true meeting of independent creatures."[6] One sees here the rudiments of friendship, the recognition of the other as a you, other but alike.

Sociability is supported not only by looks that reveal but also by looks that conceal or even dissemble. One communicates *truly* with those who are akin, but it sometimes pays to show false looks to one's enemies. For example, some species of birds will feign an injury in attempts to lure an enemy away from the brood of young (Fig. 7).

Fig. 7. Deceptive manoeuvres, such as injury-feigning lameness in many birds, which is meant to distract attention from the chicks, confront the biologist with especially difficult problems. On one hand, they are exactly like the "dissimulation" of a human being; and yet they must not be regarded as being derived from a knowledge of the deceptive effect. The illustration shows a golden plover, a species in which both sexes adopt this device of feigning injury (after a drawing by Tunnicliffe).

Reprinted by permission of Faber and Faber Ltd from ANIMAL FORMS AND PATTERNS by Adolf Portmann, Fig. 109, p. 199.

Though entirely a matter of innate reaction, this dissembling is never-theless a clear attempt at deception aimed at the eye of the beholder. The false is, in the first instance, not a matter of logic or will, and hence not unique to man. Understood as the taking of something for what it is not, a mis-take, it can be and is communicated by inarti-culate looks alone.

In considering animal sociability, Portmann acknowledges and emphasizes the contributions of social life to the preservation both of individuals and the group, and ultimately of the species. Many socially relevant aspects of the appearance and behavior of individuals are in-terpreted in terms of their survival value. Yet Portmann's sociobiol-ogy goes beyond the prevailing orthodoxy by considering the contributions of social life to *enhancing*—and not only preserving—the life of *individuals*, by providing an arena for the full development of their vitality and individuality:

> As a species not only survives and evolves further, but relatively to a particular time simply 'exists' with all its individual members' peculiar-ities; so the society of animals is not merely an evolving instrument for species preservation, but also an aspect of individual life: individuals 'make' a society, but society is also a factor of life in the service of in-dividuals.[7]

This contention can be illustrated with numerous examples, from ownership behavior of bird couples, ruling over territories with "fa-vorite places," attachment to which becomes a mark of enhanced in-dividuality, to the display of social rank among deer, based upon a recognized importance of antlers, high-ranking deer losing face when its antlers are shed.

> When it comes to fights with rivals, where the antlers have their main function, only deer with nearly the same antler strength go into action; a deer with weaker antlers does not fight but acknowledges inferiority. H. Bruhin's study (1953) of the antlers' social role yields important con-clusions: —"The position of a male with high social standing is abruptly shattered as soon as he loses his antlers. In the spring of 1951, in the Basle Zoo, I was able to watch the moment when a fallow buck α sank to a lower level (α, β, γ here describe the levels in social rank). On April 18th, at 3:45 p.m., the herd of five males and eight females were begging for food from the zoo visitors. Suddenly they were slightly startled by a playing child, so that some of them trotted off, including the α male. He happened to graze with the right side of his antlers the branch of a fir-tree lying in the enclosure. Immediately this half of his antlers fell clat-tering to the ground. Obviously upset, with tail raised, he sniffed at the piece he had just lost. Almost at the same moment the β buck realized what had happened, and attacked and pursued him vigorously. The other three yearling antler-less γ bucks took scarcely any notice of the occur-rence nor did the does. After about half an hour both the α and β bucks

had more or less calmed down and were again begging for food. But the former α buck was not tolerated at the fence by his rival, and therefore kept right at the back of the enclosure. There was only an indication of a social clash between α and γ bucks. Up to the evening the one-palmed animal carried out peculiar head movements, as were observed by Heck (1935) after the loss of antlers. On 23rd April the β buck also shed his first antlers. From this time on there was the same social ranking as had prevailed before the α animal shed his."[8]

Communication for the sake of sociality is only part of the social meaning of the look. The social group itself provides the arena in which individual animals engage in display for its own sake. The group and its environs are a theater, a stage, on which especially the high ranking actors show off before a "respectful" audience. Ultimately, the animal form is more than an organ of self-preservation or of communication. It is a means of *self-presentation,* of display simply, whereby the animal shows forth its special nature, intrinsic worth, and rank, both as one of a particular *kind* but also as a particular *one* of the kind (see Fig. 8). The look, like the animal displaying it, is, at least in part, an end in itself.

Nature, in its variety and profusion of forms, is ordered to be seen, perhaps to be appreciated, but in any case to be displayed. The aesthetic—*pace,* all you relativistic "eyes of the beholder"—is not the unique and merely artificial creation of man. So, too, regarding sociality. In living nature, man has a rich teacher, and at least a willing partner.*

Good Looks

Turning from the looks of animals to the looks of the human animal, we wonder what we can learn about human being from human appearance. Does the human look provide any clues to the peculiarly human way of being-in-the-world? What is the relation between the outside and the inside, between the patterns and activities of the surface and the deeper longings and concerns of the human soul?

What might we expect to find? On the one hand, we might be able to confirm what we suspect without examination, namely, that the human look and way are not simply on all fours with those of animals. On the other hand, we should not forget the theory of evolution, which

---

*This point was stressed by Kant, who put together his aesthetic and naturalistic teachings in the *Critique of Judgment,* comprising a critique of aesthetic judgment and a critique of teleological judgment. Kant also instructs us in the connection between beauty and taste, between that which is to be appreciated and that which is capable of appreciation. He also draws suggestive connections to morality.

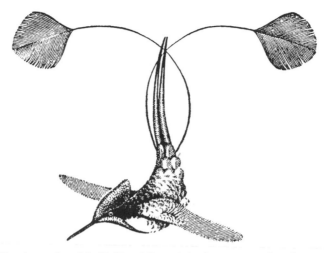

Fig. 8. The 'wonder bird' (*Loddigesia*) shows a quite fantastic formation of tail in the older males. Instead of the ten tail feathers typical of humming-birds, only four are here developed. The inner pair is partially hidden and relatively short; the outer pair form two flags crossing one another. These modifications take place only at maturity; younger males have the usual number of tail feathers. The tail coverts (those feathers which form the peacock's train) are large, one pair being particularly long. The small 'wonder bird' lives in the higher parts of Northern Peru (after Gould).

Reprinted by permission of Faber and Faber Ltd from ANIMAL FORMS AND PATTERNS by Adolf Portmann, Fig. 117, p. 219.

teaches the continuity, at least in origins, of human and animal life. Thus, we might also expect to find intensifications and elaborations, due to speech and self-consciousness, of animal looks and ways. In fact, the study of the human look may help to illuminate the look of our silent relations.

Many human appearances bear consideration. The twin looks of crying and of laughing are both rich openings to the center of humanity, the first related in some way to awareness of our mortality and vulnerability, the second to our ability to detach ourselves and rise above such serious matters of life and death. The upright posture itself is highly significant, both a cause and a mirror of the peculiarly human way of life.[9] Our present concern—nature and ethics—recommends yet another superficial phenomenon, one that involves sociality and display, that manifests our concern with the beautiful, and that points to the moral: the phenomenon of blushing.

To blush is human. In his book, *The Expression of the Emotions in Man and Animals*, Charles Darwin opens his chapter on blushing as follows: "Blushing is the most peculiar and the most human of all expressions. Monkeys redden from passion, but it would require an overwhelming amount of evidence to make us believe that any animal can blush."[10] But although uniquely human—and universal among the various races of man—blushing is not under our control. It is entirely involuntary. There is simply no way to induce, to feign, or to prevent a blush, either by physical means or by acts of will. Indeed, our attempts to banish or arrest blushing serve only to increase it. Some people blush more than others. The capacity to blush is genetically determined and the tendency to do so is inherited. Curiously, the young blush more than the old, yet infants and very young children do not blush.

Blushing is usually confined to the face, ears, and neck, though it has been observed in rare cases spreading to the chest and below, usually when attention is directed to those parts of the body. The redness of blushing is but a part of the *gestalt*, which includes motions of concealment such as the turning down and away of the face, eyes, and eyelids, and which is sometimes opposed by conscious and often vain attempts to look directly at the person whose presence caused the blush, out of a strong wish to avoid attention to the display. In some cases, these surface manifestations are accompanied by feelings of heat and tingling over the whole body and by marked confusion of the mind. Darwin reports an extreme instance of such confusion:

> A small dinner-party was given in honor of an extremely shy man, who, when he rose to return thanks, rehearsed the speech, which he had evidently learnt by heart, in absolute silence, and did not utter a single word; but he acted as if he were speaking with much emphasis. His friends, perceiving how the case stood, loudly applauded the imaginary bursts of eloquence, whenever his gestures indicated a pause, and the man never discovered that he had remained the whole time completely silent. On the contrary, he afterwards remarked to my friend, with much satisfaction, that he thought he had succeeded uncommonly well.[11]

The causes of blushing—somatic, psychic, social—are complicated, and especially so are their interactions. Even the physiological mechanism of blushing is only partially understood. The reddening is the result of dilatation of the arterioles of the face, which increases blood flow through the capillary circulation. The dilatation in turn reflects a shifting balance in the autonomic nervous system, away from the sympathetic and toward the parasympathetic components. Associated vascular disturbances may be responsible for the mental confusion, as Darwin himself suggested.

Yet this can be only part of the story, for the same mechanisms must also produce the reddenings that accompany anger or joyful excitement. The merely physiological explanations also fail to account fully for the phenomena. The vascular dilatations, with the facial reddening and mental confusion, can be artificially induced by the administration of vasodilator drugs such as amyl nitrate, but these blood vessel changes are not accompanied by the other facial manifestations of concealment nor by the inward feeling of embarrassment or shame. Blushing is more than reddening; it is but the involuntary outward bodily manifestation of a very complex psychophysical phenomenon, the understanding of which involves nothing less than understanding the relationship between bodily processes and psychic experience. Although the body seems here not to be an instrument of emotion, as the eye is of vision, we do not know whether the bodily changes are mere accompaniments, whether these or other bodily changes trigger or are triggered by the experienced feeling, or whether there is some nondualistic formulation, yet to be discovered, that can someday fully integrate the physical and psychic aspects. At least for the present, and perhaps permanently, we can only speak separately about one side or the other of this mysterious relation, either the somatic or the psychic.

Various mental states are associated with and apparently induce blushing. Among them are shyness, modesty, embarrassment over blemished looks or breaches of etiquette, and shame arising from manifest disgrace or misconduct. These very interesting feelings are hard to describe and shade over into one another. This stands to reason, for they share common roots: Our capacity for self-attention and our concern for how our appearance is regarded by others. Indeed, Darwin believed that "originally self-attention directed to personal appearance, in relation to the opinion of others was the exciting cause; the same effect being subsequently produced, through the force of association, by self-attention in relation to moral conduct."[12] We shall later have reason to question Darwin's implied view of the moral as something hidden from view, as something nonaesthetic.

For the present, we can now discern why only human beings blush, why animals and small children do not. Though blushing is involuntary and not subject to conscious control, it nevertheless depends on a high degree of self-awareness. Blushing seems to require at least the following psychic capacities: (1) a felt notion of self, as a "something" that can be viewed and judged by others and by oneself; (2) a concern, not necessarily conscious, for how one appears to others; (3) an awareness of what counts in making and keeping a good appearance, or at least in avoiding a bad one; and (4) an awareness that one is on display, the ability to "see" oneself being seen. (The presence of sight is not indispensable. The blind are capable of blushing, but under circumstances in which they sense that they are on display as objects of scrutiny and attention.)

Blushing is not only a psychophysical phenomenon; it is also emphatically social. Blushing requires the observer. Though we may feel shame before ourselves, we do not blush except under the gaze of the other, and only if the other is a you whose opinion of us matters. Very young children, indifferent to what others think of them, do not blush. Neither do they withhold their stares from the embarrassments of others, most of whom will not blush if only small children are present. Blushing requires mutual social concern and social self-awareness.

(By the way, here are some explosive questions for those who believe only in strict mechanical causation, that all is bumping and being bumped. First, how does a seeing of oneself being seen by another move us, and move us differently than a seeing of the other when his gaze is averted? Second, how is the stare of a two-year-old physically or mechanically different from that of someone who makes us blush?)

Does the look of blushing itself hint at what is within? Are the bodily manifestations that express our emotions somehow fitting and revelatory? In the case of some expressions, say, for example, laughter, it is hard to see the link. Why should the delight in the comic be expressed by gaping mouth, retracted head, and funny noises in the throat? With blushing, it seems to be otherwise. It seems right that embarrassment or shyness are expressed by facial blushing, rather than, say, by redness and tingling of the extremities. The face is not only the organ of self-expression and self-presentation, the source of our voice and transmitter of our moods; it also contains the chief organs for beholding other selves. We are to one another, especially, when face-to-face, eye-to-eye. Little children playing hide-and-seek hide their eyes and think they are invisible. Further, the face is, generally speaking, the most individuating part of the body. It is also the chief seat of beauty or ugliness. For all these reasons, the face is most *highly regarded*, both in the sense of most looked at and in the sense of most esteemed. The attention of others and our own self-attention, both wanted and unwanted, will center largely on the face. Insofar as blushing is a sign to another in response to being seen, the face is its natural home.

Is blushing a sign to another? What is its meaning or use? These are questions that can be separated from the questions of how and why this capacity evolved with the human species. Some functions may have come into being for one reason but persist for different ones, acquiring, once here, so to speak, a life of their own. And, in any case, the search for a selective advantage of a phenomenon can only follow upon the proper understanding of its significance and function. Of the meaning and use of the more complicated and inward matter of shame, with which blushing is intimately connected, I will speak later. But as to blushing itself, it seems very difficult to say.

On the one hand, blushing is often a breakdown of concealment, a drawing attention to oneself that reinforces the attention one was hoping to avoid. It is a manifestation of vulnerability, a sign of exposure. Yet, it is, at the same time, a communication. In certain situations, a blush may function as a warning to an aggressive or boorish other to desist from the activity that is causing embarrassment, for the burn of shame is slow to heal and a great cause of anger.

In other circumstances, a blush may bespeak a request for relief. Sensitive and considerate persons are themselves often embarrassed at having been the cause of another's blush, and accordingly avert their eyes and desist from an embarrassing line of conversation. In such cases, the initial red-facedness and aversion of the eyes form a submissive *gestalt*, which seems to serve as a call for a halt. It may thus be a sign of concession, perhaps analogous to the baring of the neck by the wolf defeated in a fight by his rival, which causes the victor to desist. The appearance of a blush is *always* an acknowledgment of the importance of the other. Further, the blush of shame acknowledges acceptance of the group ideal or the standard of the other. It is indeed a confession. As such it represents, albeit unwillingly and with discomfort, a coming together of the observer and the one ashamed. To blush is, in a way, to save face, by losing it. Barefaced rather than shame-faced misconduct adds insult to injury by displaying the perpetrator's indifference or contempt.

Paradoxically, the flush of blushing might be useful to the one who blushes, as a warning of his violation and an invitation to correct it. Perhaps it is to the soul what pain is to the body, also a useful evil, an unpleasant but indispensable warning of the threat to life and limb. The loss of the ability to feel pain is an invitation to all sorts of disasters. Could the same be said of the loss of the ability to blush? However this may be, we note this difference: Whereas pain (or fear of pain) have to do with survival, blushing and especially shame are concerned with higher things: with beauty, reputation, honor, and ultimately, with self-esteem.

Yet if blushing is useful and significant in any of these ways, if blushing is among the most human of all expressions, why do we try to hide it? Is the effort to conceal the blush only a further effort to conceal the fault or gaucherie that, in blushing, we failed to conceal? Or is there perhaps an additional reason why we try to avoid attention to our blushing? Isn't the problem, at least in part, that we are caught in an involuntary revelation of ourselves? Blushing is automatic. It reveals an inability to control our bodies and our self-presentation. To the disgrace that caused the blush, the blush itself is an added disgrace. We try, either by turning away or by staring down the other, to hide this added disfigurement of our form. Both in blushing and especially in covering our blush, we display our concern for our looks. Our looks are not only useful as a sign to others; we regard them, and want them to be regarded, as something of intrinsic worth, as a thing of beauty.

Thus, analogous to the looks of higher animals, we have in blushing an inborn, genetically determined, and involuntary manifestation of fundamental human capacities and concerns. We see again the twofold direction and significance of the look: first, the social—the blush as social, as communicative, as a sign or an acknowledgment of shared conventions; second, and more compelling, the aesthetic—the blush as the display of our individual concern with our own looks, our desire to look good in the eyes of others. The two aspects are again related. Sociality requires that we keep up our face, but it thereby fosters and encourages the saving and adornment of face. It invites and supports a concern with hiding the ugly and displaying the beautiful.

The beautiful is sometimes more than skin deep. Though we blush only when we are seen, though we blush usually in relation to things visible, the image of ourselves we seek to preserve is not only bodily. Our reputation and our character, ultimately our full self-esteem, are at stake. We move from good looks to looking good, from the obviously aesthetic to the apparently moral.

Looking Good

We return for a closer look at the human surface. We began looking at the human animal by noticing an involuntary and natural look that occasionally bedecks the human face. But if we were to begin with what is most striking about the human animal, we would have to notice something more pervasive, a voluntary and not exactly natural look of the rest of the body: human beings wear clothes. The *human*-look of the human animal depends upon art and convention, on the weaving of threads and the patterns of custom. Man's look is partly of his own making. It reveals that he is by nature an animal that lives by art, that he is also by nature an animal that lives by convention. Clothing our nakedness is at once *the* sign of both these dependencies.

The clothed human look has many meanings, most of them identical with, or at least analogous to, those we observed in our animal cousins. Clothing is a thicker skin, a protection against the elements, a defense against dangers underfoot and all around. Clothing may also be useful in attracting a desirable mate, different styles and amounts appealing to different types of prospective partners. Clothing can also serve to communicate to others our inner moods and our sense of ourselves. Of great importance, clothing becomes the public garb that

covers over our erotic nature, and thereby also preserves it for intimate and private expression; clothing restrains lust, but makes love
possible.* But clothing is also display, a look for the eye, both revealing and concealing. It covers over defects acquired and natural,
among the latter, especially that sign of weakness and mark of mortality, our sexuality. More positively, clothing adorns and beautifies;
it shows us forth to advantage. In all these many ways, clothes make
the man.

But man is made human by an even more fundamental covering,
often an invisible one, yet a covering of which clothing is a visible
instance. The very wearing of clothes, not to speak of styles, is a matter of convention, of customs, of law in its broadest sense. It is, in
fact, a paradigm of convention, which also cloaks and thereby seeks
to humanize and beautify the human animal. But convention dresses
deeper than the surface. It penetrates to our desires, our passions, and
our opinions. It seeks not only to make us pleasing to the eye, but also
to make us good, according to the standards of the community. These
standards, embodied in shared speeches and beliefs, cover and color
our heart and soul, much as the clothes they sanction cover our surface. With this broader view of our "exterior," we are now ready for
a look within, to the phenomena of shame.†

The phenomena of shame are complicated, delicate, and elusive.
Our speech about them is imprecise, clumsy, but near at hand. We
begin with our English word "shame." Shame, a noun: a disgrace or
dishonor. ("He bears well his shame.") Shame, a noun: the passion or
feeling of shame, a painful and vexing condition of mind, caused by
the manifestation of speeches or deeds that lead to dishonor or disgrace. Shame, a verb: to cause the feeling of shame, or to cause the
disgrace of another, as in "to put to shame."

But behind these matters stands shame as some mysterious and
native power or capacity of the soul. Shame, as the name implies, is
a power that moves to cover, to hide: shame, from the Gothic *schama*,
to cover, the root also of the German *hemd* and the French *chemise*,
shirt. Shame hides and conceals our vulnerable and disgraceful aspects, not only from others, but also from ourselves.

Having shame and feeling shame are not the same. In fact, the
emotion of feeling ashamed, an emotion in full bloom when expressed
visibly in blushing, is in some sense a breakdown of shame, understood as that power which conceals. This breakdown is almost always

---

*Kant beautifully makes this point, among others, in commenting on the shame-cov-
ering fig leaf of Adam and Eve, in his astonishing "Conjectural Beginning of Human
History."[13] The relevant passage was quoted in Chapter 11, "Thinking About the Body,"
pp. 292–293.

†My thinking here owes much to the writings of Erwin Straus and Kurt Riezler.[14]

due to the presence of an observer, whose discovery compels our own, by establishing the fact of our disgrace or misconduct beyond our ability to hide it from ourselves. He is the witness in whose eyes we pronounce ourselves guilty, now and hereafter.

Yet with growing self-consciousness, we can be our own witnesses, by becoming, as it were, double, both seer and seen. Because we can imagine ourselves as others might see us, shame can serve, prospectively and preventatively, to shape our future conduct, by promoting or preventing actions that, if omitted or committed, would demand concealment. To be ashamed to do, no less than to be ashamed of having done, is the work of shame.

To enrich our understanding of the phenomena, we consult another language. Ancient Greek has two words that we translate as shame: *aischynē* and *aidōs*. *Aischynē*, meaning both the disgrace and the feeling of shame, is cognate with the adjective *aischos*, meaning not only the shameful, but also the ugly, the base, the deformed. The shameful is the ugly and the base. It is opposed to *kalos*, the beautiful, the noble, the splendid, the fine. *Aischos* and *kalos* apply to deeds and character, as well as to looks and things.

Whereas *aischynē* links shame to the aesthetic—a connection implicit in the importance for shame of seeing and being seen—*aidōs* links shame to awe, to feelings of wonder perhaps tinged with fear, of respect, of reverence. The venerable, that to which we look up, is that before which we are ashamed. Shame and awe, the shameful and the venerable, each and both *aidōs*, are twins by birth, even though a culture may choose to exile one of them by defiling or debunking the things it once venerated. Both Greek notions, *aidōs* and *aischynē*, teach us that shame, that power which covers and negates, stands in relation to something positive, whether beautiful or venerable.

What are the causes of shame? What are the things that shame seeks to cover? The list is long, but the following items indicate the range: faults in our personal appearance, grooming, or dress; failure to share in the esteemed things that most people or those like us share (e.g., education or employment); suffering abuses leading to dishonor or reproach, such as serving another's despotism or licentiousness (e.g., as with victims of enslavement or rape); and, most especially, defects of conduct and character, ranging from unintended gaucheries or errors of tact, through failings of etiquette or manners, to acts of blatant cowardice, infidelity, injustice, and the like.

Now, in all these matters, the standards of conduct are defined by the community, by convention, albeit in most cases tacitly and through custom rather than explicitly and through written law. Indeed, as we noted at the beginning, the things deemed noble and disgraceful vary sufficiently from culture to culture that it has become the (dare I say conventional) wisdom that what is noble or disgraceful is merely a matter of convention, that there is nothing which is noble or disgraceful by nature. From this judgment, some go on to draw the further inference, that shame *itself* is an acquired and strictly conventional passion, unlike fear or anger that are acknowledged to be natural. Yet this latter conclusion, at least, is probably wrong. The very teachability of what is shameful presupposes the native capacity. Shame could not be taught to a shameless being. To be sure, convention gives the standards of conduct in relation to which one will feel shame, but the capacity to feel shame, the concern with convention, and that power which conceals are all natural and universal. The clothes of shame are in principle mutable, but the naked power that they dress is not. (The existence of some apparently shameless individuals does not destroy the thesis, any more than the existence of deaf mutes falsifies the dictum that man is an animal that talks.)

If shame is natural—and clearly, the matter needs much more investigation—then we might venture a striking, morally relevant inference: Man is by nature a moral being, in the sense that he is by nature concerned with morality. A native power that is so readily attached to conventions and beliefs regarding things noble and ugly or venerable and disgraceful implies a native concern with the moral—one that must be nurtured if it is to grow, to be sure, but one that is present on its own, if only as a seed. Man is by nature concerned with looking good, or at least with not looking bad.

Can one speculate further into man's moral nature?* What kind of good is meant, and is this good in any sense natural? Can one infer a desire to *be* good from a desire to *look* good? Or are we merely interested in *looking* or *seeming* good, as a cover for our otherwise uninhibited attempts to indulge our selfish desires, no matter how base? Do we, at our deepest center, believe that shame might serve the good of others, but that for oneself shamelessness pays? To consider these questions, we must return to the meaning and uses of shame.

The kinds of shame already discussed clearly are socially useful. Preventative shame fosters conduct that serves defined social goals, as when a general exhorts his men to avoid the disgrace of cowardice by sticking to their guns. After the fact, concealing shame seeks to conceal deviations from the group norm, and invites repentance, restitution, and reform.

But shame not only serves the community. It serves the individual as well, partly indeed by the ties it establishes and acknowledges to

---

*The late animal geneticist C. H. Waddington pursued this question from a different biological starting point. He also wanted to know why it is that the human animal "goes in for" having ethical beliefs. He emphasized the cognitive and authority-bearing capacities that make us "ethicizing beings." See his *The Ethical Animal.*[15]

the community and its conventions. We live our lives with and toward others. To be concerned with how we are to others is unavoidable because our being to ourselves is inextricably interwoven with our being to others. Our lives are not lonely projects toward solitary deaths, but lives in community, in association, in friendship. We are not simply autonomous or independent beings, nor would we, under most circumstances, be better off attempting to be such.

The community or group serves more than our needs for survival. Among its other functions, it also provides the opportunity for individuation and distinction. Even more than with the social animals, it is the arena for performance, display, achievement, and the attainment of excellence. Preventative and concealing shame reflect this personal stake in the community. *They imply a concern for self-esteem and self-respect.* Because our being is entwined with others, our self-esteem will in part depend on them. But here our shame serves ourselves. We esteem the esteem of another as a sign of our own worthiness. To lose face to a significant other is to risk losing face to oneself. At the bottom of shame, for all its conventional trappings, lies a natural concern for self-respect, a concern grounded in self-consciousness and an aspiration to be worthy.*

It goes without saying that a self-respect grounded only in the opinions of the community and of others is problematic, to say the least. First, there is the question of which opinions and of which others. Second, a mature morality and self-respect relies increasingly on internalized standards. Third, shame, for all its usefulness, is not a virtue. At best, as has long been noted, it is a useful passion in the young.[17] Mature people should not do the things that require concealment. (Parenthetically, we may hope that we have here discovered why the young blush and the old do not. Can it be that the decline of blushing with the waning of youth is due more to a decreased *need* for social shame than to a growing indifference and shamelessness or a declining docility and concern with looking good?) Still, if ethics is not a matter solely or primarily of knowledge of precepts and rules or of purity of heart and conscience, but *a matter of action*, it may never

---

*There is another dimension and function of shame, called by Erwin Straus "protective shame,"[16] which, though not crucial to the present concern, very much deserves recognition. It lends a different support to our aspiration to be worthy, but here by decreasing self-consciousness. It protects the immediacy, engagement, and spontaneity of many of our private doings and also the intimacy of our private relations from the disruptions caused by the intrusion of outside observers. Here there is nothing disgraceful to be covered, but only our vulnerability in being open and off our guard, tentative and in the midst, absorbed or lost in our own activity, or displaying the affections of the heart. To illustrate, consider the corrosive kind of self-consciousness that disrupts any intimate conversation when a third party—even a friend of both—asks if he may join the company. Shame thus stands guard at the boundary of public and private, supporting the claims of the former against the base and the ugly, protecting the needs of the latter against objectification and publicity.

outgrow openness to public view. It may never become unresponsive to praise and blame from discriminating and decent others. In this respect, at least, and with the necessary qualifications, moral self-esteem is quite properly supported by the witnessing and the esteem of others.

We have moved from shame to its more positive twin, from the negative desire to avoid disgrace to the positive desire to look *good*, not only to another but especially to oneself—and if we do not wish to be self-deluded, looking good to ourselves must be related to *doing* good. Recalling the relation between the shameful and the noble, we remember that the desire to conceal and hide is but the backside of the desire to reveal and display. The steps we have taken, beginning with the looks of animals, through shame via blushing, point toward a concern with looking good, in which good is filled in by a morality of action-on-display, in which the key terms are not holiness and sin, right or wrong, but noble and base, in which the key matter is not the purity of one's motive or the correctness of one's maxim, but the fineness of one's deed. It is a morality that appears and seeks to appear, that is fundamentally social and political in its orientation, that does not huddle in the dim light of bar and bedroom, but goes forth into the bright light of public affairs. It is a morality that does not take its bearings from extreme situations—such as murder or revolution—but rather nourishes and displays itself in the ordinary daily matters of conduct, matters that always call for self-control, courage, grace, gentleness, tact, courtesy, finesse, generosity, ambition, and fairness. It is a morality that aspires way beyond mere selective advantage and security. Far more than the moralities of rules, rights, universalizable maxims, and utility, it provides a high place for friendship, for living well with near and dear.* In short, it is a view of the moral centering on character, in which what is present in the soul is revealed on the surface, in the look-in-action. In a way, it is the moral as a species of the aesthetic. The soul sends forth the look that is described by the appreciative beholder as fine, seemly, beautiful, noble, resplendent, as in German *schön*, in Greek *kalos*. Could it not be said that such display is the truly human look of the human animal, a look that sets the standard for our kind as do the high-ranking antlers for the deer, a look that reveals its bearer as entitled to a place apart?

---

*It is thus not a morality that begins with our radical individuality or loneliness and that finds therefore the claims of individuals at odds with the claims of communities. It is no accident that Aristotle's *Ethics*, which more than any other ethical work in our tradition emphasizes the goal of nobility, is also the only one that gives concentrated attention (two of ten books) to friendship.[18] In contrast, biblical friendships are rare; David and Jonathan or Naomi and Ruth are perhaps the only examples we have. The modern liberal tradition (Hobbes and Locke) and its critics (from Marx to Heidegger and Rawls) virtually ignore the subject.

Yet before we all go running off to our mirrors in self-congratu-
latory admiration, we need to add some necessary qualifications. A
morality grounded solely in the looks may be insufficient. First, the
moral is larger than the noble. Even those thinkers, such as Aristotle,
for whom the noble was morally crucial, spoke also of justice, of that
peculiarly human virtue that acknowledges our dependencies, relations, and debts to other human beings and institutions—not only
those we covenant with, but also, and especially, parents, ancestors,
friends, cities, and gods. Next, the desire to look and be good, though
present, is often weak when set against other baser desires. Moreover,
we seem today to have enough trouble holding up the floor of morality
as to make our gaze upon the ceiling seem chimerical. We may well
ask whether the conditions of modern life are not inhospitable to the
cultivation of the noble. Yet here is where at least a part of our animal
heritage points. Why should we not aspire to follow its direction?

Coda: Looking Well

Let us re-view the terrain we have covered. We have been looking into
the meaning of the looks of animals and especially of the human animal, taking to heart the suggestion that the outside is a mirror of and
a window to the inside. We have in every case explored both the look
as communicative and useful to the group, and the group as the stage
for the display of the look: Sociality and the beautiful have been our
themes. In exploring the human look, we have gone far afield, beginning with a modest little blush and its hidden companion shame, following up their clues for a naturally grounded nobility. Yet all along,
and especially in considering the complex looks of man, we have ourselves been illustrating another one of Portmann's generalizations,
namely, that the richness of that which is to be seen develops in parallel with a richness in the capacity to see. In human beings, the full-
grown power for nobility comes hand-in-hand with the full-grown
power to see and appreciate nobility. Whether or not our sight has
been equal to the task, we have been involved in looking, conspicu-
ously so. It is fitting, in closing, to address a few remarks to looking.

To look, to see, to understand the world and ourselves in relation
to it. Here is the highest goal of science and of thought. It is an activity that goes beyond shame and of which we need not be ashamed.
It is as natural to us as our look. We are the looking animal par excellence; our eye is sharpened and suffused with intellect, and animated by a desire to understand—not least, to understand ourselves
and our place in the nature of things.

To be the blushing and the self-esteeming animal depends upon our being also the looking and reflecting (i.e., self-conscious) animal. Yet looking well, or at least thoroughly, often requires an overcoming of shame or embarrassment, for example, in the willingness to look at certain unpleasant or distasteful matters, or an overcoming of pride and vanity, for example, in the willingness to expose ourselves as ignorant. In extreme cases, perhaps, the concern for insight may be in tension with the concern for the noble and the seemly.* But, generally, although looking well is an activity that requires unabashedness, it does not require shamelessness. It knows or should know when to withhold or avert its gaze.

Our problem today is less voyeurism than blindness. We suffer less from too much looking than from too little, at least of a certain sort. Our science, which looks so masterfully into so many things, a science blessed with telescopes and microscopes, looks too little and not well at the world we experience with the naked eye. Our attempt to master nature, to predict and control appearances, needs to be leavened by a search to discover, to understand, and to appreciate the meaning and the beauty of appearances, and of the beings that appear.

Should this happen, science could also once again open our souls to that other passion linked with shame, the passion of awe. Science could help us to encounter the mysterious, rather than to explain it away. To look into the eyes of an animal and to ponder what lies within is to be struck by a kinship and a strangeness, neither of which is articulable or fathomable. It is to discover an irreducible mystery, which our current biology does not so much deny as it ignores. It is to rediscover the things on earth that are no longer dreamt of in our philosophy. And it is, finally, to restore ourselves to our proper place *within* nature's kingdom, from which we have exiled ourselves in our audacious, not to say hubristic, project to master and possess it. For in recognizing the inexpressible mystery and the power that lies beyond our control, we manage in a way to bridge the unbridgeable gap, to make a home for ourselves in this strange and wonderful world. Looking well, like looking good, is our nature's own reward.

---

*See, for example, the story of Gyges in Herodotus's *Histories,* the story of Leontius in Plato's *Republic,* and the story of the sons of Noah in Genesis.[19] Consider also Sophocles's *Oedipus Tyrannos.*

Epilogue
From Nature to Ethics

We began this book with ethical dilemmas and the challenge to ethics posed by present-day biomedical science and technology. As we end, we note that our reconsideration of living nature, toward the goal of a more natural science, has led us back to ethics, this time for a more congenial encounter between nature and ethics. The analysis of "looking good" leads to two unorthodox suggestions. First, that the ethical, rightly considered, might be integral to the natural; for if ethics belongs naturally to man, as does man to nature, then ethics, too, is part of nature. Second, and even more heretical, the natural, rightly understood, might even provide some guidance for how we are to live. One link is forged by the beautiful or the noble, itself an (albeit silent) aspiration of living nature, which becomes a self-conscious and intentional goal of conduct for human beings. Another link, correlated with the first, is natural sociality—much richer than the orthodox sociobiologists would have us believe—present widely among the animals, but extended and given articulate regulation by the "rational animal, political animal." The deliberate and conscious concern with the right and the good may be naturally peculiar to human beings, but the *contents* of these notions need not thereby be a human peculiarity or projection. They might, in fact, be genuine discoveries, available only to man but not thereby strictly of human origin. This suggestion is, of course, highly speculative and, in any case, a topic that must be left for another inquiry. For the present, it is possible only to touch briefly on the ways in which a more natural science might be useful for ethics, and, more concretely, how the notions of purposiveness or embodiment or finitude or nobility considered near the end of the book might be helpful in thinking about the bioethical dilemmas with which it began.

It is perhaps best first to disavow the intent of seeking and hope of finding precise rules of conduct deducible from even the fullest knowledge of nature—no sensible person holds that such rules can be simply "read off" from the natural record. Aristotle, who perhaps more than anyone thinks about the human good in relation to what he takes to be the natural human work, within the natural whole, and who even distinguishes something he calls "the just by nature" from "the just by convention," deduces from man's nature no rules of conduct.* "Oughts" and "ought nots" are not to be found written on cosmic stones. But there are other ethically useful contributions to be anticipated from a richer understanding of living nature.

In the first place, a more natural science, truer to life, would be useful negatively, by defending sound practices and beliefs against dangers engendered by false or partial theories. For example, a proper appreciation of organic form and animal nobility would embarrass such reductive notions as "the selfish gene," which dissolves animal wholeness and robs life of its vitality, or the equally reductive notion of "genes for altruism," dubiously extrapolated to human social life from an already shrunken animal sociobiology. Conversely, a more profound understanding of "the body" would refute the unnatural and so-called humanistic claims that intelligence or gender-specific roles or the idea of health have no biological basis and are strictly human or cultural creations. And recognizing organic purposiveness would enable us to resist as simplistic and abstract our stimulus-response (or input-output) theories of behavior, all of which deny or ignore the inner and immanent leanings and tendencies of living things.

Clarifying the natural tendencies (or aspirations or leanings or directives) that point us toward certain objectives—in addition to mere survival—would be a second and more positive contribution. A biology and psychology that recognized more than sex and survival would enable us better to discern those activities and attainments whose realization would be humanly fulfilling. What are the things human beings by nature desire: the pleasant, the beautiful, or participation in the eternal; children, freedom, distinction, or understanding? What would satisfy our deepest aspirations? Answers to these questions might inform not the prescription of rules but an ordering of lives, according to a full standard of human flourishing; such an ordering might in turn suggest which kind of social arrangements and institutions are more and less conducive to promoting these goals. Lib-

*Indeed, Aristotle explicitly denies that there are *any* universally valid rules of action, and says, quite enigmatically, that all of justice—including natural justice—is changeable (*Nicomachean Ethics* 1104a4, 1134b30).

eralism might again have an answer to those who insist that man lives by bread alone.

Turning more specifically to contemporary bioethical dilemmas, we can see how a more natural science would allow us to recognize and discourage certain dehumanizing attitudes and practices. Appreciating the meaning of our embodiment, institutionalized already in our taboos on cannibalism and incest, would lead us to oppose the buying and selling of human organs or the practice of surrogate motherhood. Understanding the virtues of mortality would show us the folly of seeking to prolong life indefinitely and the wisdom of reaffirming procreation, regeneration, and self-sacrifice against the narcissistic prejudices connected with a strictly survivalist principle of life. Recognizing our body's immanent and purposive activities toward wholeness and well-working would instruct and restrain our willful self-manipulation by illuminating which possible alterations of bodies and minds might be helpful or harmful to human well-being. And rediscovering life's inwardness, accessible through a proper attention to the revelations of animal appearance, would even encourage improved treatment of animals, by inviting firmer limits on our (unavoidable) exploitation and ownership of living things. To be sure, these implications for practice are not yet detailed enough to enable us to decide what to do in concrete cases—and happily so, for concrete cases are never well settled by abstract rules alone but only by the prudent judgment of the man on the spot. Moreover, even the fairly clear rejection of the sale of organs or surrogate motherhood or ectogenesis would not translate necessarily into legal prohibitions: The translation of sound moral intuitions into legally enforceable practices is difficult at best, all the more so in liberal democratic regimes. But the vitality of free institutions ultimately requires an upright and morally dignified citizenry, of sound beliefs and decorous practices. Liberal and democratic regimes cannot be indifferent to the attitudes and postures of their citizens regarding the dignity of human life and its place in the larger whole.

Indeed, it is with respect to such attitudes and postures that a more natural science is likely to be ethically most useful. Such a science could reawaken not only wonder and admiring delight at the given world, but also respect, awe, and gratitude: respect for the powers of living nature, awe before the mysteries of living nature, and gratitude for the unmerited—and, in the face of evolution, simply miraculous— privilege of our being here to experience wonder and delight, respect and awe. Finally, these attitudes and sentiments toward nature will nurture a truer self-respect, no longer one we simply manufacture for ourselves, but one that is ours by nature. We stand most upright when we gladly bow our heads.

Endnotes

Introduction

1. Aldous Huxley, foreword to *Brave New World*, in *Brave New World and Brave New World Revisited*, New York: Torchbooks, Harper & Row, 1965, p. xvi.
2. Jacques Monod, *Chance and Necessity*, New York: Vintage Books, 1972, p. 164.

Chapter 1
The New Biology: What Price Relieving Man's Estate?

1. Robert S. Morison, "Death: Process or Event?" and Leon R. Kass, "Death as an Event: A Commentary on Robert Morison," *Science* 173:694–702, 1971. See also Hans Jonas, "Against the Stream," in his *Philosophical Essays*, Englewood Cliffs, N.J.: Prentice Hall, 1974.
2. Robert G. Edwards, B. D. Bavister, and Patrick C. Steptoe, "Early Stages of Fertilization *in vitro* of Human Oocytes Matured *in vitro*," *Nature* 221:632–635, 1969.
3. Robert G. Edwards, Patrick C. Steptoe, and J. M. Purdy, "Fertilization and Cleavage *in vitro* of Preovulator Human Oocytes," *Nature* 227:1307–1309, 1970.
4. Herman J. Muller, "Human Evolution by Voluntary Choice of Germ Plasm," *Science* 134:643–649, 1961.

5. José M. R. Delgado, *Physical Control of the Mind: Toward a Psycho-civilized Society*, New York: Harper and Row, 1969, p. 145.

6. Ibid., p. 88.

7. Francis Bacon, *The Advancement of Learning*, Book I, Hugh G. Dick, ed., New York: Random House, 1955, p. 193.

8. C. S. Lewis, *The Abolition of Man*, New York: Macmillan, 1965, pp. 69–71.

9. Ad Hoc Committee of the Harvard Medical School, "Definition of Irreversible Coma," *Journal of the American Medical Association 205*:337–340, 1968.

10. David D. Rutstein, "The Ethical Design of Human Experiments," *Daedalus*, Spring 1969:523–541, at p. 526.

11. Eric J. Cassell, "Death and the Physician," *Commentary*, June 1969:73–79, at pp. 76–77.

12. Bentley Glass, "Science: Endless Horizons or Golden Age?" *Science 171*:23–29, 1971.

13. Hans Jonas, "Contemporary Problems in Ethics from a Jewish Perspective," *Journal of Central Conference of American Rabbis*, January 1968:27–39, at pp. 33–34. Reprinted in Daniel J. Silver, ed., *Judaism and Ethics*, New York: Ktav Publishing Company, 1970.

14. Philip Handler, ed., *Biology and the Future of Man*, New York: Oxford University Press, 1970, p. 55.

Chapter 2
Making Babies: The New Biology and the "Old" Morality

1. *The Public Interest*, Winter 1972:18–56.

2. Albert Rosenfield, *The Second Genesis: The Coming Control of Life*, Englewood Cliffs, N.J.: Prentice-Hall, 1969, p. 117.

3. Robert G. Edwards and Ruth E. Fowler, "Human Embryos in the Laboratory," *Scientific American 223*(6):45–54, December 1970, at p. 54.

4. Paul Ramsey, *Fabricated Man*, New Haven: Yale University Press, 1970, p. 113.

5. Robert G. Edwards and David J. Sharpe, "Social Values and Research in Human Embryology," *Nature 231*:87–91, at p. 89.

6. Edwards and Fowler, *op. cit.*, p. 50.

7. Patrick C. Steptoe and Robert G. Edwards, "Laparoscopic Recovery of Preovulatory Human Oocytes After Priming of Ovaries with Gonadotrophins," *The Lancet i*:683–89, April 14, 1970, at p. 683. Emphasis added.

8. Anonymous, "Controversial Test Tube Conceptions," *Medical World News*, April 4, 1969:26–32, at p. 27.

9. Donald Gould quoted in Alvin Shuster, "Human Egg Is Fertilized in Test Tube by Britons," *The New York Times*, February 15, 1969, p. 31.

10. Bentley Glass, "Science: Endless Horizons or Golden Age?" *Science 171*:23–29, 1971, p. 28.

11. Ibid.

12. Ibid.

13. James D. Watson, "Potential Consequences of Experimentation with Human Eggs," presented at the twelfth meeting of the Panel on Science and Technology, Committee on Science and Astronautics, United States House of Representatives, Washington, D.C., January 28, 1971.

14. Ramsey, *op. cit.*, pp. 71–72.

15. Editorial, "Premature Birth of Test Tube Baby," *Nature 225*:886, 1970.

16. Robert G. Edwards, Patrick C. Steptoe, and J. M. Purdy, "Fertilization and Cleavage *in vitro* of Preovulator Human Oocytes," *Nature 227*:1307–1309, 1970.

17. Genesis 4:1.

18. Robert L. Sinsheimer, "The Prospect of Designed Genetic Change," *Engineering and Science Magazine*, California Institute of Technology, April 1969. Dr. Sinsheimer has since had a change of heart, and has become one of the advocates of caution and sobriety.

19. Karl Rahner, "Experiment: Man," *Theology Digest 16*:57–69, 1968, at p. 61.

Chapter 3
Perfect Babies: Prenatal Diagnosis and the Equal Right to Life

1. *Curlender* v. *Bio-Science Laboratories*, 165 Cal. Rptr. 477 (Ct. App. 2d Dist. Div. 1, 1980). See also George J. Annas, "Righting the Wrong of 'Wrongful Life,' " *The Hastings Center Report 11*(1):8–9, February 1981.

2. Abraham Lincoln, "Fragment on Slavery" (1854) in *The Collected Works of Abraham Lincoln*, vol. II, Roy P. Basler, ed., New Brunswick, N.J.: Rutgers University Press, 1953, pp. 222–223.

3. For a discussion of the possible biological, rather than moral, price of attempts to prevent the birth of defective children, see, e.g., James V. Neel, "Ethical Issues Resulting from Prenatal Diagnosis," in *Early Diagnosis of Human Genetic Defects: Scientific and Ethical Considerations*, Maureen Harris, ed., Washington, D.C.: U.S. Government Printing Office, 1972, pp. 366–380.

4. Pearl S. Buck, foreword to *The Terrible Choice: The Abortion Dilemma*, Robert E. Cooke, ed., New York: Bantam Books, 1968, pp. ix–xi.

5. Jerome Lejeune, "On the Nature of Men," *American Journal of Human Genetics 22*:121–128, 1970.

Chapter 4
The Meaning of Life—in the Laboratory

1. Aldous Huxley, *Brave New World,* New York: Torchbooks, Harper and Row, 1965, pp. 1–3.
2. *The Federal Register,* August 8, 1975.
3. Quoted in Peter Gwynne, "Was the Birth of Louise Brown Only a Happy Accident," *Science Digest,* October 1978, pp. 7–12, at p. 9. Emphasis added.
4. For the supporting analysis of the concept of viability, see my article "Determining Death and Viability in Fetuses and Abortuses," prepared for the National Commission for the Protection of Human Subjects of Biomedical and Behavioral Research, and published in *Appendix: Research on the Fetus,* Washington, D.C.: U.S. Department of Health, Education, and Welfare, DHEW Publ. No. (OS) 76–128, 1975.
5. 410 U.S. 113, 93 S. Ct. 705 (1973).
6. See John A. Robertson, "Surrogate Mothers: Not So Novel After All," and Herbert T. Krimmel, "The Case Against Surrogate Parenting," *The Hastings Center Report 13*(5):28–39, October 1983.
7. *Science 202:* 1 and 5, October 6, 1978.
8. See Chapter 2, "Making Babies—the New Biology and the 'Old' Morality," especially pp. 44–45. See also Chapter 6, "The End of Medicine and the Pursuit of Health," especially pp. 157–164 and 177–179.

Chapter 5
Patenting Life: Science, Politics, and the Limits of Mastering Nature

1. *Diamond* v. *Chakrabarty,* 447 U.S. 303 (1980).
2. René Descartes, *Discourse on the Method,* in *The Philosophical Works of Descartes,* vol. I, Elizabeth S. Haldane and G. R. T. Ross, eds., Cambridge, England: Cambridge University Press, 1981, pp. 119–120. Emphasis added.
3. Francis Bacon, *The New Organon,* Book I, Aphorism XC, Fulton H. Anderson, ed., Indianapolis: Bobbs-Merrill, 1960, p. 89.
4. James Madison, *Federalist* No. 43, in Alexander Hamilton, James Madison, and John Jay, *The Federalist Papers,* Clinton Rossiter, ed., New York: Mentor, New American Library, 1961, pp. 271–272.
5. Ibid.
6. Ibid.
7. Abraham Lincoln, "Second Lecture on Discoveries and Inventions" (February 11, 1859), in *The Collected Works of Abraham Lincoln,* vol. III, Roy P. Basler, ed., New Brunswick, N.J.: Rutgers University Press; 1953, pp. 356–363, at p. 361.

8. Albert Henry Walker, *Text-book of the Patent Laws of the United States of America,* chapt. VI, sec. 82. New York: L. K. Strouse and Company, 1883, pp. 54–55. Emphasis added.

9. Alexis de Tocqueville, *Democracy in America,* vol. II, part I, chapt. 10, J. P. Meyer, ed., trans. by George Lawrence, New York: Doubleday, 1969, pp. 459–465.

10. Exodus 21:35–36.

Chapter 6
The End of Medicine and the Pursuit of Health

1. Claude Bernard, *An Introduction to the Study of Experimental Medicine,* trans. by Henry Copley Green, New York: Dover, 1957, pp. 1 and 67.

2. Peter Sedgwick, "Illness—Mental and Otherwise," *Hastings Center Studies* 1(3):30–31, 1973.

3. Plato, *Charmides,* 156d–157a.

4. Nedra B. Belloc and Lester Breslow, "Relationship of Physical Health Status and Health Practices," *Preventive Medicine 1:*409–421, 1972; Nedra B. Belloc, "Relationship of Health Practices and Mortality," *Preventive Medicine 2:*67–81, 1973; and Lester Breslow, "A Positive Strategy for the Nation's Health," *Journal of the American Medical Association 242:*2093–2095, 1979.

5. Philip R. Lee and Albert R. Jonsen, editorial, "The Right to Health Care," *American Review of Respiratory Disease 109:*591–592, 1974. Dr. Lee is former assistant secretary for health. Mr. Jonsen was a member of the President's Commission for the Study of Ethical Problems in Medicine and Biomedical and Behavioral Research.

6. *Disease by Disease Toward National Health Insurance?,* Washington, D.C.: Institute of Medicine/National Academy of Sciences, 1973.

7. Robert S. Morison, "Rights and Responsibilities," *The Hastings Center Report 4*(2):4, April 1974.

Chapter 7
Practicing Prudently: Ethical Dilemmas in Caring for the Ill

1. See *Tarasoff* v. *Regents of the University of California,* 551 P.2d 334, 131 Cal. Rptr. 14 (1976).

2. *Proceedings of the American Medical Association House of Delegates, 129th Annual Convention, July 20–24, 1980.* Chicago: American Medical Association, 1980, p. 207. The complete text appears in Chapter 9, p. 231.

3. "Declaration of Geneva," *World Medical Association Bulletin 2:*109–111, 1949.

4. Aristotle, *Nicomachean Ethics*, Book I, Chap. 2, 1094a28–1094b5.

5. R. B. Schiffer and Benjamin Freedman, "The Last Bed in the ICU: A Medical or a Moral Decision?" *The Hastings Center Report* 7(6):21–22, December 1977, at p. 21. Emphasis added.

6. *Proceedings of the AMA, op cit.*

7. Ibid.

8. Richard Selzer, M.D., *Mortal Lessons: Notes on the Art of Surgery*, New York: Simon and Schuster, 1974; and *Confessions of a Knife*, New York: Simon and Schuster, 1979.

9. See, e.g., Paul Ramsey, "On (Only) Caring for the Dying," in his *The Patient as Person*, New Haven: Yale University Press, 1970.

10. David L. Jackson and Stuart Younger, "Patient Autonomy and 'Death with Dignity,' " *New England Journal of Medicine 301:*404–408, August 23, 1979.

11. See Paul Ramsey, "The Indignity of 'Death with Dignity,' " as well as commentaries on this essay by Robert S. Morison ("The Dignity of the Inevitable and Necessary") and Leon R. Kass ("Averting One's Eyes, or Facing the Music?—On Dignity and Death"), all in *Death Inside Out*, Peter Steinfels and Robert M. Veatch, eds., New York: Harper and Row, 1975.

12. Natural Death Act, California Health and Safety Code, AB 3060, div. 7, pt. I, chap. 3.9, secs. 7185–7195, August 30, 1976.

13. *Superintendent of Belchertown State School et al.* v. *Joseph Saikewicz, 370* N.E. 2d 417, Supreme Judicial Court of Massachusetts, *Mass.*, 1977. See also, Paul Ramsey, "The Saikewicz Precedent: What's Good for an Incompetent Patient," *The Hastings Center Report* 8(6):36–42, December, 1978.

14. President's Commission for the Study of Ethical Problems in Medicine and Biomedical and Behavioral Research, *Deciding to Forego Life Sustaining Treatment*, Washington, D.C.: U.S. Government Printing Office, 1983.

15. 195 Cal. Rptr. 484 (Ct. App. 2d Dist. 1983).

16. Mitchell T. Rabkin, Gerald Gillerman, and Nancy R. Rice, "Orders Not to Resuscitate," *New England Journal of Medicine 295:*364–369, August 12, 1976.

17. Hans Jonas, "The Right to Die," *The Hastings Center Report* 8(4):31–36, August 1978, at p. 34.

Chapter 8
Professing Medically: The Place of Ethics in Defining Medicine

1. *American Medical Association* v. *Federal Trade Commission*, 638 F.2d 443 (1980); 101 S. Ct. 3107 (March 23, 1982; rehearing denied May 13, 1982).

2. Clark C. Havighurst, "A comment: The Antitrust Challenge to Professionalism," *Maryland Law Review 41:*30–37, 1981, at p. 34.

3. The spirit and orientation of this inquiry resembles those of Edmund D. Pellegrino. See, e.g., "Toward a Reconstruction of Medical Morality: The Primacy of the Act of Profession and the Fact of Illness," *Journal of Medicine and Philosophy* 4(1):32–56, March 1979; Edmund D. Pellegrino and David C. Tomasma, *The Philosophical Basis of Medical Practice*, New York: Oxford University Press, 1981.

4. These and subsequent lexical quotations and observations are from the *Shorter Oxford English Dictionary*.

Chapter 9
Is There a Medical Ethic?: The Hippocratic Oath and the Sources of Ethical Medicine

1. Robert Graves, *The Greek Myths*, vol. I. Harmondsworth, Middlesex, England: Penguin Books, 1955, pp. 173–178.

2. Ludwig Edelstein, "The Hippocratic Oath: Text, Translation and Interpretation," *Supplements to the Bulletin of the History of Medicine*, no. 1, Baltimore: The Johns Hopkins Press, 1943. Reprinted in *Ancient Medicine: Selected Papers of Ludwig Edelstein*, Oswei Temkin and C. Lilian Temkin, eds., Baltimore: The John Hopkins Press, 1967, pp. 3–63.

3. Temkin and Temkin, *op. cit.*, p. 4.

4. Temkin and Temkin, *op. cit.*, p. 6. Translation by Edelstein modified in the direction of greater literalness in paragraphs two, six, and seven.

5. *Proceedings of the American Medical Association House of Delegates, 129th Annual Convention, July 20–24, 1980.* Chicago: American Medical Association, 1980, p. 207.

6. Erwin Straus, "Shame as a Historiological Problem," in his *Phenomenological Psychology*, New York: Basic Books, 1966, pp. 217–224.

7. Robert M. Veatch, "Professional Medical Ethics: The Grounding of Its Principles," *The Journal of Medicine and Philosophy* 4(1):6, March 1979.

Chapter 10
Teleology, Darwinism, and the Place of Man: Beyond Chance and Necessity?

1. Francis Bacon, *The New Organon*, Book I, Aphorism XLVIII, Fulton H. Anderson, ed., Indianapolis: Bobbs-Merrill, 1960, p. 52. Emphasis added.

2. Hans Jonas, *The Phenomenon of Life: Toward a Philosophical Biology*, Chicago: The University of Chicago Press, 1982, p. 58.

3. John Dewey, *The Influence of Darwin on Philosophy and Other Essays in Contemporary Thought*, New York: Henry Holt and Company, 1910, p. 13.

4. Jonas, *op. cit.*, p. 44.

5. Asa Gray, "Charles Darwin: A Sketch," *Nature,* June 4, 1874, reprinted in Gray, *Darwiniana,* Cambridge, Mass.: Harvard University Press, 1963, p. 237.

6. Charles Darwin, Letter to Asa Gray, June 5, 1874; *Life and Letters,* vol. III, Francis Darwin, ed., 3 vols., London: John Murray, 1888, p. 189.

7. See, e.g., E. S. Russell, *The Directiveness of Organic Activities,* Cambridge, England: Cambridge University Press, 1945.

8. Aristotle, *Physics B,* 199b26–31; see also 199a20–30. For a helpful discussion of Aristotelian teleology in the light of modern biology, see Marjorie Grene, "Aristotle and Modern Biology," *Journal of the History of Ideas 33:*395–424, July–September 1972.

9. See Jacob Klein, "Aristotle, An Introduction," in *Ancients and Moderns,* Joseph Cropsey, ed., New York: Basic Books, 1964, pp. 50–69. For an excellent modern account of animal "looks" and their relation to animal being, see Adolf Portmann, *Animal Forms and Patterns,* trans. by Hella Czech, London: Faber and Faber, 1964; paperback edition, New York: Schocken Books, 1967.

10. Charles Darwin, *The Origin of Species* (First edition), Pelican Classics edition, John. W. Burrow, ed., Harmondsworth, Middlesex, England: 1968, p. 66. Emphasis added.

11. Ibid., p. 132.

12. Samuel Butler, *Life and Habit,* London: A. C. Fifield, 1878, p. 263.

13. Charles Darwin, *The Descent of Man,* New York: Modern Library Giant, Random House, 1936, pp. 441–442. Emphasis added.

14. Charles Darwin, *Life and Letters,* vol. I, Frances Darwin, ed., 3 vols, London: John Murray, 1888, pp. 312–313.

15. Charles Darwin, Letter to Asa Gray, May 22, 1860, *Life and Letters,* vol. II, p. 312.

16. Thomas Huxley, "On the Reception of the 'Origin of Species,' " in *Life and Letters,* vol. II, pp. 201–202.

17. Darwin, *The Origin of Species,* pp. 459–460.

18. Ibid., p. 114.

19. Ibid., p. 217.

20. Ibid., p. 233.

21. Ibid., p. 445.

22. Ibid., p. 220.

23. Ibid., p. 444.

24. Ibid., p. 459.

25. Ibid., p. 129.

26. Ibid., p. 133.

27. Ibid., p. 132. Emphasis added.

28. Alfred North Whitehead, *The Function of Reason,* Boston: Beacon Press, 1962, pp. 4–5.

29. Jonas, *op. cit.,* p. 106.

30. Darwin, *The Origin of Species,* p. 445.

31. Ibid., pp. 158–159.

32. Ibid., p. 459. Emphasis added.

33. Ibid., pp. 420–421.

34. Ibid., p. 421.

35. Ibid., pp. 336–337.

36. For a discussion of Darwinism and ethics, including also some comment on Darwinism and Bible, see Antony Flew, "The Philosophical Implications of Darwinism," and Leon R. Kass, "Darwinism and Ethics: A Response to Antony Flew," in *Darwin, Marx, Freud,* Arthur Caplan and Bruce Jennings, eds., New York: Plenum Press, 1984.

37. Aristotle, *Physics B,* 192b8–13. Emphasis added.

Chapter 11
Thinking About the Body

1. Harold I. Lief and Renée C. Fox, "Training for 'Detached Concern,' " in *The Psychological Basis of Medical Practice,* Harold I. Lief, Victor F. Lief, and Nina R. Lief, eds., New York: Harper and Row, 1963, pp. 12–35. See also Renée C. Fox, "The Autopsy: Its Place in the Attitude-Learning of Second-Year Medical Students," in her *Essays in Medical Sociology,* New York: John Wiley & Sons, 1979, pp. 51–77.

2. Richard Selzer, M.D., *Confessions of a Knife,* New York: Simon and Schuster, 1979, and *Mortal Lessons: Notes on the Art of Surgery,* New York: Simon and Schuster, 1974.

3. Herodotus, *Histories,* Book III, 38, trans. by George Rawlinson, New York: The Modern Library, Random House, 1942, pp. 229–230.

4. Hans Jonas, "Is God a Mathematician? The Meaning of Metabolism," in his *The Phenomenon of Life: Toward a Philosophical Biology,* Chicago: The University of Chicago Press, 1982, pp. 64–98.

5. Adolf Portmann, *Animal Forms and Patterns,* trans. by Hella Czech, London: Faber and Faber, 1964; paperback edition, New York: Schocken Books, 1967.

6. Erwin Straus, "The Upright Posture," in his *Phenomenological Psychology,* New York: Basic Books, 1966, pp. 137–165.

7. Ibid., p. 138.

8. Ibid., p. 141.

9. Ibid., p. 143.

10. Ibid., pp. 154–155.

11. Ibid., p. 162. See also Hans Jonas, "The Nobility of Sight," *The Phenomenon of Life,* pp. 135–156.

12. Ibid., p. 162.

13. Ibid., p. 162.

14. Genesis 3:7.

15. Immanuel Kant, "Conjectural Beginning of Human History," trans. by Emil Fackenheim, in *Kant on History*, Lewis White Beck, ed., Indianapolis: Bobbs-Merrill, 1963, p. 57.

16. Herodotus, *Histories*, Book I, 139, trans. by George Rawlinson, Modern Library edition, p. 77.

17. Ibid., Book I, 140, p. 77; cf. *Illiad*, I, 1–7, and XXIV.

Chapter 12
Mortality and Morality: The Virtues of Finitude

1. René Descartes, *Discourse on the Method*, in *The Philosophical Works of Descartes*, vol. I, Elizabeth S. Haldane, and G. R. T. Ross, eds., Cambridge, England: Cambridge University Press, 1981, pp. 119–120.

2. See, e.g., *Assessing Biomedical Technologies: An Inquiry into the Nature of the Process*, chapt. IV, "Retardation of Aging," Washington, D.C.: The National Research Council–National Academy of Sciences, 1973.

3. See "Retardation of Aging," *op. cit.*

4. See, e.g., Alan Harrington, *The Immortalist*, New York: Avon, 1969.

5. From "That to Philosophize Is to Learn to Die," *The Complete Essays* of Michel Montaigne, trans. by Donald M. Frame, Stanford, Calif.: Stanford University Press, 1965, p. 63.

6. See *Odyssey*, XI, 487–491, trans. by Richmond Lattimore, New York: Harper & Row, 1965.

7. Lucretius, *De Rerum Natura*, Book III, 1080, trans. by Cyril Bailey, Oxford University Press, 1947, p. 359.

8. Wallace Stevens, "Sunday Morning," *The Collected Poems of Wallace Stevens*, New York: Alfred A. Knopf, 1954, pp. 66–70.

9. *Odyssey*, V, 215–224, *op. cit.* pp. 93–94.

10. Plato, *Symposium*, 192b8–193a1, trans. by W. R. M. Lamb, Cambridge, Mass.: Loeb Classical Library, Harvard University Press, 1925, pp. 143–145. Emphasis added.

11. Aristotle, *Rhetoric*, Book II, Chapt. 13, 1389b22, trans. by Lane Cooper, Englewood Cliffs, N.J.: Prentice-Hall, 1960, p. 135. Aristotle's wonderful discussion of the times of life—the young, the old, those in their prime—should be read in its entirety.

12. *Iliad*, VI, 146–150, trans. by Richmond Lattimore, Chicago: University of Chicago Press, 1951, p. 157.

Chapter 13
Looking Good: Nature and Nobility

1. Aristotle, *Nicomachean Ethics*, 1094b14–17 (my translation).

2. Adolf Portmann, *Animal Forms and Patterns*, trans. by Hella Czech, London: Faber and Faber, 1964; paperback edition, New York: Schocken

Books, 1967; *Animals as Social Beings,* trans. by Oliver Coburn, New York: The Viking Press, 1961.

3. Portmann, *Animal Forms and Patterns,* p. 57. Emphasis added.

4. Ibid., p. 86. Emphasis added.

5. Ibid., p. 75.

6. Ibid., p. 183.

7. Portmann, *Animals as Social Beings,* p. 172.

8. Ibid., pp. 182–183.

9. The masterful essay, "The Upright Posture," by the late Erwin Straus is a model of such phenomenological reflection. The essay appears in his collection *Phenomenological Psychology,* New York: Basic Books, 1966, pp. 137–165. It was discussed in Chapter 11.

10. Charles Darwin, *The Expression of the Emotions in Man and Animals,* Chicago: University of Chicago Press, 1965, p. 309.

11. Ibid., pp. 322–323.

12. Ibid., p. 325.

13. Immanuel Kant, "Conjectual Beginning of Human History," trans. by Emil Fackenheim, in *Kant on History,* Lewis White Beck, ed., Indianapolis: Bobbs-Merrill, 1963, p. 57.

14. See, e.g., Erwin Straus, "Shame as a Historiological Problem," in *Phenomenological Psychology, op. cit.* pp. 217–224; and Kurt Riezler, "Comment on the Social Psychology of Shame," *The American Journal of Sociology* 48:457–465, January 1943.

15. C. H. Waddington, *The Ethical Animal,* Chicago: Phoenix Edition, University of Chicago Press, 1967.

16. Straus, "Shame as a Historiological Problem," *op. cit.*

17. Aristotle, *Nicomachean Ethics,* 1128b17–26.

18. Ibid., Books VIII and IX. On friendship and the noble, see especially 1169a18–1169b2 and 1169b31–1170a13.

19. Herodotus, *Histories,* Book I, 8–12; Plato, *Republic,* 439e–440a; Genesis 9:20–27.

Index